QUADRATIC PROGRAMMING AND AFFINE VARIATIONAL INEQUALITIES

A Qualitative Study

Nonconvex Optimization and Its Applications

VOLUME 78

QUADRATIC PROGRAMMING AND AFFINE VARIATIONAL INEQUALITIES

A Qualitative Study

By

GUE MYUNG LEE
Pukyong National University, Republic of Korea

NGUYEN NANG TAM
Hanoi Pedagogical Institute No. 2, Vietnam

NGUYEN DONG YEN
Vietnamese Academy of Science and Technology, Vietnam

 Springer

Library of Congress Cataloging-in-Publication Data

A C.I.P. record for this book is available from the Library of Congress.

ISBN 0-387-24277-5   e-ISBN 0-387-24278-3   Printed on acid-free paper.

Printed in the United States of America.

9 8 7 6 5 4 3 2 1   SPIN 11375562

springeronline.com

Contents

Preface

Quadratic programs and affine variational inequalities represent two fundamental, closely-related classes of problems in the theories of mathematical programming and variational inequalities, respectively. This book develops a unified theory on qualitative aspects of nonconvex quadratic programming and affine variational inequalities. The first seven chapters introduce the reader step-by-step to the central issues concerning a quadratic program or an affine variational inequality, such as the solution existence, necessary and sufficient conditions for a point to belong to the solution set, and properties of the solution set. The subsequent two chapters discuss briefly two concrete models (linear fractional vector optimization and the traffic equilibrium problem) whose analysis can benefit a lot from using the results on quadratic programs and affine variational inequalities. There are six chapters devoted to the study of continuity and/or differentiability properties of the characteristic maps and functions in quadratic programs and in affine variational inequalities where all the components of the problem data are subject to perturbation. Quadratic programs and affine variational inequalities under linear perturbations are studied in three other chapters. One special feature of the presentation is that when a certain property of a characteristic map or function is investigated, we always try first to establish necessary conditions for it to hold, then we go on to study whether the obtained necessary conditions are sufficient ones. This helps to clarify the structures of the two classes of problems under consideration. The qualitative results can be used for dealing with algorithms and applications related to quadratic programming problems and affine variational inequalities.

This book can be useful for postgraduate students in applied mathematics and for researchers in the field of nonlinear programming and equilibrium problems. It can be used for some advanced courses on nonconvex quadratic programming and affine variational inequalities.

Among many references in the field discussed in this monograph, we would like to mention the following well-known books: *"Linear and Combinatorial Programming"* by K. G. Murty (1976), *"Non-Linear Parametric Optimization"* by B. Bank, J. Guddat, D. Klatte, B. Kummer and K. Tammer (1982), and *"The Linear Complementarity Problem"* by R. W. Cottle, J.-S. Pang and R. E. Stone (1992).

As for prerequisites, the reader is expected to be familiar with the basic facts of Linear Algebra, Functional Analysis, and Convex Analysis.

We started writing this book in Pusan (Korea) and completed our writing in Hanoi (Vietnam). This book would not be possible without the financial support from the Korea Research Foundation (Grant KRF 2000-015-DP0044), the Korean Science and Engineering Foundation (through the APEC Postdoctoral Fellowship Program and the Brain Pool Program), the National Program in Basic Sciences (Vietnam).

We would like to ask the international publishers who have published some of our research papers in their journals or proceedings volumes for letting us to use a re-edited form of these papers for this book. We thank them a lot for their kind permission.

We would like to express our sincere thanks to the following experts for their kind help or generous encouragement at different times in our research related to this book: Prof. Y. J. Cho, Dr. N. H. Dien, Prof. P. H. Dien, Prof. F. Giannessi, Prof. J. S. Jung, Prof. P. Q. Khanh, Prof. D. S. Kim, Prof. J. K. Kim, Prof. S. Kum, Prof. M. Kwapisz, Prof. B. S. Lee, Prof. D. T. Luc, Prof. K. Malanowski, Prof. C. Malivert, Prof. A. Maugeri, Prof. L. D. Muu, Prof. A. Nowakowski, Prof. S. Park, Prof. J.-P. Penot, Prof. V. N. Phat, Prof. H. X. Phu, Dr. T. D. Phuong, Prof. B. Ricceri, Prof. P. H. Sach, Prof. N. K. Son, Prof. M. Studniarski, Prof. M. Théra, Prof. T. D. Van. Also, it is our pleasant duty to thank Mr. N. Q. Huy for his efficient cooperation in polishing some arguments in the proof of Theorem 8.1.

The late Professor W. Oettli had a great influence on our research on quadratic programs and affine variational inequalities. We always remember him with sympathy and gratefulness.

We would like to thank Professor P. M. Pardalos for supporting our plan of writing this monograph.

This book is dedicated to our parents. We thank our families for patience and encouragement.

Any comment on this book will be accepted with sincere thanks.

May 2004

Gue Myung Lee, Nguyen Nang Tam, and Nguyen Dong Yen

Notations and Abbreviations

N	the set of the positive integers
R	the real line
$\overline{R}$	the extended real line
R^n	the n-dimensional Euclidean space
R^n_+	the nonnegative orthant in R^n
$\emptyset$	the empty set
x^T	the transpose of vector x
$\|x\|$	the norm of vector x
$\langle x, y \rangle$	the scalar product of x and y
A^T	the transpose of matrix A
$\mathrm{rank} A$	the rank of matrix A
$\|A\|$	the norm of matrix A
$R^{m \times n}$	the set of the $m \times n$-matrices
$R^{n \times n}_S$	the set of the symmetric $n \times n$-matrices
$\det A$	the determinant of a square matrix A
E	the unit matrix in $R^{n \times n}$
$B(x, \delta)$	the open ball centered at x with radius δ
$\bar{B}(x, \delta)$	the closed ball centered at x with radius δ
$\bar{B}_{R^n}$	the closed unit ball in R^n
$\mathrm{int}\Omega$	the interior of Ω
$\overline{\Omega}$	the closure of Ω
$\mathrm{bd}\Omega$	the boundary of Ω
$\mathrm{co}\Omega$	the convex hull of Ω
$\mathrm{dist}(x, \Omega)$	the distance from x to Ω
$\mathrm{cone} M$	the cone generated by M
$\mathrm{ri}\Delta$	the relative interior of a convex set Δ
$\mathrm{aff}\Delta$	the affine hull of Δ
$\mathrm{extr}\Delta$	the set of the extreme points of Δ
$0^+\Delta$	the recession cone of Δ
$T_\Delta(\bar{x})$	the tangent cone to Δ at $\bar{x}$
$N_\Delta(\bar{x})$	the normal cone to Δ at $\bar{x}$
$M^\perp$	the linear subspace of R^n orthogonal to $M \subset R^n$
$\mathrm{Pr}_K(\cdot)$ or $\mathrm{P}_K(\cdot)$	the metric projection from R^n onto a closed convex subset $K \subset R^n$

$\mathrm{dom}\, f$	the effective domain of function f
$f'(\bar{x}; v)$	the directional derivative of f at $\bar{x}$ in direction v
$f^0(\bar{x}; v)$	the Clarke generalized directional derivative of f at $\bar{x}$ in direction v
$\partial f(\bar{x})$	the subdifferential of a convex function f at $\bar{x}$, or the Clarke generalized gradient of a locally Lipschitz function f at $\bar{x}$
$\nabla f(\bar{x})$	the gradient of f at $\bar{x}$
$\nabla^2 f(\bar{x})$	the Hessian matrix of f at $\bar{x}$
$\mathrm{Sol}(P)$	the solution set of problem (P)
$\mathrm{loc}(P)$	the local solution set of problem (P)
$S(P)$	the KKT point set of problem (P)
$v(P)$	the optimal value of problem (P)
QP	quadratic programming
$QP(D, A, c, b)$	quadratic program defined by matrices D, A and vectors c, b
$S(D, A, c, b)$	the KKT point set of a quadratic program
$\mathrm{Sol}(D, A, c, b)$	the solution set of a quadratic program
$\mathrm{Sol}(c, b)$	the solution set of a quadratic program
$\varphi(D, A, c, b)$ or $\varphi(c, b)$	the optimal value function of a quadratic program
VI	variational inequality
$\mathrm{loc}(D, A, c, b)$	the local-solution set of a quadratic program
$\mathrm{VI}(\phi, \Delta)$	the VI defined by operator ϕ and set Δ
AVI	affine variational inequality
$\mathrm{Sol}(\mathrm{VI}(\phi, \Delta))$	the solution set of $\mathrm{VI}(\phi, \Delta)$
$\mathrm{AVI}(M, q, \Delta)$	the AVI defined by matrix M, vector q, and set Δ
$\mathrm{Sol}(\mathrm{AVI}(M, q, \Delta))$	the solution set of $\mathrm{AVI}(M, q, \Delta)$
$\mathrm{Sol}(M, A, q, b)$	the solution set of $\mathrm{AVI}(M, q, \Delta)$ where $\Delta = \{x : Ax \geq b\}$
LCP	linear complementarity
$\mathrm{LCP}(M, q)$	the LCP problem defined by matrix M and vector q
$\mathrm{Sol}(M, q)$	the solution set of $\mathrm{LCP}(M, q)$

NCP	nonlinear complementarity
NCP(ϕ, Δ)	the NCP problem defined by ϕ and Δ
LFVO	linear fractional vector optimization
VVI	vector variational inequality
Sol(VP)	the efficient solution set
	of the LFVO problem (VP)
Sol(VP)^w	the weakly efficient solution set
	of the LFVO problem (VP)
lsc	lower semicontinuous
lsc property	lower semicontinuity property
usc	upper semicontinuous
usc property	upper semicontinuity property
$OD-$pair	origin-destination pair
plq	piecewise linear-quadratic
PVI	parametric variational inequality

Chapter 1

Quadratic Programming Problems

Quadratic programming problems constitute a special class of non-linear mathematical programming problems. This chapter presents some preliminaries related to mathematical programming problems including the quadratic programming problems. The subsequent three chapters will provide a detailed exposition of the basic facts on quadratic programming problems, such as the solution existence, first-order optimality conditions, second-order optimality conditions, and properties of the solution sets.

1.1 Mathematical Programming Problems

Many practical and theoretical problems can be modeled in the form

$$(P) \qquad \text{Minimize } f(x) \quad \text{subject to} \quad x \in \Delta,$$

where $f : R^n \to \overline{R}$ is a given function, $\Delta \subset R^n$ is a given subset. Here and subsequently, $\overline{R} = [-\infty, +\infty] = R \cup \{-\infty\} \cup \{+\infty\}$ denotes the extended real line, R^n stands for the n-dimensional Euclidean space with the norm

$$\|x\| = \left(\sum_{i=1}^{n} x_i^2 \right)^{1/2}$$

for all $x = (x_1, \ldots, x_n) \in R^n$ and the scalar product

$$\langle x, y \rangle = \sum_{i=1}^{n} x_i y_i = x^T y$$

for all $x = (x_1, \ldots, x_n)$, $y = (y_1, \ldots, y_n) \in R^n$. Here and subsequently, the apex T denotes the matrix transposition. In the text, vectors are expressed as rows of real numbers; while in the matrix computations they are understood as columns of real numbers. The open ball in R^n centered at x with radius $\delta > 0$ is denoted by $B(x, \delta)$. The corresponding closed ball is denoted by $\bar{B}(x, \delta)$. Thus

$$B(x, \delta) = \{y \in R^n : \|y - x\| < \delta\}, \ \bar{B}(x, \delta) = \{y \in R^n : \|y - x\| \leq \delta\}.$$

The unit ball $\bar{B}(0, 1)$ will be frequently denoted by $\bar{B}_{R^n}$. For a set $\Omega \subset R^n$, the notations $\mathrm{int}\Omega$, $\bar{\Omega}$ and $\mathrm{bd}\Omega$, respectively, are used to denote the topological *interior*, the topological *closure* and the *boundary* of Ω. Thus $\bar{\Omega}$ is the smallest closed subset in R^n containing Ω, and

$$\mathrm{int}\Omega = \{x \in \Omega : \exists \varepsilon > 0 \text{ s.t. } B(x, \varepsilon) \subset \Omega\}, \quad \mathrm{bd}\Omega = \bar{\Omega} \setminus (\mathrm{int}\Omega).$$

We say that $U \subset R^n$ is a *neighborhood* of $x \in R^n$ if there exists $\varepsilon > 0$ such that $B(x, \varepsilon) \subset U$. Sometimes instead of (P) we write the following

$$\min\{f(x) : x \in \Delta\}.$$

Definition 1.1. We call (P) a *mathematical programming problem*. We call f the *objective function* and Δ the *constraint set* (also the *feasible region*) of (P). Elements of Δ are said to be the *feasible vectors* of (P). If $\Delta = R^n$ then we say that (P) is an unconstrained problem. Otherwise (P) is called a constrained problem.

Definition 1.2 (cf. Rockafellar and Wets (1998), p. 4) A feasible vector $\bar{x} \in \Delta$ is called a (global) *solution* of (P) if $f(\bar{x}) \neq +\infty$ and $f(x) \geq f(\bar{x})$ for all $x \in \Delta$. We say that $\bar{x} \in \Delta$ is a *local solution* of (P) if $f(\bar{x}) \neq +\infty$ and there exists a neighborhood U of $\bar{x}$ such that

$$f(x) \geq f(\bar{x}) \quad \text{for all} \quad x \in \Delta \cap U. \tag{1.1}$$

The set of all the solutions (resp., the local solutions) of (P) is denoted by $\mathrm{Sol}(P)$ (resp., $\mathrm{loc}(P)$). We say that two mathematical programming problems are *equivalent* if the solution set of the first problem coincides with that of the second one.

Definition 1.3. The *optimal value* $v(P)$ of (P) is defined by setting

$$v(P) = \inf\{f(x) : x \in \Delta\}. \tag{1.2}$$

If $\Delta = \emptyset$ then, by convention, $v(P) = +\infty$.

Remark 1.1. It is clear that $\mathrm{Sol}(P) \subset \mathrm{loc}(P)$. It is also obvious that

$$\mathrm{Sol}(P) = \{x \in \Delta : f(x) \neq +\infty, \ f(x) = v(P)\}.$$

Remark 1.2. It may happen that $\mathrm{loc}(P) \setminus \mathrm{Sol}(P) \neq \emptyset$. For example, if we choose $\Delta = [-1, +\infty)$ and $f(x) = 2x^3 - 3x^2 + 1$ then $\bar{x} = 1$ is a local solution of (P) which is not a global solution.

Remark 1.3. Instead of the minimization problem (P), one may encounter with the following maximization problem

$$(P_1) \qquad \text{Maximize} \quad f(x) \quad \text{subject to} \quad x \in \Delta.$$

A point $\bar{x} \in \Delta$ is said to be a (global) solution of (P_1) if $f(\bar{x}) \neq -\infty$ and $f(x) \leq f(\bar{x})$ for all $x \in \Delta$. We say that $\bar{x} \in \Delta$ is a local solution of (P_1) if $f(\bar{x}) \neq -\infty$ and there exists a neighborhood U of $\bar{x}$ such that $f(x) \leq f(\bar{x})$ for all $x \in \Delta \cap U$. It is clear that $\bar{x}$ is a solution (resp., a local solution) of (P_1) if and only if $\bar{x}$ is a solution (resp., a local solution) of the following minimization problem

$$\text{Minimize} \quad -f(x) \quad \text{subject to} \quad x \in \Delta.$$

Thus any maximization problem of the form (P_1) can be reduced to a minimization problem of the form (P).

Remark 1.4. Even in the case $v(P)$ is a finite real number, it may happen that $\mathrm{Sol}(P) = \emptyset$. For example, if $\Delta = [1, +\infty) \subset R$ and

$$f(x) = \begin{cases} \dfrac{1}{|x|} & \text{for } x \neq 0 \\ +\infty & \text{for } x = 0 \end{cases}$$

then $v(P) = 0$, while $\mathrm{Sol}(P) = \emptyset$.

There are different ways to classify mathematical programming problems:

- Convex vs. Nonconvex
- Smooth vs. Nonsmooth
- Linear vs. Nonlinear.

1.2 Convex Programs and Nonconvex Programs

Definition 1.4. We say that $\Delta \subset R^n$ is a *convex set* if

$$tx + (1-t)y \in \Delta \quad \text{for every } x \in \Delta,\, y \in \Delta \text{ and } t \in (0,1). \quad (1.3)$$

The smallest convex set containing a set $\Omega \subset R^n$ is called the *convex hull* of Ω and it is denoted by $\mathrm{co}\,\Omega$.

Definition 1.5. A function $f : R^n \to \overline{R}$ is said to be *convex* if its epigraph

$$\mathrm{epi} f := \{(x,\alpha) : x \in R^n,\ \alpha \in R,\ \alpha \geq f(x)\} \quad (1.4)$$

is a convex subset of the product space $R^n \times R$. A convex function f is said to be *proper* if $f(x) < +\infty$ for at least one $x \in R^n$ and $f(x) > -\infty$ for all $x \in R^n$. A function $f : R^n \to \overline{R}$ is said to be *concave* if the function $-f$ defined by the formula $(-f)(x) = -f(x)$ is convex.

By the usual convention (see Rockafellar (1970), p. 24),

$$\begin{aligned}
&\alpha + (+\infty) = (+\infty) + \alpha = +\infty \quad \text{for } -\infty < \alpha \leq +\infty, \\
&\alpha + (-\infty) = (-\infty) + \alpha = -\infty \quad \text{for } -\infty \leq \alpha < +\infty, \\
&\alpha(+\infty) = (+\infty)\alpha = +\infty,\ \ \alpha(-\infty) = (-\infty)\alpha = -\infty, \\
&\quad \text{for } 0 < \alpha \leq +\infty, \\
&\alpha(+\infty) = (+\infty)\alpha = -\infty,\ \ \alpha(-\infty) = (-\infty)\alpha = +\infty, \\
&\quad \text{for } -\infty \leq \alpha < 0, \\
&0(+\infty) = (+\infty)0 = 0 = 0(-\infty) = (-\infty)0, \\
&-(-\infty) = +\infty,\ \inf \emptyset = +\infty,\ \sup \emptyset = -\infty.
\end{aligned}$$

The combinations $(+\infty) + (-\infty)$ and $(-\infty) + (+\infty)$ have no meaning and will be avoided.

Note that a function $f : R^n \to R \cup \{+\infty\}$ is convex if and only if

$$f(tx + (1-t)y) \leq tf(x) + (1-t)f(y),\ \forall x,\, y \in R^n,\ \forall t \in (0,1). \quad (1.5)$$

Indeed, by definition, f is convex if and only if the set $\mathrm{epi} f$ defined in (1.4) is convex. This means that

$$t(x,\alpha) + (1-t)(y,\beta) \in \mathrm{epi} f$$

for all $t \in (0,1)$ and for all $x, y \in R^n$, $\alpha, \beta \in R$ satisfying $\alpha \geq f(x)$, $\beta \geq f(y)$. It is a simple matter to show that the latter is equivalent to (1.5).

More generally, a function $f : R^n \to R \cup \{+\infty\}$ is convex if and only if

$$f(\lambda_1 x_1 + \cdots + \lambda_k x_k) \leq \lambda_1 f(x_1) + \cdots + \lambda_k f(x_k) \quad \text{(Jensen's Inequality)}$$

whenever $x_1, \ldots, x_k \in R^n$ and $\lambda_1 \geq 0, \ldots, \lambda_k \geq 0$, $\lambda_1 + \cdots + \lambda_k = 1$. (See Rockafellar (1970), Theorem 4.3).

Definition 1.6. We say that (P) is a *convex program* (a convex mathematical programming problem) if Δ is a convex set and f is a convex function.

Proposition 1.1. *If (P) is a convex program then*

$$\text{Sol}(P) = \text{loc}(P). \tag{1.6}$$

Proof. It suffices to show that $\text{loc}(P) \subset \text{Sol}(P)$ whenever (P) is a convex program. Let $\bar{x} \in \text{loc}(P)$ and let U be a neighborhood of $\bar{x}$ such that (1.1) holds. If $\bar{x} \notin \text{Sol}(P)$ then there must exist $\hat{x} \in \Delta$ such that $f(\hat{x}) < f(\bar{x})$. Since $f(\bar{x}) \neq +\infty$, this implies that $f(\hat{x}) \in R \cup \{-\infty\}$.

We first consider the case $f(\hat{x}) \neq -\infty$. For any $t \in (0,1)$, we have

$$\begin{aligned} f(t\hat{x} + (1-t)\bar{x}) \ &\leq tf(\hat{x}) + (1-t)f(\bar{x}) \\ &< tf(\bar{x}) + (1-t)f(\bar{x}) = f(\bar{x}). \end{aligned} \tag{1.7}$$

Since $t\hat{x} + (1-t)\bar{x} = \bar{x} + t(\hat{x} - \bar{x})$ belongs to $\Delta \cap U$ for sufficiently small $t \in (0,1)$, (1.7) contradicts (1.1).

We now consider the case $f(\hat{x}) = -\infty$. Fix any $t \in (0,1)$. For every $\alpha \in R$, since $(\hat{x}, \alpha) \in \text{epi} f$ and $(\bar{x}, f(\bar{x})) \in \text{epi} f$, we have

$$t(\hat{x}, \alpha) + (1-t)(\bar{x}, f(\bar{x})) \in \text{epi} f.$$

Hence $f(t\hat{x} + (1-t)\bar{x}) \leq t\alpha + (1-t)f(\bar{x})$ for all $\alpha \in R$. This implies that $f(t\hat{x} + (1-t)\bar{x}) = -\infty$. Since the last equality is valid for all $t \in (0,1)$ and $t\hat{x} + (1-t)\bar{x} \in \Delta \cap U$ if $t \in (0,1)$ is sufficiently small, (1.1) cannot hold. We have arrived at a contradiction. $\square$

Definition 1.7. If Δ is nonconvex (= not convex) or f is nonconvex then we say that (P) is a *nonconvex program* (a nonconvex mathematical programming problem).

Example 1.1. Consider the problem

$$\min\{f(x) = (x_1 - c_1)^2 + (x_2 - c_2)^2 : x \in \Delta\}, \qquad (1.8)$$

where $\Delta = \{x = (x_1, x_2) : x_1 \geq 0\} \cup \{x = (x_1, x_2) : x_2 \geq 0\}$ and $c = (c_1, c_2) = (-2, -1)$. Note that f is convex, while Δ is nonconvex. It is clear that (1.8) is equivalent to the following problem

$$\min\{\|x - c\| : x \in \Delta\}. \qquad (1.9)$$

On can easily verify that the solution set of (1.8) and (1.9) consists of only one point $(-2, 0)$, and the local solution set contains two points: $(-2, 0)$ and $(0, -1)$.

Example 1.2. Let $f_1(x) = -x + 2$, $f_2(x) = x + \dfrac{3}{2}$, $x \in R$. Define $f(x) = \min\{f_1(x), \ f_2(x)\}$ and choose $\Delta = [0, 2] \subset R$. For these f and Δ, we have

$$\mathrm{Sol}(P) = \{2\}, \quad \mathrm{loc}(P) = \{0, \ 2\}.$$

Note that in this example f is a nonconvex function, while Δ is a convex set.

Convex functions have many nice properties. For example, a convex function is continuous at any interior point of its effective domain and it is directionally differentiable at any point in the domain.

Definition 1.8. For a function $f : R^n \to \overline{R}$, the set

$$\mathrm{dom}\, f := \{x \in R^n : -\infty < f(x) < +\infty\} \qquad (1.10)$$

is called the *effective domain* of f. For a point $\bar{x} \in \mathrm{dom}\, f$ and a vector $v \in R^n$, if the limit

$$f'(\bar{x}; v) := \lim_{t \downarrow 0} \frac{f(\bar{x} + tv) - f(\bar{x})}{t} \qquad (1.11)$$

(which may have the values $+\infty$ and $-\infty$) exists then f is said to be *directionally differentiable* at $\bar{x}$ in direction v and the value $f'(\bar{x}; v)$ is called the *directional derivative* of f at $\bar{x}$ in direction v. If $f'(\bar{x}; v)$ exists for all $v \in R^n$ then f is said to be directionally differentiable at $\bar{x}$.

In the next two theorems, $f : R^n \to R \cup \{+\infty\}$ is a proper convex function.

Theorem 1.1. (See Rockafellar (1970), Theorem 10.1) *If $\bar{x} \in R^n$ and $\delta > 0$ are such that the open ball $B(\bar{x}, \delta)$ is contained in $\mathrm{dom} f$, then the restriction of f to $B(\bar{x}, \delta)$ is a continuous real function.*

Theorem 1.2. (See Rockafellar (1970), Theorem 23.1) *If $\bar{x} \in \mathrm{dom} f$ then for any $v \in R^n$ the limit*

$$f'(\bar{x}; v) := \lim_{t \downarrow 0} \frac{f(\bar{x} + tv) - f(\bar{x})}{t}$$

exists, and one has

$$f'(\bar{x}; v) = \inf_{t > 0} \frac{f(\bar{x} + tv) - f(\bar{x})}{t}.$$

Definition 1.9. The *normal cone* $N_\Delta(\bar{x})$ to a convex set $\Delta \subset R^n$ at a point $\bar{x} \in R^n$ is defined by the formula

$$N_\Delta(\bar{x}) = \begin{cases} \{x^* \in R^n \ : \ \langle x^*, x - \bar{x} \rangle \leq 0 \text{ for all } x \in \Delta\} & \text{if } \bar{x} \in \Delta \\ \emptyset & \text{if } \bar{x} \notin \Delta. \end{cases}$$

(1.12)

Definition 1.10. The *subdifferential* $\partial f(\bar{x})$ of a convex function $f : R^n \to \overline{R}$ at a point $\bar{x} \in R^n$ is defined by setting

$$\partial f(\bar{x}) = \{x^* \in R^n \ : \ f(\bar{x}) + \langle x^*, x - \bar{x} \rangle \leq f(x) \text{ for every } x \in R^n\}.$$

(1.13)

Definition 1.11. A subset $M \subset R^n$ is called an *affine set* if $tx + (1 - t)y \in M$ for every $x \in M$, $y \in M$ and $t \in R$. For a convex set $\Delta \subset R^n$, the *affine hull* $\mathrm{aff}\Delta$ of Δ is the smallest affine set containing Δ. The *relative interior* of Δ is defined by the formula

$$\mathrm{ri}\Delta = \{x \in \Delta \ : \ \exists \delta > 0 \text{ such that } B(x, \delta) \cap \mathrm{aff}\Delta \subset \Delta\}.$$

The following statement describes the relation between the directional derivative and the subdifferential of convex functions.

Theorem 1.3. (See Rockafellar (1970), Theorem 23.4) *Let f be a proper convex function on R^n. If $x \notin \mathrm{dom} f$ then $\partial f(x)$ is empty. If $x \in \mathrm{ri}(\mathrm{dom} f)$ then $\partial f(x)$ is nonempty and*

$$f'(x; v) = \sup\{\langle x^*, v \rangle \ : \ x^* \in \partial f(x)\}, \quad \forall v \in R^n.$$

Besides, $\partial f(x)$ is a nonempty bounded set if and only if

$$x \in \mathrm{int}(\mathrm{dom} f),$$

in which case $f'(x; v)$ is finite for every $v \in R^n$.

The following result is called the Moreau-Rockafellar Theorem.

Theorem 1.4. (See Rockafellar (1970), Theorem 23.8) *Let $f = f_1 + \cdots + f_k$, where $f_1, \ldots, f_k$ are proper convex functions on R^n. If*

$$\bigcap_{i=1}^{k} \mathrm{ri}(\mathrm{dom} f_i) \neq \emptyset$$

then

$$\partial f(x) = \partial f_1(x) + \cdots + \partial f_k(x), \quad \forall x \in R^n.$$

First-order necessary and sufficient optimality conditions for convex programs can be stated as follows.

Theorem 1.5. (See Rockafellar (1970), Theorem 27.4) *Suppose that f is a proper convex function on R^n and $\Delta \subset R^n$ is a nonempty convex set. If the inclusion*

$$0 \in \partial f(\bar{x}) + N_\Delta(\bar{x}) \tag{1.14}$$

holds for some $\bar{x} \in R^n$, then $\bar{x}$ is a solution of (P). Conversely, if

$$\mathrm{ri}(\mathrm{dom} f) \cap \mathrm{ri}\Delta \neq \emptyset \tag{1.15}$$

then (1.14) is a necessary and sufficient condition for $\bar{x} \in R^n$ to be a solution of (P). In particular, if $\Delta = R^n$ then $\bar{x}$ is a solution of (P) if and only if $0 \in \partial f(\bar{x})$.

Inclusion (1.14) means that there exist $x^* \in \partial f(\bar{x})$ and $u^* \in N_\Delta(\bar{x})$ such that $0 = x^* + u^*$. Note that (1.15) is a regularity condition for convex programs of the type (P).

The facts stated in Proposition 1.1 and Theorem 1.5 are the most characteristic properties of convex mathematical programming problems.

Theorem 1.5 can be used for solving effectively many convex programs. For illustration, let us consider the following example.

Example 1.3. (The Fermat point)   Let A, B, C be three points in the two-dimensional space R^2 with the coordinates

$$a = (a_1, a_2), \quad b = (b_1, b_2), \quad c = (c_1, c_2),$$

respectively. Assume that there exists no straight line containing all the three points. The problem consists of finding a point M in R^2

with the coordinates $\bar{x} = (\bar{x}_1, \bar{x}_2)$ such that the sum of the distances from M to A, B and C is minimal. This amounts to saying that $\bar{x}$ is a solution of the following unconstrained convex program:

$$\min\{f(x) := \|x - a\| + \|x - b\| + \|x - c\| : x \in R^2\}. \qquad (1.16)$$

In Lemma 1.1 below it will be proved that problem (1.16) has solutions and the solution set is a singleton. Note that $f = f_1 + f_2 + f_3$, where $f_1(x) = \|x - a\|$, $f_2(x) = \|x - b\|$, $f_3(x) = \|x - c\|$. By Theorem 1.5, $\bar{x}$ is a solution of (1.16) if and only if $0 \in \partial f(\bar{x})$. As $\mathrm{dom} f_i = R^2$ $(i = 1, 2, 3)$, using Theorem 1.4 we can write the last inclusion in the following equivalent form

$$0 \in \partial f_1(\bar{x}) + \partial f_2(\bar{x}) + \partial f_3(\bar{x}). \qquad (1.17)$$

We first consider the case where $\bar{x}$ coincides with one of the three vectors a, b, c. Let $\bar{x} = a$, i.e. $M \equiv A$. In this case,

$$\partial f_1(\bar{x}) = \bar{B}_{R^2}, \quad \partial f_2(\bar{x}) = \left\{ \frac{a - b}{\|a - b\|} \right\}, \quad \partial f_3(\bar{x}) = \left\{ \frac{a - c}{\|a - c\|} \right\}.$$

Hence (1.17) is equivalent to saying that there exists $u^* \in \bar{B}_{R^2}$ such that

$$0 = u^* - v^* - w^*, \qquad (1.18)$$

where $v^* := (b - a)/\|b - a\|$, $w^* := (c - a)/\|c - a\|$. From (1.18) it follows that

$$\begin{aligned} 1 \geq \|u^*\|^2 &= \langle u^*, u^* \rangle \\ &= \langle v^* + w^*, v^* + w^* \rangle \\ &= \|v^*\|^2 + \|w^*\|^2 + 2\langle v^*, w^* \rangle. \end{aligned}$$

As $\|v^*\| = 1$ and $\|w^*\| = 1$, this yields $\langle v^*, w^* \rangle \leq -\frac{1}{2}$. Denoting by α the geometric angle between the vectors v^* and w^* (which is equal to angle $\hat{A}$ of the triangle ABC), we deduce from the last inequality that

$$\cos \alpha = \frac{\langle v^*, w^* \rangle}{\|v^*\| \|w^*\|} = \langle v^*, w^* \rangle \leq -\frac{1}{2}.$$

Hence

$$\frac{2\pi}{3} \leq \alpha < \pi. \qquad (1.19)$$

(The case $\alpha = \pi$ is excluded because there exists no straight line containing A, B and C.) It is easy to show that (1.19) implies that

$\tilde{u}^* := v^* + w^*$ belongs to $\bar{B}_{R^2}$. Thus (1.19) is equivalent to (1.17). This means that (1.19) holds if and only if $\bar{x} = a$ is a solution of (1.16).

We now turn to the case where $\bar{x} \neq a$, $\bar{x} \neq b$ and $\bar{x} \neq c$, i.e. M does not coincide with anyone from the three vertexes A, B, C of the triangle ABC. In this case, as

$$\partial f_1(\bar{x}) = \left\{ \frac{\bar{x} - a}{\|\bar{x} - a\|} \right\}, \ \partial f_2(\bar{x}) = \left\{ \frac{\bar{x} - b}{\|\bar{x} - b\|} \right\}, \ \partial f_3(\bar{x}) = \left\{ \frac{\bar{x} - c}{\|\bar{x} - c\|} \right\},$$

(1.17) is equivalent to the equality

$$0 = u^* + v^* + w^*, \qquad (1.20)$$

where $u^* := (a - \bar{x})/\|a - \bar{x}\|$, $v^* := (b - \bar{x})/\|b - \bar{x}\|$ and $w^* := (c - \bar{x})/\|c - \bar{x}\|$. By (1.20),

$$\begin{aligned} 1 = \|u^*\|^2 &= \langle u^*, u^* \rangle \\ &= \langle -v^* - w^*, -v^* - w^* \rangle \\ &= \|v^*\|^2 + \|w^*\|^2 + 2\langle v^*, w^* \rangle. \end{aligned}$$

Since $\|v^*\| = 1$ and $\|w^*\| = 1$, this implies that $\langle v^*, w^* \rangle = -\frac{1}{2}$. Hence the geometric angle α between v^* and w^* is $2\pi/3$. Similarly, we deduce from (1.20) that the geometric angle β (resp., γ) between u^* and w^* (resp., between u^* and v^*) is equal to $2\pi/3$. (Geometrically, we have shown that M sees the edges BC, AC and AB of the triangle ABC under the same angle $120°$.) It is easily seen that if

$$\alpha = \beta = \gamma = \frac{2\pi}{3}$$

then (1.20) is satisfied; hence (1.17) is valid and $\bar{x}$ is a solution of (1.16).

Summarizing all the above in the language of Euclidean Geometry, we have the following conclusions:

(i) If one of the three angles of the triangle ABC, say $\hat{A}$, is larger than or equal to $120°$, then $M \equiv A$ is the unique solution of our problem.

(ii) If all the three angles of the triangle ABC are smaller than $120°$, then the unique solution of our problem is the point M seeing the edges BC, AC and AB of the triangle ABC

under the same angle $120°$. (This special point M is called the *Fermat point* or the *Torricelli point* (see Weisstein (1999)). It can be proved that the Fermat point belongs to the interior of the triangle ABC.)

If the necessary and sufficient optimality condition stated in Theorem 1.5 yields a unique point $\bar{x}$ which can be expressed explicitly via the data of the optimization problem (see, for instance, the situation in Example 1.6 below) then the problem has solutions and the solution set is a singleton. In the other case, information about the solution existence and uniqueness can be obtained by analyzing furthermore the structure of the problem under consideration.

For the illustrative problem described in Example 1.3, the following statement is valid.

Lemma 1.1. *Let* $a = (a_1, a_2)$, $b = (b_1, b_2)$, $c = (c_1, c_2)$ *be given points in* R^2 *such that there exists no straight line containing all the three points. Then problem (1.16) has solutions and the solution set is a singleton.*

Proof. In order to show that (1.16) has solutions, we observe that

$$f(x) \geq 3\|x\| - \|a\| - \|b\| - \|c\|.$$

Therefore $\lim\limits_{\|x\| \to +\infty} f(x) = +\infty$. Fix any $z \in R^2$ and put $\gamma = f(z)$. Let $\varrho \in [\|z\|, +\infty)$ be such that

$$f(x) > \gamma \quad \text{for every } x \in R^2 \setminus \bar{B}(0, \varrho).$$

By the Weierstrass Theorem, the restriction of the continuous function $f(x)$ on the compact set $\bar{B}(0, \varrho)$ achieves minimum at some point $\bar{x} \in \bar{B}(0, \varrho)$, that is $f(\bar{x}) \leq f(y)$ for every $y \in \bar{B}(0, \varrho)$. Since

$$f(\bar{x}) \leq f(z) = \gamma < f(x) \quad \text{for all } x \in R^2 \setminus \bar{B}(0, \varrho),$$

it follows that $\bar{x}$ is a solution of (1.16).

We now prove that $f(x)$ is a *strictly convex function*, that is

$$f(tx + (1-t)y)) < tf(x) + (1-t)f(y)$$

for all x, y in R^2 with $x \neq y$ and for all $t \in (0,1)$. Given any $x = (x_1, x_2)$, $y = (y_1, y_2)$ in R^2 with $x \neq y$ and $t \in (0,1)$, we consider the following vector systems

$$\{x - a, y - a\}, \quad \{x - b, y - b\}, \quad \{x - c, y - c\}. \tag{1.21}$$

We claim that at least one of the three systems is linearly independent. Suppose the claim were false. Then we would have

$$\det \begin{pmatrix} x_1 - a_1 & y_1 - a_1 \\ x_2 - a_2 & y_2 - a_2 \end{pmatrix} = 0, \quad \det \begin{pmatrix} x_1 - b_1 & y_1 - b_1 \\ x_2 - b_2 & y_2 - b_2 \end{pmatrix} = 0,$$

$$\det \begin{pmatrix} x_1 - c_1 & y_1 - c_1 \\ x_2 - c_2 & y_2 - c_2 \end{pmatrix} = 0,$$

where $\det Z$ denotes the *determinant* of a square matrix Z. These equalities imply that

$$\begin{array}{ll} (x_1 - y_1)a_2 - (x_2 - y_2)a_1 &= x_1 y_2 - x_2 y_1, \\ (x_1 - y_1)b_2 - (x_2 - y_2)b_1 &= x_1 y_2 - x_2 y_1, \qquad (1.22) \\ (x_1 - y_1)c_2 - (x_2 - y_2)c_1 &= x_1 y_2 - x_2 y_1. \end{array}$$

Since $x \neq y$, we have $(x_1 - y_1)^2 + (x_2 - y_2)^2 \neq 0$. So the set

$$L := \{z = (z_1, z_2) \in R^2 : (x_1 - y_1)z_2 - (x_2 - y_2)z_1 = x_1 y_2 - x_2 y_1\}$$

is a straight line in R^2. By (1.22), L contains all the points a, b, c. This contradicts our assumption. We have thus proved that at least one of the three vector systems in (1.21) is linearly independent. Without loss of generality, we can assume that the system $\{x-a, y-a\}$ is linearly independent. Then the system $\{t(x-a), (1-t)(y-a)\}$ is also linearly independent. This implies that

$$\|t(x - a) + (1 - t)(y - a)\| < t\|x - a\| + (1 - t)\|y - a\|.$$

So we have

$$\begin{aligned} f(tx + (1-t)y) &= \|tx + (1-t)y - a\| + \|tx + (1-t)y - b\| \\ &\quad + \|tx + (1-t)y - c\| \\ &= \|t(x - a) + (1-t)(y - a)\| \\ &\quad + \|t(x - b) + (1-t)(y - b)\| \\ &\quad + \|t(x - c) + (1-t)(y - c)\| \\ &< t\|x - a\| + (1-t)\|y - a\| \\ &\quad + t\|x - b\| + (1-t)\|y - b\| \\ &\quad + t\|x - c\| + (1-t)\|y - c\| \\ &= tf(x) + (1-t)f(y). \end{aligned}$$

The strict convexity of f has been established. From this property it follows immediately that (1.16) cannot have more than one solution.

Indeed, if there were two different solutions x and y of the problem, then by the strict convexity of f we would have

$$f\left(\frac{1}{2}x + \frac{1}{2}y\right) < \frac{1}{2}f(x) + \frac{1}{2}f(y) = f(x).$$

This contradicts the fact that x is a solution of (1.16). The proof of the lemma is complete. □

Remark 1.5. It follows from the above results that (1.16) admits a unique solution belonging to the convex hull of the set $\{a, b, c\}$. Hence (1.16) is equivalent to the following constrained convex program

$$\min\{\|x - a\| + \|x - b\| + \|x - c\| : x \in \text{co}\{a, b, c\}\}.$$

In problem (P), if Δ is the solution set of a system of inequalities and equalities then first-order optimality conditions can be written in a form involving some *Lagrange multipliers*.

Let us consider problem (P) under the assumptions that $f : R^n \rightarrow R$ is a convex function and

$$\Delta = \{x \in R^n : g_1(x) \leq 0, \ldots, g_m(x) \leq 0, h_1(x) = 0, \ldots, h_s(x) = 0\}, \tag{1.23}$$

where $g_i : R^n \rightarrow R$ for $i = 1, \ldots, m$ is a convex function, $h_j : R^n \rightarrow R$ for $j = 1, \ldots, s$ is an *affine function*, i.e. there exist $a_j \in R^n$ and $\alpha_j \in R$ such that $h_j(x) = \langle a_j, x \rangle + \alpha_j$ for every $x \in R^n$. It is admitted that the equality constraints (resp., the equality constraints) can be absent in (1.23). For abbreviation, we use the formal writing $m = 0$ (resp., $s = 0$) to indicate that all the inequality constraints (resp., all the equality constraints) in (1.23) are absent.

Theorem 1.6. (Kuhn-Tucker Theorem for convex programs; see Rockafellar (1970), p. 283) *Let (P) be a convex program where Δ is given by (1.23). Let the above assumptions on f, g_i ($i = 1, \ldots, m$) and h_j ($j = 1, \ldots s$) be satisfied. Assume that there exists a vector $z \in R^n$ such that*

$$g_i(z) < 0 \text{ for } i = 1, \ldots, m \quad \text{and} \quad h_j(z) = 0 \text{ for } j = 1, \ldots, s. \tag{1.24}$$

Then $\bar{x}$ is a solution of (P) if and only if there exist $m + s$ real numbers $\lambda_1, \ldots, \lambda_m, \mu_1, \ldots, \mu_s$, which are called the Langrange multipliers corresponding to $\bar{x}$, such that the following Kuhn-Tucker conditions are fulfilled:

(a) $\lambda_i \geq 0$, $g_i(\bar{x}) \leq 0$ *and* $\lambda_i f_i(\bar{x}) = 0$ *for* $i = 1, \ldots, m$,

(b) $h_j(\bar{x}) = 0$ *for* $j = 1, \ldots, s$,

(c) $0 \in \partial f(\bar{x}) + \sum_{i=1}^{m} \lambda_i \partial g_i(\bar{x}) + \sum_{j=1}^{s} \mu_j a_j$.

Note that (1.24) is a *constraint qualification* for convex programs. If $s = 0$ then it becomes

$\exists z \in R^n$ s.t. $g_i(z) < 0$ for $i = 1, \ldots, m$.   (The Slater condition)

If $m = 0$ then (1.24) is equivalent to the requirement that Δ is nonempty. Actually, in that case condition (1.24) can be omitted in the formulation of Theorem 1.6.

1.3   Smooth Programs and Nonsmooth Programs

For brevity, if $f : R^n \to R$ is a continuously Fréchet differentiable function then we shall say that f is a C^1-*function*. Similarly, if f is twice continuously Fréchet differentiable function then we shall say that f is a C^2-*function*. The vector

$$\nabla f(\bar{x}) = \begin{pmatrix} \dfrac{\partial f(\bar{x})}{\partial x_1} \\ \vdots \\ \dfrac{\partial f(\bar{x})}{\partial x_n} \end{pmatrix},$$

where $\dfrac{\partial f(\bar{x})}{\partial x_i}$ for $i = 1, \ldots, n$ denotes the partial derivative of f at $\bar{x}$ with respect to x_i, is called the *gradient* of f at $\bar{x}$. The matrix

$$\nabla^2 f(\bar{x}) = \begin{pmatrix} \dfrac{\partial^2 f(\bar{x})}{\partial x_1 \partial x_1} & \dfrac{\partial^2 f(\bar{x})}{\partial x_1 \partial x_2} & \cdots & \dfrac{\partial^2 f(\bar{x})}{\partial x_1 \partial x_n} \\ \dfrac{\partial^2 f(\bar{x})}{\partial x_2 \partial x_1} & \dfrac{\partial^2 f(\bar{x})}{\partial x_2 \partial x_2} & \cdots & \dfrac{\partial^2 f(\bar{x})}{\partial x_2 \partial x_n} \\ \dfrac{\partial^2 f(\bar{x})}{\partial x_n \partial x_1} & \dfrac{\partial^2 f(\bar{x})}{\partial x_n \partial x_2} & \cdots & \dfrac{\partial^2 f(\bar{x})}{\partial x_n \partial x_n} \end{pmatrix},$$

where $\dfrac{\partial^2 f(\bar{x})}{\partial x_j \partial x_i}$ denotes the second-order partial derivative of f at $\bar{x}$ w.r.t. x_j and x_i, is called the *Hessian matrix* of f at $\bar{x}$. It is well-known that if f is a C^1-function on R^n then f is directionally differentiable on R^n (see Definition 1.8) and

$$f'(\bar{x}; v) = \nabla f(\bar{x})v = \sum_{i=1}^{n} \frac{\partial f(\bar{x})}{\partial x_i} v_i,$$

for every $\bar{x} \in R^n$ and $v = (v_1, \ldots, v_n) \in R^n$.

Definition 1.12. We say that (P) is a *smooth program* (a smooth mathematical programming problem) if $f : R^n \to R$ is a C^1-function and Δ can be represented in the form (1.23) where $g_i : R^n \to R$ $(i = 1, \ldots, m)$ and $h_j : R^n \to R$ $(j = 1, \ldots, s)$ are C^1-functions. Otherwise, (P) is called a *nonsmooth program*.

We have considered problem (1.16) of finding the Fermat point. It is an example of nonsmooth programs. Function $f(x)$ in (1.16) is not a C^1-function. However, it is a *Lipschitz function* because

$$|f(x) - f(y)| \leq 3\|x - y\| \quad \text{for all } x, y \text{ in } R^2.$$

Definition 1.13. A function $f : R^n \to R$ is said to be a *locally Lipschitz* near $\bar{x} \in R^n$ if there exist a constant $\ell \geq 0$ and a neighborhood U of $\bar{x}$ such that

$$|f(x') - f(x)| \leq \ell\|x' - x\| \quad \text{for all } x, x' \text{ in } U.$$

If f is locally Lipschitz near every point in R^n then f is said to be a *locally Lipschitz function* on R^n. If f is locally Lipschitz near $\bar{x}$ then the *generalized directional derivative* of f at $\bar{x}$ in direction $v \in R^n$ is defined by

$$
\begin{aligned}
f^0(\bar{x}; v) \quad &:= \limsup_{x \to \bar{x}, \, t \downarrow 0} \frac{f(x + tv) - f(x))}{t} \\
&= \sup\{\xi \in R : \exists \text{ sequences } x_k \to \bar{x} \text{ and } t_k \to 0+ \\
&\qquad \text{such that } \xi = \lim_{k \to +\infty} \frac{f(x_k + t_k v) - f(x_k)}{t_k}\}.
\end{aligned}
$$

The *Clarke generalized gradient* of f at $\bar{x}$ is given by

$$\partial f(\bar{x}) := \{x^* \in R^n : f^0(\bar{x}; v) \geq \langle x^*, v \rangle \text{ for all } v \in R^n\}.$$

Theorem 1.7. (See Clarke (1983), Propositions 2.1.2, 2.2.4, 2.2.6 and 2.2.7) *Let $f : R^n \to R$ be a real function. Then the following assertions hold:*

(a) *If f is locally Lipschitz near $\bar{x} \in R^n$ then*

$$f^0(\bar{x}; v) = \max\{\langle x^*, v\rangle \,:\, x^* \in \partial f(\bar{x})\}$$

for every $v \in R^n$.

(b) *If f is a C^1-function then f is a locally Lipschitz function and $\partial f(\bar{x}) = \{\nabla f(\bar{x})\}$, $f^0(\bar{x}; v) = \langle \nabla f(\bar{x}), v\rangle$ for all $\bar{x} \in R^n$ and $v \in R^n$.*

(c) *If f is convex then f is a locally Lipschitz function and, for every $\bar{x} \in R^n$, the Clarke generalized gradient $\partial f(\bar{x})$ coincides with the subdifferential of f at $\bar{x}$ defined by formula (1.13). Besides, $f^0(\bar{x}; v) = f'(\bar{x}; v)$ for every $v \in R^n$.*

As concerning the above assertion (c), we note that the directional derivative $f'(\bar{x}; v)$ exists according to Theorem 1.2.

Definition 1.14. Let $C \subset R^n$ be a nonempty subset. The *Clarke tangent cone* $T_C(x)$ to C at $x \in C$ is the set of all $v \in R^n$ satisfying $d^0_C(x; v) = 0$, where $d^0_C(x; v)$ denotes the generalized directional derivative of the Lipschitzian function $d_C(z) := \inf\{\|y - z\| \,:\, y \in C\}$ at x in direction v. The *Clarke normal cone* $N_C(x)$ to C at x is defined as the *dual cone* of $T_C(x)$, i.e.

$$N_C(x) = \{x^* \in R^n \,:\, \langle x^*, v\rangle \leq 0 \quad \text{for all } v \in T_C(x)\}.$$

Theorem 1.8. (See Clarke (1983), Propositions 2.4.3, 2.4.4 and 2.4.5) *For any nonempty subset $C \subset R^n$ and any point $x \in C$, the following assertions hold:*

(a) $N_C(x) = \overline{\left\{ \cup_{t \geq 0} \, t\partial d_C(x) \right\}}.$

(b) *If C is convex then $N_C(x)$ coincides with the normal cone to C at x defined by formula (1.12), and $T_C(x)$ coincides with the topological closure of the set $\mathrm{cone}(C - x) := \{tz \,:\, t \geq 0, \, z \in C - x\}$.*

(c) *The inclusion $v \in T_C(x)$ is valid if and only if, for every sequence x_k in C converging to x and sequence t_k in $(0, +\infty)$ converging to 0, there exists a sequence v_k in R^n converging to v such that $x_k + t_k v_k \in C$ for all k.*

We now consider problem (P) under the assumptions that $f : R^n \to R$ is a locally Lipschitz function and

$$\Delta = \{x \in C : g_1(x) \le 0, \ldots, g_m(x) \le 0, \ h_1(x) = 0, \ldots, h_s(x) = 0\}, \tag{1.25}$$

where $C \subset R^n$ is a nonempty subset, $g_i : R^n \to R$ $(i = 1, \ldots, m)$ and $h_j : R^n \to R$ $(j = 1, \ldots, s)$ are locally Lipschitz functions.

Theorem 1.9. (See Clarke (1983), Theorem 6.1.1 and Remark 6.1.2) *If $\bar{x}$ is a local solution of (P) then there exist $m + s + 1$ real numbers $\lambda_0 \ge 0$, $\lambda_1 \ge 0, \ldots, \lambda_m \ge 0$, $\mu_1, \ldots, \mu_s$, not all zero, such that*

$$0 \in \lambda_0 \partial f(\bar{x}) + \sum_{i=1}^{m} \lambda_i \partial g_i(\bar{x}) + \sum_{j=1}^{s} \mu_j \partial h_j(\bar{x}) + N_C(\bar{x}) \tag{1.26}$$

and

$$\lambda_i g_i(\bar{x}) = 0 \quad \text{for all } i = 1, 2, \ldots, m. \tag{1.27}$$

The preceding theorem expresses the first-order necessary optimality condition for a class of nonsmooth programs in the *Fritz-John form*. Under some suitable *constraint qualifications*, the multiplier λ_0 corresponding to the objective function f is positive. In that case, dividing both sides of the inclusion in (1.26) and the equalities in (1.27) by λ_0 and setting $\widetilde{\lambda}_i = \lambda_i / \lambda_0$ for $i = 1, \ldots, m$, $\widetilde{\mu}_j = \mu_j / \lambda_0$ for $j = 1, \ldots, s$, we obtain

$$0 \in \partial f(\bar{x}) + \sum_{i=1}^{m} \widetilde{\lambda}_i \partial g_i(\bar{x}) + \sum_{j=1}^{s} \widetilde{\mu}_j \partial h_j(\bar{x}) + N_C(\bar{x}) \tag{1.28}$$

and

$$\widetilde{\lambda}_i g_i(\bar{x}) = 0 \quad \text{for all } i = 1, 2, \ldots, m. \tag{1.29}$$

Similarly as in the case of convex programs (see Theorem 1.6), if (1.28) and (1.29) are fulfilled then the numbers $\widetilde{\lambda}_1 \ge 0, \ldots, \widetilde{\lambda}_m \ge 0$, $\widetilde{\mu}_1 \in R, \ldots, \widetilde{\mu}_s \in R$ are called the *Lagrange multipliers* corresponding to $\bar{x}$.

It is a simple matter to obtain the following two *Lagrange multiplier rules* from Theorem 1.9. (See Clarke (1983), pp. 234–236).

Corollary 1.1. *If $\bar{x}$ is a local solution of (P) and if the constraint qualification*

$$
\begin{cases}
\left[0 \in \sum_{i=1}^{m} \lambda_i \partial g_i(\bar{x}) + \sum_{j=1}^{s} \mu_j \partial h_j(\bar{x}) + N_C(\bar{x}), \right. \\
\lambda_1 \geq 0, \ldots, \lambda_m \geq 0, \ \mu_1 \in R, \ldots, \mu_s \in R; \\
\left. \lambda_i g_i(\bar{x}) = 0 \ \text{for } i = 1, \ldots, m \right] \\
\implies \left[\lambda_1 = \ldots = \lambda_m = 0, \ \mu_1 = \ldots = \mu_s = 0 \right]
\end{cases}
$$

holds, then there exist Lagrange multipliers $\lambda_1 \geq 0, \ldots, \lambda_m \geq 0$, $\mu_1 \in R, \ldots, \mu_s \in R$ such that $\lambda_i g_i(\bar{x}) = 0$ for $i = 1, 2, \ldots, m$, and

$$
0 \in \partial f(\bar{x}) + \sum_{i=1}^{m} \lambda_i \partial g_i(\bar{x}) + \sum_{j=1}^{s} \mu_j \partial h_j(\bar{x}) + N_C(\bar{x}).
$$

Corollary 1.2. *Assume that $\bar{x}$ is a local solution of a smooth program (P) where Δ is given by formula (1.23). If the following Mangasarian-Fromovitz constraint qualification*

$$
\begin{cases}
\text{The vectors } \{\nabla h_j(\bar{x}) : j = 1, \ldots, s\} \text{ are linearly independent,} \\
\text{and there exists } v \in R^n \text{ such that } \langle \nabla h_j(\bar{x}), v \rangle = 0 \\
\text{for } j = 1, \ldots, s, \text{ and } \langle \nabla g_i(\bar{x}), v \rangle < 0 \\
\text{for every } i = 1, \ldots, m \text{ satisfying } g_i(\bar{x}) = 0
\end{cases}
$$

is satisfied, then there exist Lagrange multipliers $\lambda_1 \geq 0, \ldots, \lambda_m \geq 0$, $\mu_1 \in R, \ldots, \mu_s \in R$ such that $\lambda_i g_i(\bar{x}) = 0$ for $i = 1, 2, \ldots, m$, and

$$
0 = \nabla f(\bar{x}) + \sum_{i=1}^{m} \lambda_i \nabla g_i(\bar{x}) + \sum_{j=1}^{s} \mu_j \nabla h_j(\bar{x}).
$$

From Theorem 1.9 we can derive the basic Lagrange multiplier rule for convex programs stated in Theorem 1.6. Indeed, suppose that the assumptions of Theorem 1.6 are satisfied and $\bar{x}$ is a solution of (P). Consider separately the following two cases: (i) The vectors $\{a_j : j = 1, \ldots, s\}$ are linearly independent; (ii) The vectors $\{a_j : j = 1, \ldots, s\}$ are linearly dependent. In the first case, Theorem 1.9 shows that there exist real numbers $\lambda_0 \geq 0$, $\lambda_1 \geq 0, \ldots, \lambda_m \geq 0$, $\mu_1, \ldots, \mu_s$, not all zero, such that (1.26) and (1.27) are satisfied. Condition (1.24) forces $\lambda_0 > 0$. Hence there exist Lagrange multipliers satisfying the Kuhn-Tucker conditions. In the second case, when $a_j = 0$ for $j = 1, \ldots, s$, we can obtain the desired result; when $a_j \neq 0$ for some $j = 1, \ldots, s$, we choose a maximal linearly independent subsystem, say $\{a_1, \ldots, a_k\}$, of the vector system $\{a_1, \ldots, a_s\}$. Then we consider the problem

$$\min\{f(x) \,:\, g_1(x) \leq 0, \ldots, g_m(x) \leq 0, h_1(x) = 0, \ldots, h_k(x) = 0\}. \tag{1.30}$$

It is easy to show that the constraint set of this problem coincides with Δ. Hence $\bar{x}$ is a solution of (1.30). Applying Theorem 1.9 to problem (1.30) and using condition (1.24) we can find a set of Lagrange multipliers satisfying the Kuhn-Tucker conditions.

1.4 Linear Programs and Nonlinear Programs

Definition 1.15. A subset $\Delta \subset R^n$ is called a *polyhedral convex set* if Δ can be represented as the intersection of finitely many closed half spaces of R^n; that is, there exist nonzero vectors $a_1, \ldots, a_m \in R^n$ and real numbers $\beta_1, \ldots, \beta_m$ such that

$$\Delta = \{x \in R^n : \langle a_i, x \rangle \geq \beta_i \text{ for } i = 1, \ldots, m\}. \tag{1.31}$$

In other words, Δ is the solution set of a system of finitely many linear inequalities. (We admit that the intersection any *empty family* of closed half spaces of R^n is R^n. Hence $\Delta = R^n$ is also a polyhedral convex set.) A point $x \in \Delta$ is called an *extreme point* of Δ if there is no way to express x in the form $x = ty + (1-t)z$ where $y \in \Delta$, $z \in \Delta$, $y \neq z$, and $t \in (0,1)$. The set of all the extreme points of Δ is denoted by $\text{extr}\Delta$.

Let A be the $m \times n$-matrix with the elements a_{ij} ($i = 1, \ldots, m$, $j = 1, \ldots, n$), where a_{ij} stands for the j-th component of a_i. Set $b = (\beta_1, \ldots, \beta_m) \in R^m$. Then (1.31) can be rewritten as

$$\Delta = \{x \in R^n : Ax \geq b\}.$$

Here and subsequently, for any two vectors $y = (y_1, \ldots, y_m) \in R^m$ and $z = (z_1, \ldots, z_m) \in R^m$, we write $y \geq z$ if $y_i \geq z_i$ for all $i = 1, \ldots, m$. We shall write $y > z$ if $y_i > z_i$ for all $i = 1, \ldots, m$. Since

$$\{x \in R^n : Ax = b\} = \{x \in R^n : Ax \geq b, \ (-A)x \geq -b\},$$

it follows that $\{x \in R^n : Ax = b\}$ is a polyhedral convex set.

Definition 1.16. Problem (P) is called a *linear program* (a linear programming problem) if f is an affine function and Δ is a polyhedral convex set. Otherwise, (P) is said to be a *nonlinear program*.

There are three typical forms for describing linear programs

$$\min\{f(x) = \langle c, x \rangle \ : \ x \in R^n, \ Ax \geq b\},$$
$$\min\{f(x) = \langle c, x \rangle \ : \ x \in R^n, \ Ax = b, \quad x \geq 0\},$$
$$\min\{f(x) = \langle c, x \rangle \ : \ x \in R^n, \ Ax \geq b, \ Cx = d\}$$

which are called the *standard form*, the *canonical form* and the *general form*, respectively. Here $A \in R^{m \times n}$, $C \in R^{s \times n}$ are given matrices, $c \in R^n$, $b \in R^m$ and $d \in R^s$ are given vectors.

Example 1.4. Consider the following linear program of the standard form:

$$\min \left\{ x_1 + \frac{1}{2} x_2 : x = (x_1, x_2), \ x_1 + x_2 \geq 1, \ x_1 \geq 0, \ x_2 \geq 0 \right\}.$$

It is easy to check that $\text{Sol}(P) = \{(0, 1)\}$. Note that the constraint set

$$\Delta = \{x \in R^2 : x_1 + x_2 \geq 1, \ x_1 \geq 0, \ x_2 \geq 0\}$$

has two extreme points, namely $\text{extr}\Delta = \{(1, 0), \ (0, 1)\}$. One of these points is the solution of our problem.

Definition 1.17. The *dual problems* of linear programs of the standard, canonical and general forms, respectively, are the following linear programs:

$$\max\{\langle b, y \rangle : y \in R^m, \ A^T y = c, \ y \geq 0\},$$
$$\max\{\langle b, y \rangle : y \in R^m, \ A^T y \leq c\},$$
$$\max\{\langle b, y \rangle + \langle d, z \rangle : (y, z) \in R^m \times R^s, \ A^T y + C^T z = c, \ y \geq 0\}.$$

Of course, linear programs are convex mathematical programming problems. Hence they enjoy all the properties of the class of convex programs. Besides, linear programs have many other special properties.

Theorem 1.10. (See Dantzig (1963)) *Let (P) be a linear program in one of the three typical forms. The following properties hold true:*

(i) *If the constraint set is nonempty and if $v(P) > -\infty$, then $\text{Sol}(P)$ is a nonempty polyhedral convex set.*

(ii) *If both the sets* extrΔ *and* Sol(P) *are nonempty, then the intersection* extr$\Delta \cap$ Sol(P) *is also nonempty.*

(iii) *If* rank$A = n$ *and the set* $\Delta := \{x \in R^n : Ax = b, \ x \geq 0\}$ *is nonempty, then* Δ *must have an extreme point.*

(iv) *The optimal value* $v(P)$ *of* (P) *and the optimal value* $v(P')$ *of the dual problem* (P') *of* (P) *are equal, provided that the constraint set of at least one of these problems is nonempty.*

Note that the five problems considered in Remarks 1.2, 1.4 and Examples 1.1-1.3 are all nonlinear.

We now consider one important class of nonlinear programs, which contains the class of linear programs as a special subclass.

1.5 Quadratic Programs

Definition 1.18. We say that $f : R^n \to R$ is a *linear-quadratic function* if there exist a matrix $D \in R^{n \times n}$, a vector $c \in R^n$ and a real number α such that

$$
\begin{aligned}
f(x) &= \frac{1}{2} x^T D x + c^T x + \alpha \\
&= \frac{1}{2} \langle x, Dx \rangle + \langle c, x \rangle + \alpha
\end{aligned}
\tag{1.32}
$$

for all $x \in R^n$.

If

$$
D = \begin{pmatrix} d_{11} & \cdots & d_{1n} \\ \cdots & \cdots & \cdots \\ d_{n1} & \cdots & d_{nn} \end{pmatrix}, \quad c = \begin{pmatrix} c_1 \\ \vdots \\ c_n \end{pmatrix}, \quad x = \begin{pmatrix} x_1 \\ \vdots \\ x_n \end{pmatrix},
$$

then (1.32) means that

$$
f(x) = \frac{1}{2} \left(\sum_{j=1}^{n} \sum_{i=1}^{n} d_{ij} x_i x_j \right) + \sum_{i=1}^{n} c_i x_i + \alpha.
$$

Since $x^T D x = \frac{1}{2} x^T (D + D^T) x$ for every $x \in R^n$, representation (1.32) remains valid if we replace D by the symmetric matrix $\frac{1}{2}(D + D^T)$. For this reason, we will assume that the square matrix in the representation of a linear-quadratic function is symmetric. The space of the symmetric $n \times n$-matrices will be denoted by $R_S^{n \times n}$.

Definition 1.19. Problem (P) is called a *linear-quadratic mathematical programming problem* (or a *quadratic program*, for brevity) if f is a linear-quadratic function and Δ is a polyhedral convex set.

In (1.32), if D is the zero matrix then f is an affine function. Thus the class of linear programs is a subclass of the class of quadratic programs. In general, quadratic programs are *nonconvex* mathematical programming problems.

Example 1.5. The following quadratic program is nonconvex:

$$\min\{f(x) = x_1^2 - x_2^2 : x = (x_1, x_2), \ 1 \leq x_1 \leq 3, \ 1 \leq x_2 \leq 3\}.$$

It is obvious that f is a nonconvex function. One can verify that $\text{Sol}(P) = \{(1,3)\}$ and $v(P) = -8$.

It is clear that if we delete the constant α in the representation (1.32) of f then we do not change the solution set of the problem $\min\{f(x) : x \in \Delta\}$, where $\Delta \subset R^n$ is a polyhedral convex set. Therefore, instead of (1.32) we will usually use the simplified form $f(x) = \frac{1}{2}x^T D x + c^T x$ of the objective function.

Modifying the terminology used for linear programs, we call the following forms of quadratic programs

$$\min\left\{\frac{1}{2}x^T D x + c^T x : x \in R^n, \ A x \geq b\right\},$$

$$\min\left\{\frac{1}{2}x^T D x + c^T x : x \in R^n, \ A x \geq b, \ x \geq 0\right\},$$

$$\min\left\{\frac{1}{2}x^T D x + c^T x : x \in R^n, \ A x \geq b, \ C x = d\right\}$$

the *standard form*, the *canonical form* and the *general form*, respectively. (The meaning of A, C, b and d is the same as in the description of the typical forms of linear programs.) Note that the representation of the constraint set of canonical quadratic programs is slightly different from that of canonical linear programs. The above definition of canonical quadratic programs is adopted because quadratic programs of this type have a very tight connection with *linear complementarity problems* (see, for instance, Murty (1976) and Cottle et al. (1992)). In Chapter 5 we will clarify this point. The relation between the general quadratic programs and *affine variational inequalities* will be studied in the same chapter.

Definition 1.20. A matrix $D \in R^{n \times n}$ is said to be *positive definite* (resp., *negative definite*) if $v^T D v > 0$ (resp., $v^T D v < 0$) for every $v \in R^n \setminus \{0\}$. If $v^T D v \geq 0$ (resp., $v^T D v \leq 0$) for every $v \in R^n$ then D is said to be *positive semidefinite* (resp., *negative semidefinite*).

Proposition 1.2. Let $f(x) = \frac{1}{2} x^T D x + c^T x + \alpha$ where $D \in R_S^{n \times n}$, $c \in R^n$ and $\alpha \in R$. If D is a positive semidefinite matrix, then f is a convex function.

Proof. Since $x \mapsto c^T x + \alpha$ is a convex function and the sum of two convex functions is a convex function, it suffices to show that $f_1(x) := x^T D x$ is a convex function. As D is a positive semidefinite matrix, for every $u \in R^n$ and $v \in R^n$ we have

$$0 \leq (u - v)^T D (u - v) = u^T D u - 2 v^T D u + v^T D v.$$

This implies that

$$v^T D v \leq u^T D u - 2 v^T D (u - v). \tag{1.33}$$

Given any $x \in R^n$, $y \in R^n$ and $t \in (0, 1)$, we set $z = tx + (1 - t)y$. Taking account of (1.33) we have

$$z^T D z \leq y^T D y - 2 z^T D (y - z),$$
$$z^T D z \leq x^T D x - 2 z^T D (x - z).$$

Since $y - z = t(y - x)$ and $x - z = (1 - t)(x - y)$, from the last two inequalities we deduce that

$$(1 - t) z^T D z + t z^T D z \leq (1 - t) y^T D y + t x^T D x,$$

hence

$$f_1(tx + (1 - t)y) = f_1(z) \leq t f_1(x) + (1 - t) f(y).$$

Thus f_1 is a convex function. $\square$

If D is negative semidefinite, then the function f given by (1.32) is *concave*, i.e.

$$f(tx + (1 - t)y) \geq t f(x) + (1 - t) f(y)$$

for every $x \in R^n$, $y \in R^n$ and $t \in (0, 1)$. In the case where matrix D is neither assumed to be positive semidefinite nor assumed to be negative semidefinite, we say that $f(x) = \frac{1}{2}x^T D x + c^T x$, where $c \in R^n$, is an *indefinite* linear-quadratic function. Quadratic programming problems with indefinite linear-quadratic objective functions are called *indefinite* quadratic programs.

Remark 1.6. It is clear that if f is given by (1.32), where $D \in R_S^{n \times n}$, then $\nabla^2 f(x) = D$ for every $x \in R^n$. Therefore, the fact stated in Proposition 1.2 is a direct consequence of the following theorem (see Rockafellar (1970), Theorem 4.5): *"If $f : R^n \to R$ is a C^2-function and if the Hessian matrix $\nabla^2 f(x)$ is positive semidefinite for every $x \in R^n$, then f is a convex function."*

By using Proposition 1.2 one can verify whether a given quadratic program is convex or not.

Let us consider a simple example of convex quadratic programs.

Example 1.6. Given k points $a_1, a_2, \ldots, a_k$ in R^n, we want to find a point $x \in R^n$ at which the sum

$$f(x) := \|x - a_1\|^2 + \cdots + \|x - a_k\|^2$$

attains its minimal value. Observe that

$$\begin{aligned} f(x) &= \sum_{i=1}^{k}(x - a_i)^T(x - a_i) \\ &= k x^T x - 2\left(\sum_{i=1}^{k} a_i\right)^T x + \sum_{i=1}^{k} a_i^T a_i \end{aligned}$$

is a convex linear-quadratic function. By Theorem 1.5, $\bar{x}$ is a solution of our problem if and only if $\nabla f(\bar{x}) = 0$. Since

$$\nabla f(\bar{x}) = 2k\bar{x} - 2\sum_{i=1}^{k} a_i,$$

one can write the condition $0 = \nabla f(\bar{x})$ equivalently as

$$\bar{x} = \frac{1}{k}\sum_{i=1}^{k} a_i.$$

Thus $\bar{x} = \dfrac{1}{k}\displaystyle\sum_{i=1}^{k} a_i$ is the unique solution of our problem. That special point $\bar{x}$ is called the *barycenter* of the system $\{a_1, a_2, \ldots, a_k\}$.

Observe that there is a simple algorithm for constructing the barycenter. Namely, first we define $z_1 = \frac{1}{2}a_1 + \frac{1}{2}a_2$. Then we put

$$z_i = \frac{i}{i+1}z_{i-1} + \frac{1}{i+1}a_{i+1} \quad \text{for every } i \geq 2.$$

By induction it is not difficult to show that $\bar{x} := z_{k-1}$ is the barycenter of the system $\{a_1, a_2, \ldots, a_k\}$. For performing a sequential construction of the barycenter of a system of points in R^2, it is convenient to use the following equivalent vector form of the formula defining z_i:

$$\overrightarrow{z_{i-1}z_i} = \frac{1}{i+1}\overrightarrow{z_{i-1}a_i} \quad \text{(for every } i \geq 2\text{)}.$$

The following geometrical example leads to a (nonconvex) quadratic program of the general form.

Example 1.7. Let $\Delta = \{x \in R^n \ : \ Ax \geq b, \ Cx = d\}$, where $A \in R^{m \times n}$, $C \in R^{s \times n}$, $b \in R^m$ and $d \in R^s$. (The equality $Cx = d$ can be absent in that formula. Likewise, the inequality $Ax \geq b$ can be absent too.) Let α_i $(i = 1, \ldots, n)$, $\tilde{\alpha}_i$ $(i = 1, \ldots, n)$, β and $\tilde{\beta}$ be a family of $2n + 2$ real numbers satisfying the conditions

$$\sum_{i=1}^{n} \alpha_i^2 = 1, \quad \sum_{i=1}^{n} \tilde{\alpha}_i^2 = 1. \tag{1.34}$$

Note that

$$M = \{x \in R^n \ : \ \sum_{i=1}^{n} \alpha_i x_i + \beta = 0\}$$

and

$$\widetilde{M} = \{x \in R^n \ : \ \sum_{i=1}^{n} \tilde{\alpha}_i x_i + \tilde{\beta} = 0\}$$

are two *hyperplanes* in R^n. The task is to find $x \in \Delta$ such that the function

$$f(x) = (\text{dist}(x, M))^2 - (\text{dist}(x, \widetilde{M}))^2,$$

where $\text{dist}(x, \Omega) = \inf\{\|x - z\| \ : \ z \in \Omega\}$ is the *distance* from x to a subset $\Omega \subset R^n$, achieves its minimum. We have

$$\text{dist}(x, M) = |\sum_{i=1}^{n} \alpha_i x_i + \beta|. \tag{1.35}$$

In order to prove this formula we consider the following convex program

$$\min\{\varphi(z) = \|x - z\|^2 \ : \ z \in M\}. \tag{1.36}$$

By Theorem 1.6, $\bar{z} = (\bar{z}_1, \ldots, \bar{z}_n) \in M$ is a solution of (1.36) if and only if there exists $\mu \in R$ such that

$$0 \in \partial\varphi(\bar{z}) + \mu(\alpha_1, \ldots, \alpha_n).$$

Since $\partial\varphi(\bar{z}) = \{\nabla\varphi(\bar{z})\} = \{-2(x - \bar{z})\}$, this inclusion is valid if and only if

$$2(x - \bar{z}) = \mu(\alpha_1, \ldots, \alpha_n).$$

This implies $\bar{z} = x - \dfrac{\mu}{2}\alpha$, where $\alpha := (\alpha_1, \ldots \alpha_n)$. As $\bar{z} \in M$, we must have

$$0 = \langle \alpha, \bar{z} \rangle + \beta = \langle \alpha, x \rangle - \frac{\mu}{2}\langle \alpha, \alpha \rangle + \beta.$$

Taking account of (1.34), we obtain $\mu = 2(\langle \alpha, x \rangle + \beta)$. Therefore

$$\begin{aligned}
(\operatorname{dist}(x, M))^2 = \|x - \bar{z}\|^2 &= \|x - (x - \tfrac{\mu}{2}\alpha)\|^2 \\
&= \left(\frac{\mu}{2}\right)^2 \langle \alpha, \alpha \rangle \\
&= (\langle \alpha, x \rangle + \beta)^2,
\end{aligned}$$

hence (1.35) holds. Similarly,

$$\operatorname{dist}(x, \widetilde{M}) = |\sum_{i=1}^{n} \widetilde{\alpha}_i x_i + \widetilde{\beta}|.$$

Consequently,

$$\begin{aligned}
f(x) &= \left(\sum_{i=1}^{n} \alpha_i x_i + \beta\right)^2 - \left(\sum_{i=1}^{n} \widetilde{\alpha}_i x_i + \widetilde{\beta}\right)^2 \\
&= \sum_{j=1}^{n}\sum_{i=1}^{n}(\alpha_i\alpha_j - \widetilde{\alpha}_i\widetilde{\alpha}_j)x_i x_j \\
&\quad + 2\sum_{i=1}^{n}(\beta\alpha_i - \widetilde{\beta}\widetilde{\alpha}_i)x_i + (\beta^2 - \widetilde{\beta}^2).
\end{aligned}$$

From this we conclude that $f(x)$ is a linear-quadratic function; so the optimization problem under consideration is a quadratic program of the general form.

It is easy to verify that if we choose

$$A = \begin{pmatrix} 1 & 0 \\ 0 & 1 \\ -1 & 0 \\ 0 & -1 \end{pmatrix}, \quad b = \begin{pmatrix} 1 \\ 1 \\ -3 \\ -3 \end{pmatrix},$$

$$\alpha = (1,0), \quad \beta = 0, \quad \widetilde{\alpha} = (0,1), \quad \widetilde{\beta} = 0,$$

then the preceding problem, where the equation $Cx = d$ is absent, becomes the one discussed in Example 1.5. In this case, we have $M = \{x = (x_1, x_2) : x_1 = 0, x_2 \in R\}$, $\widetilde{M} = \{x = (x_1, x_2) : x_1 \in R, x_2 = 0\}$, dist$(x, M) = |x_1|$ and dist$(x, \widetilde{M}) = |x_2|$.

1.6 Commentaries

Mathematical Programming is one important branch of Optimization Theory. Other branches with many interesting problems and results are known under the names The Calculus of Variations and Optimal Control Theory. Of course, the informal division of Optimization Theory into such branches is only for convenience. In fact, research problems and methods of the three branches are actively interacted. The classical work of Ioffe and Tihomirov (1979) is an excellent textbook addressing all the three branches of Optimization Theory.

Convex Analysis and Convex Programming theory can be studied by using the books of Rockafellar (1970) and of Ioffe and Tihomirov (1979). The book of Mangasarian (1969) gives a nice introductory course to Mathematical Programming.

Linear Programming can be learned by using the books of Dantzig (1963), Murty (1976), and many other nice books.

Nonsmooth Analysis and Nonsmooth Optimization can be learned by the books of Clarke (1983), Rockafellar and Wets (1998), and many other excellent books. In addition to these books, one can study Mordukhovich (1988, 1993, 1994) to be familiar with a powerful approach to Nonsmooth Analysis and Nonsmooth Optimization which has been developed intensively in recent years.

Some theoretical results on (Nonconvex) Quadratic Programming are available in the books of Murty (1976), Bank et al. (1982), Cottle et al. (1992), and other books. The next three chapters of this book are intended to cover the basic facts on (Nonconvex) Quadratic Programming, such as the solution existence, necessary and sufficient optimality conditions, and structure of the solution sets. Eight other chapters (Chapters 10-17) establish various results on stability and sensitivity of parametric quadratic programs.

The geometric problem described in Example 1.3 is called Fermat's problem or Steiner's problem. It was proposed by Fermat to

Torricelli. Torricelli's solution was published in 1659 by his pupil
Viviani (see Weisstein (1999), p. 623).

Chapter 2

Existence Theorems for Quadratic Programs

In this chapter we shall discuss the Frank-Wolfe Theorem and the Eaves Theorem, which are two fundamental existence theorems for quadratic programming problems.

2.1 The Frank-Wolfe Theorem

Consider a quadratic program of the standard form

$$\begin{cases} \text{Minimize } f(x) := \dfrac{1}{2}x^T Dx + c^T x \\ \text{subject to } x \in R^n, \ Ax \geq b, \end{cases} \tag{2.1}$$

where $D \in R_S^{n \times n}$, $A \in R^{m \times n}$, $c \in R^n$ and $b \in R^m$. For the constraint set and the optimal value of (2.1) we shall use the following abbreviations:

$$\Delta(A, b) = \{x \in R^n : Ax \geq b\},$$
$$\bar{\theta} = \inf\{f(x) : x \in \Delta(A, b)\}.$$

If $\Delta(A, b) = \emptyset$ then $\bar{\theta} = +\infty$ by convention. If $\Delta(A, b) \neq \emptyset$ then there are two situations: (i) $\bar{\theta} \in R$, (ii) $\bar{\theta} = -\infty$. If (ii) occurs then, surely, (2.1) has no solutions. It is natural to ask: *Whether the problem always has solutions when (i) occurs?*

Note that optimization problems with non-quadratic objective functions may have no solutions even in the case the optimal value is finite. For example, the problem $\min \left\{ \dfrac{1}{x} : x \in R, \ x \geq 1 \right\}$ has no

solutions, while the optimal value $\bar{\theta} = \inf\left\{\dfrac{1}{x} : x \in R, \ x \geq 1\right\} = 0$
is finite.

The following result was published by Frank and Wolfe in 1956.

Theorem 2.1. (The Frank-Wolfe Theorem; See Frank and Wolfe (1956), p. 108) *If $\bar{\theta} = \inf\{f(x) : x \in \Delta(A,b)\}$ is a finite real number then problem* (2.1) *has a solution.*

Proof. We shall follow the analytical proof proposed by Blum and Oettli (1972). The assumption $\bar{\theta} \in R$ implies that $\Delta(A,b) \neq \emptyset$. Select a point $x^0 \in \Delta(A,b)$. Let $\rho > 0$ be given arbitrarily. Define

$$\Delta_\rho = \Delta(A,b) \cap \bar{B}(x^0, \rho).$$

Note that Δ_ρ is a convex, nonempty, compact set. Consider the following problem

$$\min\{f(x) : x \in \Delta_\rho\}. \tag{2.2}$$

By the Weierstrass Theorem, there exists some $y \in \Delta_\rho$ such that $f(y) = q_\rho := \min\{f(x) : x \in \Delta_\rho\}$. Since the solution set of (2.2) is nonempty and compact, there exists $y_\rho \in \Delta_\rho$ such that

$$\|y_\rho - x^0\| = \min\{\|y - x^0\| : y \in \Delta_\rho, \ f(y) = q_\rho\}.$$

We claim that there exists $\hat{\rho} > 0$ such that

$$\|y_\rho - x^0\| < \rho \quad \text{for all } \rho \geq \hat{\rho}. \tag{2.3}$$

Indeed, if the claim were false then we would find an increasing sequence $\rho_k \to +\infty$ such that for every k there exists $y_{\rho_k} \in \Delta_{\rho_k}$ such that

$$f(y_{\rho_k}) = q_{\rho_k}, \quad \|y_{\rho_k} - x^0\| = \rho_k. \tag{2.4}$$

For simplicity of notation, we write y^k instead of y_{ρ_k}. Since $y^k \in \Delta(A,b)$, we must have $A_i y^k \geq b_i$ for $i = 1, \ldots, m$, where A_i denotes the i-th row of A and b_i denotes the i-th component of b. For $i = 1$, since the sequence $\{A_1 y^k\}$ is bounded below, one can choose a subsequence $\{k'\} \subset \{k\}$ such that $\lim\limits_{k' \to \infty} A_1 y^{k'}$ exists. (It may happen that $\lim\limits_{k' \to \infty} A_1 y^{k'} = +\infty$.) Without restriction of generality we can assume that $\{k'\} \equiv \{k\}$, that is the sequence $\{A_1 y^k\}$ itself is convergent. Similarly, for $i = 2$ there exists a subsequence $\{k'\} \subset \{k\}$ such that $\lim\limits_{k' \to \infty} A_2 y^{k'}$ exists. Without loss of generality we can assume that $\{k'\} \equiv \{k\}$. Continue the process until $i = m$ to find a subsequence $\{k'\} \subset \{k\}$ such that all the limits

$$\lim_{k' \to \infty} A_i y^{k'} \quad (i = 1, \ldots, m)$$

exist. For simplicity of notation, we will assume that $\{k'\} \equiv \{k\}$. Let $I = \{1, \ldots, m\}$, $I_0 = \{i \in I : \lim_{k \to \infty} A_i y^k = b_i\}$ and

$$I_1 = I \setminus I_0 = \{i \in I : \lim_{k \to \infty} A_i y^k > b_i\}.$$

Of course, there exists $\varepsilon > 0$ such that

$$\lim_{k \to \infty} A_i y^k \geq b_i + \varepsilon \quad \text{for every } i \in I_1.$$

By (2.4), $\|(y^k - x^0)/\rho_k\| = 1$ for every k. Since the unit sphere in R^n is a compact set, there is no loss of generality in assuming that the sequence

$$\left\{ \frac{y^k - x^0}{\rho_k} \right\}$$

converges to some $\bar{v} \in R^n$ as $k \to \infty$. Clearly, $\|\bar{v}\| = 1$. As $\rho_k \to +\infty$, for every $i \in I_0$ we have

$$
\begin{aligned}
0 &= \lim_{k \to \infty} (A_i y^k - b_i) \\
&= \lim_{k \to \infty} \left(\frac{A_i y^k - b_i}{\rho_k} \right) \\
&= \lim_{k \to \infty} \left(A_i \frac{y^k - x^0}{\rho_k} \right) + \lim_{k \to \infty} \left(\frac{A_i x^0 - b_i}{\rho_k} \right) = A_i \bar{v}.
\end{aligned}
$$

Similarly, for every $i \in I_1$ we have

$$
\begin{aligned}
0 &\leq \liminf_{k \to \infty} \left(\frac{A_i y^k - b_i}{\rho_k} \right) \\
&= \liminf_{k \to \infty} \left(\frac{A_i y^k - A_i x^0}{\rho_k} + \frac{A_i x^0 - b_i}{\rho_k} \right) \\
&= \lim_{k \to \infty} \left(A_i \frac{y^k - x^0}{\rho_k} \right) + \lim_{k \to \infty} \left(\frac{A_i x^0 - b_i}{\rho_k} \right) = A_i \bar{v}.
\end{aligned}
$$

Therefore

$$A_i \bar{v} = 0 \quad \text{for every } i \in I_0, \quad A_i \bar{v} \geq 0 \quad \text{for every } i \in I_1. \quad (2.5)$$

From this we can conclude that $\bar{v}$ is a direction of recession of the polyhedral convex set $\Delta(A, b)$. Recall (Rockafellar (1970), p. 61) that a nonzero vector $v \in R^n$ is said to be a *direction of recession* of a nonempty convex set $\Omega \subset R^n$ if

$$x + tv \in \Omega \quad \text{for every } t \geq 0 \text{ and } x \in \Omega.$$

Recall also that the set composed by $0 \in R^n$ and all the directions $v \in R^n$ satisfying the last condition, is called the *recession cone* of Ω. In our case, from (2.5) we deduce immediately that

$$y + t\bar{v} \in \Delta(A, b) \quad \text{for every } t \geq 0 \text{ and } y \in \Delta(A, b). \qquad (2.6)$$

Since

$$
\begin{aligned}
f(y^k) = f(y_{\rho_k}) &= q_{\rho_k} \\
&= \min\{f(x) : x \in \Delta_{\rho_k}\} \\
&= \min\{f(x) : x \in \Delta(A, b) \cap \bar{B}(x_0, \rho_k)\}
\end{aligned}
$$

and the increasing sequence $\{\rho_k\}$ converges to $+\infty$, we see that the sequence $\{f(y^k)\}$ is non-increasing and $f(y^k) \to \bar{\theta}$. Consequently, for k sufficiently large, we have

$$\bar{\theta} - 1 \leq f(y^k) \leq \bar{\theta} + 1.$$

Using the formula of f we can rewrite these inequalities as follows

$$
\begin{aligned}
\bar{\theta} - 1 \; \leq \; &\frac{1}{2}(y^k - x^0)^T D(y^k - x^0) + c^T(y^k - x^0) \\
&+ (x^0)^T D(y^k - x^0) + \frac{1}{2}(x^0)^T D x^0 + c^T x^0 \leq \bar{\theta} + 1.
\end{aligned}
$$

Dividing these expressions by ρ_k^2 and taking the limits as $k \to \infty$, we get $0 \leq \frac{1}{2}\bar{v}^T D\bar{v} \leq 0$. Hence

$$\bar{v}^T D \bar{v} = 0. \qquad (2.7)$$

By (2.6),

$$y^k + t\bar{v} \in \Delta(A, b) \quad \text{for every } t \geq 0 \text{ and } k \in N,$$

where N stands for the set of the positive integers. On account of (2.7), we have

$$
\begin{aligned}
f(y^k + t\bar{v}) &= \frac{1}{2}(y^k + t\bar{v})^T D(y^k + t\bar{v}) + c^T(y^k + t\bar{v}) \\
&= \frac{1}{2}(y^k)^T D y^k + c^T y^k + t((y^k)^T D\bar{v} + c^T \bar{v}).
\end{aligned}
$$

Note that
$$(y^k)^T D\bar{v} + c^T \bar{v} \geq 0 \quad \text{for every } k \in N. \tag{2.8}$$

Indeed, if (2.8) were false then we would have $f(y^k + t\bar{v}) \to -\infty$ as $t \to +\infty$, which contradicts the assumption $\bar{\theta} \in R$.

Since $\langle \bar{v}, \bar{v} \rangle = 1$ and $\dfrac{y^k - x^0}{\rho_k} \to \bar{v}$, there exists $k_1 \in N$ such that $\left\langle \dfrac{y^k - x^0}{\rho_k}, \bar{v} \right\rangle > 0$ for all $k \geq k_1$. For any fixed index $k \geq k_1$, we have $\langle y^k - x^0, \bar{v} \rangle > 0$. Therefore

$$\|y^k - x^0 - t\bar{v}\|^2 = \|y^k - x^0\|^2 - 2t\langle y^k - x^0, \bar{v}\rangle + t^2\|\bar{v}\|^2 < \|y^k - x^0\|^2 \tag{2.9}$$

for $t > 0$ small enough. By (2.5),

$$A_i(y^k - t\bar{v}) = A_i y^k \geq b_i \quad \text{for all } i \in I_0.$$

Since $\lim\limits_{k\to\infty} A_i y^k \geq b_i + \varepsilon$ for every $i \in I_1$, there exists $k_2 \in N$, $k_2 \geq k_1$, such that $A_i y^k \geq b_i + \dfrac{\varepsilon}{2}$ for every $k \geq k_2$ and $i \in I_1$. Fix an index $k \geq k_2$ and choose $\delta_k > 0$ as small as $tA_i\bar{v} \leq \dfrac{\varepsilon}{2}$ for every $i \in I_1$ and $t \in (0, \delta_k)$. (Of course, this choice is made only in the case $I_1 \neq \emptyset$.) Then we have

$$A_i(y^k - t\bar{v}) \geq b_i + \frac{\varepsilon}{2} - tA_i\bar{v} \geq b_i$$

for all $i \in I_1$ and $t \in (0, \delta_k)$. From what has already been proved, it may be concluded that

$$y^k - t\bar{v} \in \Delta(A, b) \quad \text{for all } t \in (0, \delta_k).$$

Combining this with (2.9) we see that $y^k - t\bar{v} \in \Delta(A, b)$ and

$$\|(y^k - t\bar{v}) - x^0\| = \|y^k - x^0 - t\bar{v}\| < \|y^k - x^0\| = \rho_k \tag{2.10}$$

for all $t \in (0, \delta_k)$ small enough. By (2.7) and (2.8), we have

$$f(y^k - t\bar{v}) = f(y^k) - t((y^k)^T D\bar{v} + c^T\bar{v}) \leq f(y^k).$$

So $y^k - t\bar{v}$ is a solution of the problem

$$\min\{f(x) : x \in \Delta_{\rho_k}\}. \tag{2.11}$$

From the inequality $\|(y^k - t\bar{v}) - x^0\| < \|y^k - x^0\|$ in (2.10) it follows that y^k cannot be a solution of (2.11) with the minimal distance to x^0, a contradiction.

We have shown that there exists $\hat{\rho} > 0$ such that (2.3) holds. We proceed to show that

$$\text{there exists } \rho \geq \hat{\rho} \text{ such that } q_\rho = \bar{\theta}. \tag{2.12}$$

As $q_\rho = \min\{f(x) : x \in \Delta_\rho\}$, it is easily seen that the conclusion of the theorem follows from (2.12). In order to obtain (2.12), we assume on the contrary that

$$q_\rho > \bar{\theta} \quad \text{for all } \rho \geq \hat{\rho}. \tag{2.13}$$

Note that $q_\rho \geq q_{\rho'}$ whenever $\rho' \geq \rho$. Note also that $q_\rho \to \bar{\theta}$ as $\rho \to +\infty$. Hence from (2.13) it follows that there exist $\rho_i \in (\hat{\rho}, +\infty)$ ($i = 1, 2$) such that $\rho_1 < \rho_2$ and $q_{\rho_1} > q_{\rho_2}$. Since $\rho_2 > \hat{\rho}$, by (2.3) we have

$$\|y_{\rho_2} - x^0\| < \rho_2.$$

Since $q_{\rho_1} > q_{\rho_2}$, we must have $\rho_1 < \|y_{\rho_2} - x^0\|$. (Indeed, if $\rho_1 \geq \|y_{\rho_2} - x^0\|$ then $y_{\rho_2} \in \Delta_{\rho_1}$ and $f(y_{\rho_2}) = q_{\rho_2} < q_{\rho_1} = f(y_{\rho_1})$. This contradicts the choice of y_{ρ_1}.) Setting $\rho_3 = \|y_{\rho_2} - x^0\|$ we have $\rho_1 < \rho_3 < \rho_2$. Since $\rho_3 > \hat{\rho}$ and $\rho_2 > \hat{\rho}$, from (2.3) it follows that

$$\|y_{\rho_3} - x^0\| < \rho_3 = \|y_{\rho_2} - x^0\| < \rho_2. \tag{2.14}$$

Since $\rho_2 > \rho_3$, we have

$$q_{\rho_3} = f(y_{\rho_3}) \geq f(y_{\rho_2}) = q_{\rho_2}.$$

If $f(y_{\rho_3}) = f(y_{\rho_2})$ then from (2.14) we see that y_{ρ_3} is a feasible vector of the problem

$$\min\{f(x) : x \in \Delta_{\rho_2}\} \tag{2.15}$$

at which the objective function attains its optimal value $q_{\rho_2} = f(y_{\rho_2})$. Hence y_{ρ_3} is a solution of (2.15). By (2.14),

$$\|y_{\rho_3} - x^0\| < \|y_{\rho_2} - x^0\|.$$

This implies that y_{ρ_2} cannot be a solution of (2.15) with the minimal distance to x^0, a contradiction. So we must have $f(y_{\rho_3}) > f(y_{\rho_2})$. Since $\|y_{\rho_2} - x^0\| = \rho_3$, we deduce that y_{ρ_2} is a feasible vector of the

problem $\min\{f(x) : x \in \Delta_{\rho_3}\}$. Then the inequality $f(y_{\rho_3}) > f(y_{\rho_2})$ contradicts the fact that y_{ρ_3} is a solution of this optimization problem. We have established property (2.12). The proof is complete.
□

In Theorem 2.1, it is assumed that f is a linear-quadratic function and Δ is a polyhedral convex set. From Definition 1.15 it follows immediately that for any polyhedral convex set $\Delta \subset R^n$ there exists an integer $m \in N$, a matrix $A \in R^{m \times n}$ and a vector $b \in R^m$ such that $\Delta = \{x \in R^n : Ax \geq b\}$. This means that the Frank-Wolfe Theorem can be stated as follows: *"If a linear-quadratic function is bounded from below on a nonempty polyhedral convex set, then the problem of minimizing this function on the set must have a solution."*

If f is a linear-quadratic function but Δ is not assumed to be a polyhedral convex set, then the conclusion of Theorem 2.1 may not hold.

Example 2.1. Let $f(x) = x_1$ for every $x = (x_1, x_2) \in R^2$. Let $\Delta = \{x = (x_1, x_2) \in R^2 : x_1 x_2 \geq 1, \ x_1 \geq 0, \ x_2 \geq 0\}$. We have $\bar{\theta} := \inf\{f(x) : x \in \Delta\} = 0$, but the problem $\min\{f(x) : x \in \Delta\}$ has no solutions.

If Δ is a polyhedral convex set but f is not assumed to be a linear-quadratic function, then the conclusion of Theorem 2.1 may not hold. In the following example, f is a polynomial function of degree 4 of the variables x_1 and x_2.

Example 2.2. (See Frank and Wolfe (1956), p. 109) Let $f(x) = x_1^2 + (1 - x_1 x_2)^2$ for every $x = (x_1, x_2) \in R^2$. Let $\Delta = \{x = (x_1, x_2) \in R^2 : x_1 \geq 0, \ x_2 \geq 0\}$. Observe that $f(x) \geq 0$ for every $x \in R^2$. Choosing $x^k := \left(\dfrac{1}{k}, 1 + k\right)$, $k \in N$, we have

$$f(x^k) = \frac{2}{k^2} \to 0 \quad \text{as } k \to \infty.$$

This implies that

$$\bar{\theta} := \inf\{f(x) : x \in \Delta\} = 0 = \inf\{f(x) : x \in R^2\}.$$

It is a simple matter to show that both the problems $\min\{f(x) : x \in \Delta\}$ and $\min\{f(x) : x \in R^2\}$ have no solutions.

In Frank and Wolfe (1956), the authors informed that Irving Kaplansky has pointed out that the problem of minimizing a polynomial function of degree greater than 2 on a nonempty polyhedral

convex set may not have solutions even in the case the function is
bounded from below on the set.

Given a linear-quadratic function and a polyhedral convex set,
verifying whether the function is bounded from below on the set is
a rather difficult task. In the next section we will discuss another
fundamental existence theorem for quadratic programming which
gives us a tool for dealing with the task.

2.2   The Eaves Theorem

The following result was published by Eaves in 1971.

Theorem 2.2. (The Eaves Theorem; See Eaves (1971), Theorem
3 and Corollary 4, p. 702) *Problem* (2.1) *has solutions if and only
if the following three conditions are satisfied:*

(i) $\Delta(A, b)$ *is nonempty;*

(ii) *If* $v \in R^n$ *and* $Av \geq 0$ *then* $v^T Dv \geq 0$;

(iii) *If* $v \in R^n$ *and* $x \in R^n$ *are such that* $Av \geq 0$, $v^T Dv = 0$ *and*
$Ax \geq b$, *then* $(Dx + c)^T v \geq 0$.

Proof. *Necessity:* Suppose that (2.1) has a solution $\bar{x}$. Since $\bar{x} \in$
$\Delta(A, b)$, condition (i) is satisfied. Given any $v \in R^n$ with $Av \geq$
0, since $A(\bar{x} + tv) = A\bar{x} + tAv \geq b$ for every $t \geq 0$, we have
$\bar{x} + tv \in \Delta(A, b)$ for every $t \geq 0$. Hence $f(\bar{x} + tv) \geq f(\bar{x})$ for
every $t \geq 0$. It follows that $\frac{1}{2} t^2 v^T Dv + t(D\bar{x} + c)^T v \geq 0$ for every
$t \geq 0$, hence $v^T Dv \geq 0$. This shows that condition (ii) is satisfied.
We now suppose that there are given any $v \in R^n$ and $x \in R^n$
with the properties that $Av \geq 0$, $v^T Dv = 0$ and $Ax \geq b$. Since
$x + tv \in \Delta(A, b)$ for every $t \geq 0$ and $\bar{x}$ is a solution of (2.1), we
have $f(x + tv) \geq f(\bar{x})$ for every $t \geq 0$. From this and the condition
$v^T Dv = 0$ we deduce that $t(Dx + c)^T v + \frac{1}{2} x^T Dx + c^T x \geq f(\bar{x})$ for
every $t \geq 0$. This implies that $(Dx + c)^T v \geq 0$. We have thus shown
that condition (iii) is satisfied.

Sufficiency: Assume that conditions (i), (ii) and (iii) are satis-
fied. Define $\bar{\theta} = \inf\{f(x) : x \in R^n, Ax \geq b\}$. As $\Delta(A, b) \neq \emptyset$, we
have $\bar{\theta} \neq +\infty$. If $\bar{\theta} \in R$ then the assertion of the theorem follows
from the Frank-Wolfe Theorem. Hence we only need to show that

the situation $\bar{\theta} = -\infty$ cannot occur. To obtain a contradiction, suppose that $\bar{\theta} = -\infty$. We can now proceed analogously to the proof of Theorem 2.1.

Fix a point $x^0 \in \Delta(A, b)$. For every $\rho > 0$, define $\Delta_\rho = \Delta(A, b) \cap \bar{B}(x^0, \rho)$ and consider the minimization problem $\min\{f(x) : x \in \Delta_\rho\}$. Denote by q_ρ the optimal value of this problem. Let $y_\rho \in \Delta_\rho$ be such that $f(y_\rho) = q_\rho$ and

$$\|y_\rho - x^0\| = \min\{\|y - x^0\| : y \in \Delta_\rho, \ f(y) = q_\rho\}.$$

We claim that there exists $\hat{\rho} > 0$ such that $\|y_\rho - x^0\| < \rho$ for all $\rho \geq \hat{\rho}$. Suppose the claim were false. Then we would find an increasing sequence $\rho_k \to +\infty$ such that for every k there exists $y_{\rho_k} \in \Delta_{\rho_k}$ such that

$$f(y_{\rho_k}) = q_{\rho_k}, \quad \|y_{\rho_k} - x^0\| = \rho_k.$$

For simplicity of notation, we write y^k instead of y_{ρ_k}. Since $y^k \in \Delta(A, b)$, we must have $A_i y^k \geq b_i$ for $i = 1, \ldots, m$. Analysis similar to that in the proof of Theorem 2.1 shows that there exists a subsequence $\{k'\} \subset \{k\}$ such that all the limits

$$\lim_{k' \to \infty} A_i y^{k'} \quad (i = 1, \ldots, m)$$

exist. Without restriction of generality we can assume that $\{k'\} \equiv \{k\}$. Let $I = \{1, \ldots, m\}$, $I_0 = \{i \in I : \lim_{k \to \infty} A_i y^k = b_i\}$ and

$$I_1 = I \setminus I_0 = \{i \in I : \lim_{k \to \infty} A_i y^k > b_i\}.$$

Let $\varepsilon > 0$ be such that

$$\lim_{k \to \infty} A_i y^k \geq b_i + \varepsilon \quad \text{for every } i \in I_1.$$

Since $\|(y^k - x^0)/\rho_k\| = 1$ for every k, there is no loss of generality in assuming that the sequence

$$\left\{ \frac{y^k - x^0}{\rho_k} \right\}$$

converges to some $\bar{v} \in R^n$, $\|\bar{v}\| = 1$, as $k \to \infty$. Since $\rho_k \to +\infty$, for every $i \in I_0$ we have

$$\begin{aligned}
0 &= \lim_{k \to \infty} (A_i y^k - b_i) \\
&= \lim_{k \to \infty} \frac{A_i y^k - b_i}{\rho_k} \\
&= \lim_{k \to \infty} \left(A_i \frac{y^k - x^0}{\rho_k} \right) + \lim_{k \to \infty} \left(\frac{A_i x^0 - b_i}{\rho_k} \right) = A_i \bar{v}.
\end{aligned}$$

Similarly, for every $i \in I_1$ we have

$$
\begin{aligned}
0 \;\le\; & \liminf_{k \to \infty} \frac{A_i y^k - b_i}{\rho_k} \\
= \; & \liminf_{k \to \infty} \left(\frac{A_i y^k - A_i x^0}{\rho_k} + \frac{A_i x^0 - b_i}{\rho_k} \right) \\
= \; & \lim_{k \to \infty} \left(A_i \frac{y^k - x^0}{\rho_k} \right) + \lim_{k \to \infty} \frac{A_i x^0 - b_i}{\rho_k} = A_i \bar{v}.
\end{aligned}
$$

Therefore

$$A_i \bar{v} = 0 \quad \text{for every } i \in I_0, \quad A_i \bar{v} \ge 0 \quad \text{for every } i \in I_1.$$

From this we deduce that

$$y + t\bar{v} \in \Delta(A, b) \quad \text{for every } t \ge 0 \text{ and } y \in \Delta(A, b). \tag{2.16}$$

Since

$$
\begin{aligned}
f(y^k) = f(y_{\rho_k}) \;=\; & q_{\rho_k} \\
= \; & \min\{ f(x) : x \in \Delta_{\rho_k} \} \\
= \; & \min\{ f(x) : x \in \Delta(A, b) \cap \bar{B}(x_0, \rho_k) \}
\end{aligned}
$$

and the increasing sequence $\{\rho_k\}$ converges to $+\infty$, we see that the sequence $\{f(y^k)\}$ is non-increasing and $f(y^k) \to \bar{\theta} = -\infty$. Hence $f(y^k) < 0$ for all k sufficiently large. Using the formula of f we can rewrite the last inequality as follows

$$
\begin{aligned}
& \frac{1}{2}(y^k - x^0)^T D(y^k - x^0) + c^T(y^k - x^0) \\
& \quad + (x^0)^T D(y^k - x^0) + \frac{1}{2}(x^0)^T D x^0 + c^T x^0 < 0.
\end{aligned}
$$

Dividing this inequality by ρ_k^2 and taking the limits as $k \to \infty$, we get $\bar{v}^T D \bar{v} \le 0$. Since $A\bar{v} \ge 0$, from condition (ii) it follows that $\bar{v}^T D \bar{v} \ge 0$. Hence

$$\bar{v}^T D \bar{v} = 0. \tag{2.17}$$

By (2.16),

$$y^k + t\bar{v} \in \Delta(A, b) \quad \text{for every } t \ge 0 \text{ and } k \in N.$$

By virtue of (2.17), we have

$$
\begin{aligned}
f(y^k + t\bar{v}) \;=\; & \frac{1}{2}(y^k + t\bar{v})^T D(y^k + t\bar{v}) + c^T(y^k + t\bar{v}) \\
= \; & \frac{1}{2}(y^k)^T D y^k + c^T y^k + t((y^k)^T D \bar{v} + c^T \bar{v}).
\end{aligned}
$$

Since $y^k \in \Delta(A, b)$, $A\bar{v} \geq 0$ and $\bar{v}^T D\bar{v} = 0$, by condition (iii) we have

$$(y^k)^T D\bar{v} + c^T \bar{v} = (Dy^k + c)^T \bar{v} \geq 0 \quad \text{for every } k \in N. \quad (2.18)$$

Since $\langle \bar{v}, \bar{v} \rangle = 1$ and $\dfrac{y^k - x^0}{\rho_k} \to \bar{v}$, there exists $k_1 \in N$ such that $\left\langle \dfrac{y^k - x^0}{\rho_k}, \bar{v} \right\rangle > 0$ for all $k \geq k_1$. For any fixed index $k \geq k_1$, we have $\langle y^k - x^0, \bar{v} \rangle > 0$. Therefore

$$\|y^k - x^0 - t\bar{v}\|^2 = \|y^k - x^0\|^2 - 2t\langle y^k - x^0, \bar{v} \rangle + t^2 \|\bar{v}\|^2 < \|y^k - x^0\|^2 \quad (2.19)$$

for $t > 0$ small enough. We have

$$A_i(y^k - t\bar{v}) = A_i y^k \geq b_i \quad \text{for all } i \in I_0.$$

Since $\lim\limits_{k \to \infty} A_i y^k \geq b_i + \varepsilon$ for every $i \in I_1$, there exists $k_2 \in N$, $k_2 \geq k_1$, such that $A_i y^k \geq b_i + \dfrac{\varepsilon}{2}$ for every $k \geq k_2$ and $i \in I_1$. Fix an index $k \geq k_2$ and choose $\delta_k > 0$ as small as $t A_i \bar{v} \leq \dfrac{\varepsilon}{2}$ for every $i \in I_1$ and $t \in (0, \delta_k)$. Then we have

$$A_i(y^k - t\bar{v}) \geq b_i + \frac{\varepsilon}{2} - t A_i \bar{v} \geq b_i$$

for all $i \in I_1$ and $t \in (0, \delta_k)$. From what has already been proved, we deduce that

$$y^k - t\bar{v} \in \Delta(A, b) \quad \text{for all } t \in (0, \delta_k).$$

Combining this with (2.19) we see that $y^k - t\bar{v} \in \Delta(A, b)$ and

$$\|(y^k - t\bar{v}) - x^0\| = \|y^k - x^0 - t\bar{v}\| < \|y^k - x^0\| = \rho_k \quad (2.20)$$

for all $t \in (0, \delta_k)$ small enough. By (2.17) and (2.18), we have

$$f(y^k - t\bar{v}) = f(y^k) - t((y^k)^T D\bar{v} + c^T \bar{v}) \leq f(y^k).$$

So $y^k - t\bar{v}$ is a solution of the problem

$$\min\{f(x) : x \in \Delta_{\rho_k}\}. \quad (2.21)$$

From the inequality $\|(y^k - t\bar{v}) - x^0\| < \|y^k - x^0\|$ in (2.20) it follows that y^k cannot be a solution of (2.21) with the minimal distance to x^0, a contradiction.

We have shown that there exists $\hat{\rho} > 0$ such that

$$\|y_\rho - x^0\| < \rho \quad \text{for all } \rho \geq \hat{\rho}. \tag{2.22}$$

Note that $q_\rho \geq q_{\rho'}$ whenever $\rho' \geq \rho$. Note also that $q_\rho \to \bar{\theta} = -\infty$ as $\rho \to +\infty$. Hence there must exist $\rho_i \in (\hat{\rho}, +\infty)$ $(i = 1, 2)$ such that $\rho_1 < \rho_2$ and $q_{\rho_1} > q_{\rho_2}$. Since $\rho_2 > \hat{\rho}$, by (2.22) we have

$$\|y_{\rho_2} - x^0\| < \rho_2.$$

Since $q_{\rho_1} > q_{\rho_2}$, we must have $\rho_1 < \|y_{\rho_2} - x^0\|$. (Indeed, if $\rho_1 \geq \|y_{\rho_2} - x^0\|$ then $y_{\rho_2} \in \Delta_{\rho_1}$ and $f(y_{\rho_2}) = q_{\rho_2} < q_{\rho_1} = f(y_{\rho_1})$. This contradicts the choice of y_{ρ_1}.) Setting $\rho_3 = \|y_{\rho_2} - x^0\|$ we have $\rho_1 < \rho_3 < \rho_2$. Since $\rho_3 > \hat{\rho}$ and $\rho_2 > \hat{\rho}$, from (2.22) it follows that

$$\|y_{\rho_3} - x^0\| < \rho_3 = \|y_{\rho_2} - x^0\| < \rho_2. \tag{2.23}$$

Since $\rho_2 > \rho_3$, we have

$$q_{\rho_3} = f(y_{\rho_3}) \geq f(y_{\rho_2}) = q_{\rho_2}.$$

If $f(y_{\rho_3}) = f(y_{\rho_2})$ then from (2.23) we see that y_{ρ_3} is a feasible vector of the problem

$$\min\{f(x) : x \in \Delta_{\rho_2}\} \tag{2.24}$$

at which the objective function attains its optimal value $q_{\rho_2} = f(y_{\rho_2})$. Hence y_{ρ_3} is a solution of (2.24). By (2.23),

$$\|y_{\rho_3} - x^0\| < \|y_{\rho_2} - x^0\|.$$

This implies that y_{ρ_2} cannot be a solution of (2.24) with the minimal distance to x^0, a contradiction. So we must have $f(y_{\rho_3}) > f(y_{\rho_2})$. Since $\|y_{\rho_2} - x^0\| = \rho_3$, we deduce that y_{ρ_2} is a feasible vector of the problem $\min\{f(x) : x \in \Delta_{\rho_3}\}$. Then the inequality $f(y_{\rho_3}) > f(y_{\rho_2})$ shows that y_{ρ_3} cannot be a solution of this optimization problem, a contradiction. The proof is complete. $\square$

Here are several important consequences of the Eaves Theorem.

Corollary 2.1. *Assume that D is a positive semidefinite matrix. Then problem (2.1) has solutions if and only if $\Delta(A, b)$ is nonempty and the following condition is satisfied:*

$$\left(v \in R^n, \ x \in R^n, \ Av \geq 0, \ v^T Dv = 0, \ Ax \geq b\right) \Rightarrow (Dx + c)^T v \geq 0. \tag{2.25}$$

Proof. Note that condition (ii) in Theorem 2.2 is satisfied because, by our assumption, $v^T D v \geq 0$ for every $v \in R^n$. Therefore the conclusion follows from Theorem 2.2.   □

Corollary 2.2. *Assume that D is a negative semidefinite matrix. Then problem (2.1) has solutions if and only if $\Delta(A, b)$ is nonempty and the following conditions are satisfied:*

(i) $(v \in R^n,\ Av \geq 0) \Rightarrow v^T D v = 0$;

(ii) $(v \in R^n,\ x \in R^n,\ Av \geq 0,\ Ax \geq b) \Rightarrow (Dx + c)^T v \geq 0$.

Proof. Since $v^T D v \leq 0$ for every $v \in R^n$ by our assumption, we see that condition (ii) in Theorem 2.2 is can be rewritten as condition (i) in this corollary. Besides, since $Av \geq 0$ implies $v^T D v = 0$, condition (iii) in Theorem 2.2 can be rewritten as condition (ii) in this corollary. Therefore the conclusion follows from Theorem 2.2. □

Corollary 2.3. *If D is a positive definite matrix, then problem (2.1) has solutions if and only if $\Delta(A, b)$ is nonempty.*

Proof. Since D is a positive definite matrix, the equality $v^T D v = 0$ implies that $v = 0$. Then the assertion follows from Corollary 2.1 because condition (2.25) is satisfied.   □

Corollary 2.4. *If D is a negative definite matrix, then problem (2.1) has solutions if and only if $\Delta(A, b)$ is nonempty and compact.*

Proof. Since D is a negative definite matrix, conditions (i) and (ii) in Corollary 2.2 are fulfilled if and only if the set $L := \{v \in R^n : Av \geq 0\}$ contains just one element $v = 0$. Since L is the recession cone of the polyhedral convex set $\Delta(A, b) = \{x \in R^n : Ax \geq b\}$ (see Rockafellar (1970), p. 62), the condition $L = \{0\}$ is equivalent to the compactness of $\Delta(A, b)$ (see Rockafellar (1970), Theorem 8.4). Hence the assertion follows from Corollary 2.2.   □

Theorem 2.2 allows one to verify the existence of solutions of a quadratic program of the form (2.1) through analyzing its data set $\{D,\ A,\ c,\ b\}$. If any one from the three conditions (i), (ii) and (iii) in the theorem is violated, then the problem cannot have solutions.

Formally, the Eaves Theorem formulated above allows one to deal only with quadratic programs of the standard form. It is a simple matter to derive existence results for quadratic programs of the canonical and the general forms from Theorem 2.2.

Corollary 2.5. *Let* $D \in R_S^{n \times n}$, $A \in R^{m \times n}$, $c \in R^n$ *and* $b \in R^m$. *The quadratic program*

$$\min \left\{ \frac{1}{2} x^T D x + c^T x \ : \ x \in R^n, \ A x \geq b, \ x \geq 0 \right\} \qquad (2.26)$$

has solutions if and only if the following three conditions are satisfied:

(i) *The constraint set* $\{x \in R^n \ : \ Ax \geq b, \ x \geq 0\}$ *is nonempty;*

(ii) *If* $v \in R^n$, $Av \geq 0$ *and* $v \geq 0$, *then* $v^T D v \geq 0$;

(iii) *If* $v \in R^n$ *and* $x \in R^n$ *are such that* $Av \geq 0$, $v \geq 0$, $v^T D v = 0$, $Ax \geq b$ *and* $x \geq 0$, *then* $(Dx + c)^T v \geq 0$.

Proof. Define $\widetilde{A} \in R^{(m+n) \times n}$ and $\widetilde{b} \in R^{m+n}$ by setting

$$\widetilde{A} = \begin{pmatrix} A \\ E \end{pmatrix}, \quad \widetilde{b} = \begin{pmatrix} b \\ 0 \end{pmatrix},$$

where E denotes the unit matrix in $R^{n \times n}$ and 0 stands for the zero vector in R^n. It is clear that (2.26) can be rewritten in the following form

$$\min \left\{ \frac{1}{2} x^T D x + c^T x \ : \ x \in R^n, \ \widetilde{A} x \geq \widetilde{b} \right\}.$$

Applying Theorem 2.2 to this quadratic program we obtain the desired result. $\square$

Corollary 2.6. *Let* $D \in R_S^{n \times n}$, $A \in R^{m \times n}$, $C \in R^{s \times n}$, $c \in R^n$, $b \in R^m$ *and* $d \in R^s$. *The quadratic program*

$$\min \left\{ \frac{1}{2} x^T D x + c^T x \ : \ x \in R^n, \ A x \geq b, \ C x = d \right\} \qquad (2.27)$$

has solutions if and only if the following three conditions are satisfied:

(i) *The constraint set* $\{x \in R^n \ : \ Ax \geq b, \ Cx = d\}$ *is nonempty;*

(ii) *If* $v \in R^n$, $Av \geq 0$ *and* $Cv = 0$, *then* $v^T D v \geq 0$;

(iii) *If* $v \in R^n$ *and* $x \in R^n$ *are such that* $Av \geq 0$, $Cv = 0$, $v^T D v = 0$, $Ax \geq b$ *and* $Cx = d$, *then* $(Dx + c)^T v \geq 0$.

Proof. Define $\widetilde{A} \in R^{(m+2s) \times n}$ and $\widetilde{b} \in R^{m+2s}$ by setting

$$\widetilde{A} = \begin{pmatrix} A \\ C \\ -C \end{pmatrix}, \quad \widetilde{b} = \begin{pmatrix} b \\ d \\ -d \end{pmatrix}.$$

It is clear that (2.27) can be rewritten in following form

$$\min \left\{ \frac{1}{2} x^T D x + c^T x \ : \ x \in R^n, \ \widetilde{A} x \geq \widetilde{b} \right\}.$$

Applying Theorem 2.2 to this quadratic program we obtain the desired result. □

2.3 Commentaries

The Frank-Wolfe Theorem has various applications. For example, in (Cottle et al. (1992), Chapter 3) it has been used as the main tool for obtaining many existence results for linear complementarity problems. Usually, existence theorems for optimization problems and for variational problems give only *sufficient conditions* to assure that the problem under consideration has solutions. Here we see that the Frank-Wolfe Theorem and the Eaves Theorem provide some criteria (both the necessary and sufficient conditions) for the solution existence. This is possible because the quadratic programs have a relatively simple structure.

We realized the importance of the Eaves Theorem when we studied a paper by Klatte (1985) and applied the theorem to investigate the lower semicontinuity of the solution map in parametric quadratic programs (see Chapter 15 of this book), the directional differentiability and the *piecewise linear-quadratic property* of the optimal value function in quadratic programs under linear perturbations (see Chapter 16 of this book). We believe that the theorem is really very useful and important.

The proof of the Frank-Wolfe Theorem given in Section 2.1 follows exactly the scheme proposed in Blum and Oettli (1972). For the convenience of the reader, all the arguments of Blum and Oettli are described in detail. Two other proofs of the theorem can be found in Frank and Wolfe (1956) and Eaves (1971).

The proof of the Eaves Theorem given in Section 2.2 is rather different from the original proof given in Eaves (1971). The repetition of one part of the arguments used for proving the Frank-Wolfe

Theorem is intended to show the close interrelations between the two existence theorems.

Chapter 3

Necessary and Sufficient Optimality Conditions for Quadratic Programs

This chapter is devoted to a discussion on first-order optimality conditions and second-order optimality conditions for quadratic programming problems.

3.1 First-Order Optimality Conditions

In this section we will establish first-order necessary and sufficient optimality conditions for quadratic programs. Second-order necessary and sufficient optimality conditions for these problems will be obtained in the next section.

The first assertion of the following proposition states the *Fermat rule*, which is a basic first-order necessary optimality condition for mathematical programming problems, for quadratic programs. The second assertion states the so-called *first-order sufficient optimality condition* for quadratic programs and its consequence.

Theorem 3.1. *Let $\bar{x}$ be a feasible vector of the quadratic program*

$$\min \left\{ f(x) = \frac{1}{2} x^T D x + c^T x : x \in \Delta \right\}, \tag{3.1}$$

where $D \in R_S^{n \times n}$, $c \in R^n$, and $\Delta \subset R^n$ is a polyhedral convex set.

(i) *If $\bar{x}$ is a local solution of this problem, then*

$$\langle D\bar{x} + c, x - \bar{x} \rangle \geq 0 \quad \text{for every } x \in \Delta. \tag{3.2}$$

(ii) *If*
$$\langle D\bar{x} + c, x - \bar{x} \rangle > 0 \quad \text{for every } x \in \Delta \setminus \{\bar{x}\}, \qquad (3.3)$$

then $\bar{x}$ is a local solution of (3.1) and, moreover, there exist $\varepsilon > 0$ and $\varrho > 0$ such that

$$f(x) - f(\bar{x}) \geq \varrho \|x - \bar{x}\| \quad \text{for every } x \in \Delta \cap B(\bar{x}, \varepsilon). \quad (3.4)$$

Proof. (i) Let $\bar{x} \in \Delta$ be a local solution of (3.1). Choose $\mu > 0$ so that

$$f(y) \geq f(\bar{x}), \quad \forall y \in \Delta \cap B(\bar{x}, \mu).$$

Given any $x \in \Delta \setminus \{\bar{x}\}$, we observe that there exists $\delta > 0$ such that $\bar{x} + t(x - \bar{x})$ belongs to $\Delta \cap B(\bar{x}, \mu)$ whenever $t \in (0, \delta)$. Therefore

$$
\begin{aligned}
0 \leq \lim_{t \downarrow 0} \frac{f(\bar{x} + t(x - \bar{x})) - f(\bar{x})}{t} &= f'(\bar{x}; x - \bar{x}) \\
&= \langle \nabla f(\bar{x}), x - \bar{x} \rangle \\
&= \langle D\bar{x} + c, x - \bar{x} \rangle.
\end{aligned}
$$

Property (3.2) has been established.

(ii) It suffices to show that if (3.3) holds then there exist $\varepsilon > 0$ and $\varrho > 0$ such that (3.4) is satisfied. To obtain a contradiction, suppose that (3.3) holds but for every $\varepsilon > 0$ and $\varrho > 0$ there exists $x \in \Delta \cap B(\bar{x}, \varepsilon)$ such that $f(x) - f(\bar{x}) < \varrho \|x - \bar{x}\|$. Then there exists a sequence $\{x^k\}$ in R^n such that, for every $k \in N$, we have $x^k \in \Delta \cap B(\bar{x}, \frac{1}{k})$ and $f(x^k) - f(\bar{x}) < \frac{1}{k}\|x^k - \bar{x}\|$. There is no loss of generality in assuming that the sequence of unit vectors $\{(x^k - \bar{x})/\|x^k - \bar{x}\|\}$ converges to some unit vector $\bar{v} \in R^n$. Since

$$
\begin{aligned}
f(x^k) - f(\bar{x}) &= \frac{1}{2}(x^k - \bar{x})^T D(x^k - \bar{x}) + c^T(x^k - \bar{x}) \\
&\quad + (x^k - \bar{x})^T D\bar{x} < \frac{1}{k}\|x^k - \bar{x}\|.
\end{aligned}
$$

Dividing the last inequality by $\|x^k - \bar{x}\|$ and taking the limits as $k \to \infty$, we obtain

$$(D\bar{x} + c)^T \bar{v} \leq 0. \qquad (3.5)$$

Since Δ is a polyhedral convex set, there exist $m \in N$, $a_1, \ldots, a_m$ in R^n and $\beta_1, \ldots, \beta_m$ in R such that Δ has the representation (1.31). Let $I_0 = \{i : \langle a_i, \bar{x} \rangle = \beta_i\}$, $I_1 = \{i : \langle a_i, \bar{x} \rangle > \beta_i\}$. For each $i \in I_0$, we have $\langle a_i, x^k - \bar{x} \rangle = \langle a_i, x^k \rangle - \beta_i \geq 0$. Therefore $\langle a_i, \bar{v} \rangle = \lim_{k \to \infty} \langle a_i, (x^k - \bar{x})/\|x^k - \bar{x}\| \rangle \geq 0$. Obviously, there exists $\delta_1 > 0$ such that $\langle a_i, \bar{x} + t\bar{v} \rangle > \beta_i$ for every $i \in I_1$ and $t \in (0, \delta_1)$. Consequently, $\bar{x} + t\bar{v} \in \Delta$ for every $t \in (0, \delta_1)$. Substituting $x = \bar{x} + t\bar{v}$, where t is a value from $(0, \delta_1)$, into (3.3) gives $(D\bar{x} + c)^T \bar{v} > 0$, which contradicts (3.5). The proof is complete. $\square$

For obtaining a first-order necessary optimality condition for quadratic programs in the form of a *Lagrange multiplier rule*, we shall need the following basic result concerning linear inequalities.

Theorem 3.2. (Farkas' Lemma; See Rockafellar (1970), p. 200) *Let $a_0, a_1, \ldots, a_k$ be vectors from R^n. The inequality $\langle a_0, x \rangle \leq 0$ is a consequence of the system*

$$\langle a_i, x \rangle \leq 0, \quad i = 1, 2, \ldots, k,$$

if and only if there exist nonnegative real numbers $\lambda_1, \ldots, \lambda_k$ such that

$$\sum_{i=1}^{k} \lambda_i a_i = a_0.$$

Theorem 3.3. (See, for instance, Cottle et al. (1992), p. 118) *If $\bar{x} \in R^n$ is a local solution of problem (2.1) then there exists $\lambda = (\lambda_1, \ldots, \lambda_m) \in R^m$ such that*

$$\begin{cases} D\bar{x} - A^T \lambda + c = 0, \\ A\bar{x} - b \geq 0, \quad \lambda \geq 0, \\ \lambda^T (A\bar{x} - b) = 0. \end{cases} \tag{3.6}$$

Proof. Denote by A_i the i-th row of A, and set $a_i = A_i^T$. Denote by b_i the i-th component of vector b. Set $\Delta = \Delta(A, b) = \{x \in R^n : Ax \geq b\}$. Let $\bar{x}$ be a local solution of (2.1). By Theorem 3.1(i), property (3.2) holds. Set $I = \{1, \ldots, m\}$, $I_0 = \{i \in I : \langle a_i, \bar{x} \rangle = b_i\}$ and $I_1 = \{i \in I : \langle a_i, \bar{x} \rangle > b_i\}$. For any $v \in R^n$ satisfying

$$\langle a_i, v \rangle \geq 0 \quad \text{for every } i \in I_0,$$

analysis similar to that in the proof of Theorem 3.1(ii) shows that there exists $\delta_1 > 0$ such that $\langle a_i, \bar{x} + tv \rangle \geq b_i$ for every $i \in I$ and $t \in (0, \delta_1)$. Substituting $x = \bar{x} + tv$, where t is a value from $(0, \delta_1)$, to (3.2) yields $\langle D\bar{x} + c, v \rangle \geq 0$. We have thus shown that

$$\langle -D\bar{x} - c, v \rangle \leq 0$$

for any $v \in R^n$ satisfying

$$\langle -a_i, v \rangle \leq 0 \quad \text{for every } i \in I_0.$$

By Theorem 3.2, there exist nonnegative real numbers λ_i ($i \in I_0$) such that

$$\sum_{i \in I_0} \lambda_i(-a_i) = -D\bar{x} - c. \tag{3.7}$$

Put $\lambda_i = 0$ for all $i \in I_1$ and $\lambda = (\lambda_1, \ldots, \lambda_m)$. Since $a_i = A_i^T$ for every $i \in I$, from (3.7) we obtain the first equality in (3.6). Since $\bar{x} \in \Delta(A, b)$ and $\lambda_i(A_i\bar{x} - b_i) = 0$ for each $i \in I$, the other conditions in (3.6) are satisfied too. The proof is complete. $\square$

From Theorem 3.3 we can derive the following Lagrange multiplier rules for quadratic programs of the canonical and the general forms.

Corollary 3.1. (See, for instance, Murty (1972)) *If $\bar{x}$ is a local solution of problem (2.26), then there exist $\lambda = (\lambda_1, \ldots, \lambda_m) \in R^m$ such that*

$$\begin{cases} D\bar{x} - A^T\lambda + c \geq 0, \\ A\bar{x} - b \geq 0, \quad \bar{x} \geq 0, \quad \lambda \geq 0, \\ \bar{x}^T(D\bar{x} - A^T\lambda + c) + \lambda^T(A\bar{x} - b) = 0. \end{cases} \tag{3.8}$$

Proof. Define matrix $\tilde{A} \in R^{(m+n) \times n}$ and vector $\tilde{b} \in R^{m+n}$ as in the proof of Corollary 2.5 and note that problem (2.26) can be rewritten in the form

$$\min \left\{ \frac{1}{2}x^T D x + c^T x : x \in R^n, \ \tilde{A}x \geq \tilde{b} \right\}.$$

Applying Theorem 3.3 to this quadratic program we deduce that there exists $\tilde{\lambda} = (\lambda_1, \ldots, \lambda_{m+n}) \in R^{m+n}$ such that

$$D\bar{x} - \tilde{A}^T\tilde{\lambda} + c = 0, \quad \tilde{A}\bar{x} \geq \tilde{b}, \quad \tilde{\lambda} \geq 0, \quad \tilde{\lambda}^T(\tilde{A}\bar{x} - \tilde{b}) = 0.$$

Taking $\lambda = (\lambda_1, \ldots, \lambda_m)$, we can obtain the properties stated in (3.8) from the last ones. $\square$

Corollary 3.2. *If $\bar{x}$ is a local solution of problem (2.27), then there exist $\lambda = (\lambda_1, \ldots, \lambda_m) \in R^m$ and $\mu = (\mu_1, \ldots, \mu_s) \in R^s$ such that*

$$\begin{cases} D\bar{x} - A^T\lambda - C^T\mu + c = 0, \\ A\bar{x} - b \geq 0, \quad C\bar{x} = d, \quad \lambda \geq 0, \\ \lambda^T(A\bar{x} - b) = 0. \end{cases} \tag{3.9}$$

Proof. Define $\tilde{A} \in R^{(m+2s) \times n}$ and $\tilde{b} \in R^{m+2s}$ as in the proof of Corollary 2.6 and note that problem (2.27) can be rewritten in the form

$$\min \left\{ \frac{1}{2} x^T D x + c^T x \ : \ x \in R^n, \ \tilde{A} x \geq \tilde{b} \right\}.$$

Applying Theorem 3.3 to this quadratic program we see that there exists $\tilde{\lambda} = (\lambda_1, \ldots, \lambda_{m+2n}) \in R^{m+2n}$ such that

$$D\bar{x} - \tilde{A}^T \tilde{\lambda} + c = 0, \quad \tilde{A}\bar{x} \geq \tilde{b}, \quad \tilde{\lambda} \geq 0, \quad \tilde{\lambda}^T(\tilde{A}\bar{x} - \tilde{b}) = 0.$$

Taking $\lambda = (\lambda_1, \ldots, \lambda_m)$ and $\mu = (\mu_1, \ldots, \mu_s)$, where $\mu_j = \lambda_{m+j} - \lambda_{m+s+j}$ for $j = 1, \ldots, s$, we can obtain the properties stated in (3.9) from the last ones. $\square$

Definition 3.1. If $(\bar{x}, \lambda) \in R^n \times R^m$ is such a pair that (3.6) (resp., (3.8)) holds, then we say that $(\bar{x}, \lambda)$ is a *Karush-Kuhn-Tucker pair* (KKT pair for short) of the standard quadratic program (2.1) (resp., of the canonical quadratic program (2.26)). The point $\bar{x}$ is called a *Karush-Kuhn-Tucker point* (KKT point for short), and the real numbers $\lambda_1, \ldots, \lambda_m$ are called the *Lagrange multipliers* corresponding to $\bar{x}$. Similarly, if $(\bar{x}, \lambda, \mu) \in R^n \times R^m \times R^s$ is such a triple that (3.9) is satisfied then $\bar{x}$ is called a *Karush-Kuhn-Tucker point* of the general quadratic program (2.27), and the real numbers $\lambda_1, \ldots, \lambda_m, \mu_1, \ldots, \mu_s$ are called the *Lagrange multipliers* corresponding to $\bar{x}$. Sometimes the vectors $\lambda = (\lambda_1, \ldots, \lambda_m)$ and $\mu = (\mu_1, \ldots, \mu_s)$ are also called the *Lagrange multipliers* corresponding to $\bar{x}$.

In the sequel, by abuse of notation, we will abbreviate both the KKT point sets of (2.1) and (2.26) to $S(D, A, c, b)$. Likewise, both the solution sets (resp., both the local-solution sets) of (2.1) and (2.26) are abbreviated to $\text{Sol}(D, A, c, b)$ (resp., to $\text{loc}(D, A, c, b)$).

From Theorem 3.3 and Corollary 3.1 it follows that

$$\text{Sol}(D, A, c, b) \subset \text{loc}(D, A, c, b) \subset S(D, A, c, b). \tag{3.10}$$

Later on, we will encounter with examples where the three sets figured in (3.10) are different each from others.

3.2   Second-Order Optimality Conditions

This section provides a detailed exposition of second-order necessary and sufficient optimality conditions for quadratic programming
problems. The main result in this direction was published by Majthay in 1971. It is stated in Theorem 3.4 below. The proof given
by Majthay contains one inaccurate argument (see Contesse (1980),
p. 331). The first rigorous proof of this result belongs to Contesse
(1980).

Theorem 3.4. (See Majthay (1971) and Contesse (1980), Théorème
1) *The necessary and sufficient condition for a point $\bar{x} \in R^n$ to be a
local solution of problem (2.27) is that there exists a pair of vectors*

$$(\bar{\lambda}, \bar{\mu}) = (\bar{\lambda}_1, \ldots, \bar{\lambda}_m, \bar{\mu}_1, \ldots, \bar{\mu}_s) \in R^m \times R^s$$

such that

(i) *the system*

$$\begin{cases} D\bar{x} - A^T\bar{\lambda} - C^T\bar{\mu} + c = 0, \\ A\bar{x} - b \geq 0, \quad C\bar{x} = d, \quad \bar{\lambda} \geq 0, \\ \bar{\lambda}^T(A\bar{x} - b) = 0. \end{cases} \qquad (3.11)$$

  is satisfied, and

(ii) *if $v \in R^n \setminus \{0\}$ is such that $A_{I_1}v = 0$, $A_{I_2}v \geq 0$, $Cv = 0$, where*

$$I_1 = \{i : A_i\bar{x} = b_i, \ \bar{\lambda}_i > 0\}, \quad I_2 = \{i : A_i\bar{x} = b_i, \ \bar{\lambda}_i = 0\}, \qquad (3.12)$$

  then $v^T Dv \geq 0$.

Consider problem (2.27) and set

$$\Delta = \{x \in R^n : Ax \geq b, \ Cx = d\}.$$

Denote the objective function of (2.27) by $f(x)$. Let the symbol
$(\nabla f(x))^\perp$ stand for the linear subspace of R^n orthogonal to $\nabla f(x)$,
that is

$$(\nabla f(x))^\perp = \{v \in R^n : \langle \nabla f(x), v \rangle = 0\}.$$

For proving Theorem 3.4 we shall need the following fact.

Lemma 3.1. *Let* $\bar{x} \in R^n$, $\bar{\lambda} \in R^m$ *and* $\bar{\mu} \in R^s$ *be such that the system* (3.11) *is satisfied. Let* I_1 *and* I_2 *be as in* (3.12). *Then*

$$\{v \in R^n : A_{I_1}v = 0, \ A_{I_2}v \geq 0, \ Cv = 0\}$$
$$= \{v \in R^n : A_{I_0}v \geq 0, \ Cv = 0\} \cap (\nabla f(\bar{x}))^{\perp}$$
$$= T_{\Delta}(\bar{x}) \cap (\nabla f(\bar{x}))^{\perp},$$

where $I_0 := I_1 \cup I_2 = \{i : A_i\bar{x} = b_i\}$, *and* $T_{\Delta}(\bar{x})$ *denotes the tangent cone to* Δ *at* $\bar{x}$.

Proof. Using Theorem 1.8 (b), it is a simple matter to show that

$$T_{\Delta}(\bar{x}) = \{v \in R^n : A_{I_0}v \geq 0, \ Cv = 0\}. \tag{3.13}$$

So it suffices to prove the first equality in the assertion of the lemma. Suppose that $v \in R^n$, $A_{I_1}v = 0$, $A_{I_2}v \geq 0$, $Cv = 0$. Define $I = \{1, 2, \ldots, m\}$. By (3.11) we have

$$\begin{aligned}
\langle \nabla f(\bar{x}), v \rangle = (D\bar{x} + c)^T v &= (A^T\bar{\lambda} + C^T\bar{\mu})^T v \\
&= \bar{\lambda}^T Av + \bar{\mu}^T \underbrace{Cv}_{=0} \\
&= \bar{\lambda}^T_{I_1} \underbrace{A_{I_1}v}_{=0} + \bar{\lambda}^T_{I_2} \underbrace{A_{I_2}v}_{=0} + \bar{\lambda}^T_{I\setminus I_0} \underbrace{A_{I\setminus I_0}v}_{=0} \\
&= 0.
\end{aligned}$$

Hence $v \in (\nabla f(\bar{x}))^{\perp}$. It follows that

$$\{v \in R^n : A_{I_1}v = 0, \ A_{I_2}v \geq 0, \ Cv = 0\}$$
$$\subset \{v \in R^n : A_{I_0}v \geq 0, \ Cv = 0\} \cap (\nabla f(\bar{x}))^{\perp}.$$

To obtain the reverse inclusion, suppose that $v \in R^n$, $A_{I_0}v \geq 0$, $Cv = 0$, $v \in (\nabla f(\bar{x}))^{\perp}$. We need only to show that $A_{I_1}v = 0$. From (3.11) we deduce that

$$\begin{aligned}
0 = \langle \nabla f(\bar{x}), v \rangle &= (D\bar{x} + c)^T v \\
&= (A^T\bar{\lambda} + C^T\bar{\mu})^T v \\
&= \bar{\lambda}^T Av \\
&= \underbrace{\bar{\lambda}^T_{I_1}}_{>0} \underbrace{A_{I_1}v}_{\geq 0} + \bar{\lambda}^T_{I_2} \underbrace{A_{I_2}v}_{=0}.
\end{aligned}$$

Hence $A_{I_1}v = 0$, and the proof is complete. $\square$

Note that Theorem 3.4 can be reformulated in the following equivalent form which does not require the use of Lagrange multipliers.

Theorem 3.5. (See Cottle et al. (1992), p. 116) *The necessary and sufficient condition for a point* $\bar{x} \in R^n$ *to be a local solution of problem* (2.27) *is that the next two properties are valid:*

(i) $\langle \nabla f(\bar{x}), v \rangle = (D\bar{x} + c)^T v \geq 0$ *for every* $v \in T_\Delta(\bar{x}) = \{v \in R^n :$
$A_{I_0} v \geq 0, \; Cv = 0\}$, *where* $I_0 = \{i : A_i \bar{x} = b_i\}$;

(ii) $v^T Dv \geq 0$ *for every* $v \in T_\Delta(\bar{x}) \cap (\nabla f(\bar{x}))^\perp$, *where* $(\nabla f(\bar{x}))^\perp =$
$\{v \in R^n : \langle \nabla f(\bar{x}), v \rangle = 0\}$.

The fact that the first property is equivalent to the existence of a pair $(\bar{\lambda}, \bar{\mu}) \in R^m \times R^s$ satisfying system (3.11) can be established by using the Farkas Lemma (see Theorem 3.2) and some arguments similar to those in the proof of Lemma 3.1. The equivalence between property (ii) in Theorem 3.5 and property (ii) in Theorem 3.4, which is formulated via a Lagrange multipliers set $(\lambda, \bar{\mu}) \in R^m \times R^s$, follows from Lemma 3.1. Hence Theorem 3.5 is an equivalent form of Theorem 3.4.

Proof of Theorem 3.4.

Necessity: Suppose that $\bar{x}$ is a local solution of (2.27). Then there exists $\varepsilon > 0$ such that

$$f(x) - f(\bar{x}) \geq 0 \quad \forall x \in \Delta \cap B(\bar{x}, \varepsilon). \tag{3.14}$$

According to Corollary 3.2, there exists $(\bar{\lambda}, \bar{\mu}) \in R^m \times R^n$ such that condition (i) is satisfied. Let I_1 and I_2 be defined as in (3.12). Suppose that property (ii) were false. Then we could find $\bar{v} \in R^n \setminus \{0\}$ such that

$$A_{I_1} \bar{v} = 0, \quad A_{I_2} \bar{v} \geq 0, \quad C\bar{v} = 0, \quad \bar{v}^T D\bar{v} < 0.$$

By Lemma 3.1, $(D\bar{x} + c)^T \bar{v} = \langle \nabla f(\bar{x}), \bar{v} \rangle = 0$. Consequently, for each $t \in (0, 1)$ we have

$$f(\bar{x} + t\bar{v}) - f(\bar{x}) = t(D\bar{x} + c)^T \bar{v} + \frac{t^2}{2} \bar{v}^T D\bar{v} = \frac{1}{2} t^2 \bar{v}^T D\bar{v} < 0.$$

As $\bar{x} + t\bar{v} \in \Delta \cap B(\bar{x}, \varepsilon)$ for all $t \in (0, 1)$ sufficiently small, the last fact contradicts (3.14). Thus (ii) must hold true.

Sufficiency: Suppose that $\bar{x} \in R^n$ is such that there exists $(\bar{\lambda}, \bar{\mu}) \in R^m \times R^n$ such that conditions (i) and (ii) are satisfied. We shall prove that $\bar{x}$ is a local solution of (2.27). The main idea of the proof is to decompose the tangent cone $T_\Delta(\bar{x})$ into the sum of a subspace and a pointed polyhedral convex cone. Set $I = \{1, 2, \ldots, m\}$, $I_0 = \{i \in I : A_i \bar{x} = b_i\}$, and observe that $I_0 = I_1 \cup I_2$. Define

$$M = \{v \in R^n : A_{I_0} v = 0, \; Cv = 0\},$$
$$M^\perp = \{v \in R^n : \langle v, u \rangle = 0 \; \forall u \in M\}.$$

Let

$$K = \{v \in M^\perp : v = z - u \text{ for some } z \in T_\Delta(\bar{x}) \text{ and } u \in M\}$$
$$= \mathrm{Pr}_{M^\perp}(T_\Delta(\bar{x})),$$

where $\mathrm{Pr}_{M^\perp}(\cdot)$ denotes the orthogonal projection of R^n onto $M^\perp$. Since $K = \mathrm{Pr}_{M^\perp}(T_\Delta(\bar{x}))$ and $M \subset T_\Delta(\bar{x})$, it follows that

$$T_\Delta(\bar{x}) = M + K. \tag{3.15}$$

We have

$$K = M^\perp \cap T_\Delta(\bar{x}). \tag{3.16}$$

Indeed, if $v \in K$ then $v \in M^\perp$ and $v = z - u$ for some $z \in T_\Delta(\bar{x})$ and $u \in M$. Hence

$$A_{I_0} u = 0, \quad Cu = 0, \quad A_{I_0} z \geq 0, \quad Cz = 0.$$

Therefore

$$A_{I_0} v = A_{I_0} z - A_{I_0} u = A_{I_0} z \geq 0,$$
$$Cv = Cz - Cu = 0.$$

So $v \in M^\perp \cap T_\Delta(\bar{x})$. It follows that $K \subset M^\perp \cap T_\Delta(\bar{x})$. For proving the reverse inclusion, it suffices to note that each $v \in M^\perp \cap T_\Delta(\bar{x})$ admits the representation $v = v - 0$ where $v \in T_\Delta(\bar{x})$, $0 \in M$.

We next show that K is a pointed polyhedral convex cone. Recall that a cone $K \subset R^n$ is said to be *pointed* if $K \cap (-K) = \{0\}$. From (3.16) it follows that K is a polyhedral convex cone. So we need only to show that K is pointed. If it were true that K is not pointed, there would be $v \in K \setminus \{0\}$ such that $-v \in K$. On one hand, from (3.13) and (3.16) it follows that

$$A_{I_0} v \geq 0, \quad Cv = 0, \quad A_{I_0}(-v) \geq 0.$$

This implies that $A_{I_0} v = 0$ and $Cv = 0$. So $v \in M$. On the other hand, since $v \in K$, by (3.16) we have $v \in M^\perp$. Thus $v \in M \cap M^\perp = \{0\}$, a contradiction.

Define $K_0 = \{v \in K : \langle \nabla f(\bar{x}), v \rangle = 0\}$. Since K is a pointed polyhedral convex cone, we see that K_0 is also a pointed polyhedral convex cone.

From (i) it follows that

$$\langle \nabla f(\bar{x}), v \rangle \geq 0 \quad \forall v \in T_\Delta(\bar{x}). \tag{3.17}$$

Indeed, let $v \in T_\Delta(\bar{x})$. By (i) and (3.13),

$$
\begin{aligned}
\langle \nabla f(\bar{x}), v \rangle = (D\bar{x} + c)^T v &= (A^T \bar{\lambda} + C^T \bar{\mu})^T v \\
&= \bar{\lambda}^T Av + \bar{\mu}^T Cv \\
&= \bar{\lambda}^T Av \\
&\geq 0.
\end{aligned}
$$

Since $M \subset T_\Delta(\bar{x})$ and $-v \in M$ whenever $v \in M$, it follows that

$$\langle \nabla f(\bar{x}), v \rangle = 0 \quad \forall v \in M. \tag{3.18}$$

From (3.17) and the definition of K_0 we have

$$\langle \nabla f(\bar{x}), v \rangle > 0 \quad \forall v \in K \setminus K_0. \tag{3.19}$$

Since K is a polyhedral convex cone, according to Theorem 19.1 in Rockafellar (1970), K is a *finitely generated cone*. The latter means that there exists a finite system of nonzero vectors $\{z^1, \ldots, z^q\}$, called the *generators* of K, such that

$$K = \{v = \sum_{j=1}^q t_j z^j : t_j \geq 0 \text{ for all } j = 1, \ldots, q\}. \tag{3.20}$$

If $K_0 \neq \{0\}$ then some of the vectors z^j $(j = 1, \ldots, q)$ must belong to K_0. To prove this, suppose, contrary to our claim, that $K_0 \neq \{0\}$ and all the generators z^j $(j = 1, \ldots, q)$ belong to $K \setminus K_0$. Let $\bar{v} = \sum_{j=1}^q t_j z^j$, where $t_j \geq 0$ for all j, be a nonzero vector from K_0. Since at least one of the values t_j must be nonzero, from (3.19) we deduce that

$$\langle \nabla f(\bar{x}), \bar{v} \rangle = \sum_{j=1}^q t_j \langle \nabla f(\bar{x}), z^j \rangle > 0.$$

This contradicts our assumption that $\bar{v} \in K_0$. If $K_0 \neq \{0\}$, there is no loss of generality in assuming that the first q_0 generators z^j $(j = 1, \ldots, q_0)$ belong to K_0, and the other generators z_j $(j = q_0 + 1, \ldots, q)$ belong to $K \setminus K_0$. Thus $\langle \nabla f(\bar{x}), z^j \rangle = 0$ for $j = 1, \ldots, q_0$, $\langle \nabla f(\bar{x}), z^j \rangle > 0$ for $j = q_0 + 1, \ldots, q$, and $q_0 \in \{1, \ldots, q\}$.

We are now in a position to prove the following claim:

CLAIM. $\bar{x}$ *is a local solution of* (2.27).

If $\bar{x}$ were not a local solution of (2.27), we would find a sequence $\{x^k\} \subset \Delta$ such that $x^k \to \bar{x}$, and

$$f(x^k) - f(\bar{x}) < 0 \quad \forall k \in N.$$

For each $k \in N$, on account of (3.15), we have

$$x^k - \bar{x} \in T_\Delta(\bar{x}) = M + K.$$

Combining this with (3.20) we deduce that there exist $t_j^k \geq 0$ $(j = 1, \ldots, q)$ and $u^k \in M$ such that

$$x^k - \bar{x} = u^k + \sum_{j=1}^{q_0} t_j^k z^j + \sum_{j=q_0+1}^{q} t_j^k z^j. \qquad (3.21)$$

Writing $v^k = \sum_{j=1}^{q_0} t_j^k z^j$ and $w^k = \sum_{j=q_0+1}^{q} t_j^k z^j$ yields

$$x^k - \bar{x} = u^k + v^k + w^k. \qquad (3.22)$$

It is understood that v^k (resp., w^k) is absent in the last representation if $K_0 = \{0\}$ (resp., $K \setminus K_0 = \emptyset$). There are two possible cases:

- *Case 1:* There exists a subsequence $\{k'\} \subset \{k\}$ such that $w^{k'} = 0$ for all k'. (If $K \setminus K_0 = \emptyset$, then w^k is vacuous for all $k \in N$. Such situation is also included in this case.)

- *Case 2:* There exists a number $k_* \in N$ such that $w^k \neq 0$ for every $k \geq k_*$.

If *Case 1* occurs, then without restriction of generality we can assume that $\{k'\} \equiv \{k\}$. Since $x^k - \bar{x} = u^k + v^k$, from (3.18) we have

$$(D\bar{x} + c)^T (x^k - \bar{x}) = \langle \nabla f(\bar{x}), x^k - \bar{x} \rangle = \langle \nabla f(\bar{x}), u^k \rangle + \langle \nabla f(\bar{x}), v^k \rangle$$
$$= 0.$$

Therefore

$$f(x^k) - f(\bar{x}) = \frac{1}{2}(x^k - \bar{x})^T D(x^k - \bar{x}) + (D\bar{x} + c)^T (x^k - \bar{x})$$
$$= \frac{1}{2}(x^k - \bar{x})^T D(x^k - \bar{x}).$$

Hence

$$(x^k - \bar{x})^T D(x^k - \bar{x}) < 0 \quad \forall k \in N. \qquad (3.23)$$

Since $x^k - \bar{x} \in T_\Delta(\bar{x})$ and $\langle \nabla f(\bar{x}), x^k - \bar{x} \rangle = 0$, we have $x^k - \bar{x} \in T_\Delta(\bar{x}) \cap (\nabla f(\bar{x}))^\perp$. Consequently, from Lemma 3.1 and from assumption (ii) we obtain $(x^k - \bar{x})^T D(x^k - \bar{x}) \geq 0$, contrary to (3.23).

If *Case 2* happens, then there is no loss of generality in assuming that $w^k \neq 0$ for all $k \in N$. For each k, since t_j^k $(j = q_0 + 1, \ldots, q)$ are nonnegative and not all zero, there must exists some $j(k) \in \{q_0 + 1, \ldots, q\}$ such that

$$t_{j(k)}^k = \max\{t_j^k : j \in \{q_0 + 1, \ldots, q\}\} > 0.$$

It is clear that there must exist an index $j_* \in \{q_0 + 1, \ldots, q\}$ and a subsequence $\{k'\} \subset \{k\}$ such that $j(k') = j_*$ for every k'. Without loss of generality we can assume that $\{k'\} \equiv \{k\}$. On account of (3.21) and (3.22), we have

$$
\begin{aligned}
f(x^k) - f(\bar{x}) &= \frac{1}{2}(x^k - \bar{x})^T D(x^k - \bar{x}) + (D\bar{x} + c)^T (x^k - \bar{x}) \\
&= \frac{1}{2}(u^k + v^k + w^k)^T D(u^k + v^k + w^k) \\
&\qquad + (D\bar{x} + c)^T (u^k + v^k + w^k) \\
&= \underbrace{\frac{1}{2}(u^k + v^k)^T D(u^k + v^k)}_{\geq 0} + (u^k + v^k)^T D w^k \\
&\quad + \sum_{j=q_0+1}^q t_j^k (D\bar{x} + c)^T z^j + \frac{1}{2}(w^k)^T D w^k \\
&\geq \sum_{j=q_0+1}^q t_j^k (u^k + v^k)^T D z^j + t_{j_*}^k (D\bar{x} + c)^T z^{j_*} \\
&\quad + \frac{1}{2}\left(\sum_{j=q_0+1}^q t_j^k z^j\right)^T D w^k.
\end{aligned}
$$

In these transformations, we have used the inequality

$$(u^k + v^k)^T D(u^k + v^k) \geq 0,$$

which is a consequence of Lemma 3.1 and condition (ii). From what has already been proved, it follows that

$$0 > \sum_{j=q_0+1}^q t_j^k (u^k + v^k)^T D z^j + t_{j_*}^k (D\bar{x} + c)^T z^{j_*} + \frac{1}{2}\left(\sum_{j=q_0+1}^q t_j^k z^j\right)^T D w^k.$$

$$(3.24)$$

Dividing (3.24) by $t_{j_*}^k$, noting that $0 \leq t_j^k / t_{j_*}^k \leq 1$ for every $j = q_0 + 1, \ldots, q$, letting $k \to \infty$ and using the following

FACT. *If $x^k \to \bar{x}$ then $u^k \to 0$, $v^k \to 0$, and $w^k \to 0$,* we get

$$0 \geq (D\bar{x} + c)^T z^{j*}, \tag{3.25}$$

a contradiction. This finishes the proof our Claim.

What is left is to show that the Fact is true. For proving it, we first observe that

$$\|x^k - \bar{x}\|^2 = \langle x^k - \bar{x}, x^k - \bar{x}\rangle = \langle u^k + v^k + w^k, u^k + v^k + w^k\rangle$$
$$= \|u^k\|^2 + \|v^k + w^k\|^2.$$

Since $\|x^k - \bar{x}\| \to 0$, it follows that $u^k \to 0$ and $v^k + w^k \to 0$. We have

$$v^k + w^k = \sum_{j=1}^{q} t_j^k z^j.$$

It suffices to prove that, for any $j \in \{1, \dots, q\}$, $t_j^k \to 0$ as $k \to \infty$. On the contrary, suppose that there exists $j_1 \in \{1, \dots, q\}$ such that the sequence $\{t_{j_1}^k\}$ does not converge to 0 as $k \to \infty$. Then there exist $\varepsilon > 0$ and a subsequence $\{k'\} \subset \{k\}$ such that $t_{j_1}^{k'} \geq \varepsilon$ for every k'. Since $\sum_{j=1}^{q} t_j^{k'} \geq t_{j_1}^{k'} \geq \varepsilon$ for every k', we can write

$$v^{k'} + w^{k'} = \sum_{j=1}^{q} t_j^{k'} z^j = \left(\sum_{j=1}^{q} t_j^{k'}\right) \sum_{j=1}^{q} \frac{t_j^{k'}}{\sum_{j=1}^{q} t_j^{k'}} z^j. \tag{3.26}$$

Replace $\{k'\}$ by a subsequence if necessary, we can assume that, for every $j \in \{1, \dots, q\}$,

$$\frac{t_j^{k'}}{\sum_{j=1}^{q} t_j^{k'}} \to \bar{\tau}_j$$

for some $\bar{\tau}_j \in [0, 1]$. It is clear that $\sum_{j=1}^{q} \bar{\tau}_j = 1$. We must have $\sum_{j=1}^{q} \bar{\tau}_j z^j \neq 0$. Indeed, if it were true that $\sum_{j=1}^{q} \bar{\tau}_j z^j \neq 0$, there would be some $j_0 \in \{1, \dots, q\}$ such that $\bar{\tau}_{j_0} \neq 0$. Then

$$\sum_{j \neq j_0} \bar{\tau}_j z^j = -\bar{\tau}_{j_0} z^{j_0}.$$

This implies that $-\bar{\tau}_{j_0} z^{j_0} \in K$, $\bar{\tau}_{j_0} z^{j_0} \in K$, $\bar{\tau}_{j_0} z^{j_0} \neq 0$. Hence the cone K is not pointed, a contradiction. We have thus proved that $\bar{z} := \sum_{j=1}^{q} \bar{\tau}_j z^j$ is a nonzero vector. If the sequence $\{\sum_{j=1}^{q} t_j^{k'}\}$ is bounded, then without loss of generality we can assume that it converges to some limit $\hat{\tau} \geq \varepsilon$. Letting $k' \to \infty$, from (3.26) we deduce that $0 = \hat{\tau}\bar{z}$, a contradiction. If the sequence $\{\sum_{j=1}^{q} t_j^{k'}\}$ is unbounded, then without loss of generality we can assume that it converges to $+\infty$. From (3.26) it follows that

$$\|v^{k'} + w^{k'}\| = \left(\sum_{j=1}^{q} t_j^{k'}\right) \left\|\sum_{j=1}^{q} \frac{t_j^{k'}}{\sum_{j=1}^{q} t_j^{k'}} z^j\right\|.$$

Letting $k' \to \infty$ we obtain $0 = +\infty \|\bar{z}\|$, an absurd. $\quad\square$

Definition 3.2. (See Mangasarian (1980), p. 201) A point $\bar{x} \in \Delta$ is called a *locally unique solution* of the problem $\min\{f(x) : x \in \Delta\}$, where $f : R^n \to R$ is a real function and $\Delta \subset R^n$ is a given subset, if there exists $\varepsilon > 0$ such that

$$f(x) > f(\bar{x}) \quad \forall x \in (\Delta \cap B(\bar{x}, \varepsilon)) \setminus \{\bar{x}\}.$$

Of course, if $\bar{x}$ is a locally unique solution of a minimization problem then it is a local solution of that problem. The converse is not true in general.

The following theorem describes the (second-order) necessary and sufficient condition for a point to be a locally unique solution of a quadratic program.

Theorem 3.6. (See Mangasarian (1980), Theorem 2.1, and Contesse (1980), Théorème 1) *The necessary and sufficient condition for a point $\bar{x} \in R^n$ to be a locally unique solution of problem (2.27) is that there exists a pair of vectors*

$$(\bar{\lambda}, \bar{\mu}) = (\bar{\lambda}_1, \ldots, \bar{\lambda}_m, \bar{\mu}_1, \ldots, \bar{\mu}_s) \in R^m \times R^s$$

such that

(i) *The system (3.11) is satisfied, and*

(ii) *If $v \in R^n \setminus \{0\}$ is such that $A_{I_1}v = 0$, $A_{I_2}v \geq 0$, $Cv = 0$, where*

$$I_1 = \{i : A_i\bar{x} = b_i, \ \bar{\lambda}_i > 0\}, \quad I_2 = \{i : A_i\bar{x} = b_i, \ \bar{\lambda}_i = 0\},$$

then $v^T Dv > 0$.

Proof. *Necessity:* Suppose that $\bar{x}$ is a locally unique solution of (2.27). Then there exists $\varepsilon > 0$ such that

$$f(x) - f(\bar{x}) > 0 \quad \forall x \in (\Delta \cap B(\bar{x}, \varepsilon)) \setminus \{\bar{x}\}. \tag{3.27}$$

According to Corollary 3.2, there exists $(\bar{\lambda}, \bar{\mu}) \in R^m \times R^n$ such that condition (i) is satisfied. Suppose that property (ii) were false. Then we could find $\bar{v} \in R^n \setminus \{0\}$ such that

$$A_{I_1}\bar{v} = 0, \quad A_{I_2}\bar{v} \geq 0, \quad C\bar{v} = 0, \quad \bar{v}^T D\bar{v} \leq 0.$$

By Lemma 3.1, $(D\bar{x} + c)^T \bar{v} = \langle \nabla f(\bar{x}), \bar{v} \rangle = 0$. Consequently, for each $t \in (0, 1)$ we have

$$f(\bar{x} + t\bar{v}) - f(\bar{x}) = t(D\bar{x} + c)^T \bar{v} + \frac{t^2}{2}\bar{v}^T D\bar{v} = \frac{1}{2}t^2\bar{v}^T D\bar{v} \leq 0.$$

As $\bar{x} + t\bar{v} \in \Delta \cap B(\bar{x}, \varepsilon)$ for all $t \in (0, 1)$ sufficiently small, the last fact contradicts (3.27). Thus (ii) must hold true.

Sufficiency: Suppose that $\bar{x} \in R^n$ is such that there exists $(\bar{\lambda}, \bar{\mu}) \in R^m \times R^n$ such that (i) and (ii) are satisfied. We shall show that $\bar{x}$ is a locally unique solution of (2.27). Set $I = \{1, 2, \ldots, m\}$, $I_0 = \{i \in I : A_i\bar{x} = b_i\}$. Let $M, M^\perp, K, K_0, z^1, \ldots, z^q$, and q_0 be defined as in the proof of Theorem 3.4. Then the properties (3.15)–(3.20) are valid.

If $\bar{x}$ were not a locally unique solution of (2.27), we would find a sequence $\{x^k\} \subset \Delta$ such that $x^k \to \bar{x}$, and

$$f(x^k) - f(\bar{x}) \leq 0 \quad \forall k \in N.$$

For each $k \in N$, on account of (3.15), we have

$$x^k - \bar{x} \in T_\Delta(\bar{x}) = M + K.$$

Combining this with (3.20) we conclude that there exist $t_j^k \geq 0$ ($j = 1, \ldots, q$) and $u^k \in M$ such that (3.21) holds. Setting $v^k = \sum_{j=1}^{q_0} t_j^k z^j$ and $w^k = \sum_{j=q_0+1}^{q} t_j^k z^j$ we have (3.22). As before, if $K_0 = \{0\}$ (resp., $K \setminus K_0 = \emptyset$) then it is understood that v^k (resp., w^k) is absent in the representation (3.22). We consider separately the following two cases:

- *Case 1:* There exists a subsequence $\{k'\} \subset \{k\}$ such that $w^{k'} = 0$ for all k'. (If $K \setminus K_0 = \emptyset$, then w^k is vacuous for all $k \in N$. Such situation is also included in this case.)

- *Case 2:* There exists a number $k_* \in N$ such that $w^k \neq 0$ for every $k \geq k_*$.

If *Case 1* occurs, then without restriction of generality we can assume that $\{k'\} \equiv \{k\}$. Arguing similarly as in the analysis of *Case 1* in the preceding proof, we obtain

$$(x^k - \bar{x})^T D(x^k - \bar{x}) \leq 0. \tag{3.28}$$

Since $x^k - \bar{x} \in T_\Delta(\bar{x})$ and $\langle \nabla f(\bar{x}), x^k - \bar{x} \rangle = 0$, we have $x^k - \bar{x} \in T_\Delta(\bar{x}) \cap (\nabla f(\bar{x}))^\perp$. Hence from Lemma 3.1 and from assumption (ii) it follows that $(x^k - \bar{x})^T D(x^k - \bar{x}) > 0$, contrary to (3.28).

If *Case 2* happens then there is no loss of generality in assuming that $w^k \neq 0$ for all $k \in N$. Construct the sequence $\{j(k)\}$ $(k \in N)$ as in the proof of Theorem 3.4. Then there must exist an index $j_* \in \{q_0+1, \ldots, q\}$ and a subsequence $\{k'\} \subset \{k\}$ such that $j(k') = j_*$ for every k'. Without loss of generality we can assume that $\{k'\} \equiv \{k\}$. Analysis similar to that in the proof of Theorem 3.4 shows that

$$0 \geq \sum_{j=q_0+1}^{q} t_j^k (u^k+v^k)^T D z^j + t_{j_*}^k (D\bar{x}+c)^T z^{j_*} + \frac{1}{2} \left(\sum_{j=q_0+1}^{q} t_j^k z^j \right)^T D w^k. \tag{3.29}$$

Dividing (3.29) by $t_{j_*}^k$, noting that $0 \leq t_j^k/t_{j_*}^k \leq 1$ for every $j = q_0 + 1, \ldots, q$, letting $k \to \infty$ and using the Fact established in the preceding proof, we get (3.25). This contradicts (3.19) because $z^{j_*} \in K \backslash K_0$. We have thus proved that $\bar{x}$ is a locally unique solution of (2.27). $\square$

Note that Theorem 3.6 can be reformulated in the following equivalent form which does not require the use of Lagrange multipliers.

Theorem 3.7. *The necessary and sufficient condition for a point $\bar{x} \in R^n$ to be a locally unique solution of problem (2.27) is that the next two properties are valid:*

(i) $\langle \nabla f(\bar{x}), v \rangle = (D\bar{x} + c)^T v \geq 0$ *for every* $v \in T_\Delta(\bar{x}) = \{v \in R^n : A_{I_0} v \geq 0, \; Cv = 0\}$, *where* $I_0 = \{i : A_i \bar{x} = b_i\}$;

(ii) $v^T Dv > 0$ *for every nonzero vector* $v \in T_\Delta(\bar{x}) \cap (\nabla f(\bar{x}))^\perp$, *where* $(\nabla f(\bar{x}))^\perp = \{v \in R^n : \langle \nabla f(\bar{x}), v \rangle = 0\}$.

As it has been noted after the formulation of Theorem 3.5, the first property is equivalent to the existence of a pair $(\bar{\lambda}, \bar{\mu}) \in R^m \times R^s$ satisfying system (3.11). The equivalence between property (ii) in Theorem 3.7 and property (ii) in Theorem 3.6, which is formulated via a Lagrange multipliers set $(\lambda, \bar{\mu}) \in R^m \times R^s$, follows from Lemma 3.1. Hence Theorem 3.7 is an equivalent form of Theorem 3.6.

It is interesting to observe that if $\bar{x}$ is a locally unique solution of a quadratic program then a property similar to (3.4) holds.

Theorem 3.8. *If $\bar{x} \in R^n$ is a locally unique solution of problem (2.27) then there exist $\varepsilon > 0$ and $\varrho > 0$ such that*

$$f(x) - f(\bar{x}) \geq \varrho\|x - \bar{x}\|^2 \quad \text{for every } x \in \Delta \cap B(\bar{x}, \varepsilon), \quad (3.30)$$

where $\Delta = \{x \in R^n : Ax \geq b, \ Cx = d\}$ is the constraint set of (2.27).

Proof. Let $\bar{x} \in R^n$ be a locally unique solution of (2.27). By Theorem 3.6, there exists a pair of vectors

$$(\bar{\lambda}, \bar{\mu}) = (\bar{\lambda}_1, \dots, \bar{\lambda}_m, \bar{\mu}_1, \dots, \bar{\mu}_s) \in R^m \times R^s$$

such that

(i)' The system (3.11) is satisfied, and

(ii)' If $v \in R^n \setminus \{0\}$ is such that $A_{I_1}v = 0$, $A_{I_2}v \geq 0$, $Cv = 0$, where

$$I_1 = \{i : A_i\bar{x} = b_i, \ \bar{\lambda}_i > 0\}, \quad I_2 = \{i : A_i\bar{x} = b_i, \ \bar{\lambda}_i = 0\},$$

then $v^T Dv > 0$.

As it has been noted in the proof of Theorem 3.4, from (i)' it follows that (3.17) is valid.

To obtain a contradiction, suppose that one cannot find any pair of positive numbers (ε, ϱ) satisfying (3.30). Then, for each $k \in N$, there exists $x^k \in \Delta$ such that $\|x^k - \bar{x}\| \leq \dfrac{1}{k}$ and

$$f(x^k) - f(\bar{x}) < \frac{1}{k}\|x^k - \bar{x}\|^2. \quad (3.31)$$

The last inequality implies that $x^k \neq \bar{x}$. Without loss of generality we can assume that the sequence $\{(x^k - \bar{x})/\|x^k - \bar{x}\|\}$ converges to some $\bar{v} \in R^n$ with $\|\bar{v}\| = 1$. By (3.31), we have

$$\frac{1}{k}\|x^k - \bar{x}\|^2 > f(x^k) - f(\bar{x}) = \frac{1}{2}(x^k - \bar{x})^T D(x^k - \bar{x}) + (D\bar{x} + c)^T(x^k - \bar{x}). \quad (3.32)$$

Dividing this expression by $\|x^k - \bar{x}\|$ and letting $k \to \infty$ we get $0 \geq (D\bar{x} + c)^T \bar{v}$. Since $x^k - \bar{x} \in T_\Delta(\bar{x})$ for every $k \in N$, we must have $\bar{v} \in T_\Delta(\bar{x})$. By (3.17), $(D\bar{x} + c)^T \bar{v} \geq 0$. Thus $\langle \nabla f(\bar{x}), \bar{v} \rangle = (D\bar{x} + c)^T \bar{v} = 0$. As $x^k - \bar{x} \in T_\Delta(\bar{x})$ for every $k \in N$, according to (3.17) we have $(D\bar{x} + c)^T (x^k - \bar{x}) \geq 0$. Combining this with (3.32) yields

$$\frac{1}{k} \|x^k - \bar{x}\|^2 > \frac{1}{2}(x^k - \bar{x})^T D(x^k - \bar{x}).$$

Dividing the last inequality by $\|x^k - \bar{x}\|^2$ and letting $k \to \infty$ we obtain $0 \geq \bar{v}^T D\bar{v}$. Since $\bar{v} \in T_\Delta(\bar{x}) \cap (\nabla f(\bar{x}))^\perp$, from Lemma 3.1 and (ii)' it follows that $\bar{v}^T D\bar{v} > 0$. We have arrived at a contradiction. The proof is complete. $\square$

3.3   Commentaries

First-order necessary and sufficient optimality conditions for (non-convex) quadratic programs are proved in several textbooks. Meanwhile, to our knowledge, the paper of Contesse (1980) is the only place where one can find a satisfactory proof of the second-order necessary and sufficient optimality condition for quadratic programs which was noted firstly by Majthay in 1971 and which has many interesting applications (see, for instance, Cottle et al. (1992) and Chapters 4, 10, 14 of this book). The reason might be that the proof is rather long and complicated. The proof described in this chapter is essentially that one of Contesse. For the benefit of the reader, we have proposed a series of minor modifications in the presentation. The formulation given in Theorem 3.4 can be used effectively in performing practical calculations to find the local solution set, while the formulation given in Theorem 3.5 is very convenient for theoretical investigations concerning the solution sets of quadratic programs (see the next chapter).

The necessary and sufficient condition for locally unique solutions of quadratic programs described in Theorem 3.6 and Theorem 3.7 is also a good criterion for the stability of the local solutions. The result formulated in Theorem 3.6 was obtained independently by Mangasarian (1980) and Contesse (1980). The proof given in this chapter follows the scheme proposed by Contesse. Another nice proof of the "Necessity" part of Theorem 3.6 can be found in Mangasarian (1980). In Mangasarian (1980) it was noted that the "Sufficiency" part of the result stated in Theorem 3.6 follows from the general second-order sufficient optimality condition for smooth mathematical programming problems established by McCormick (see McCormick (1967), Theorem 6). Actually, the studies of Majthay (1971), Mangasarian (1980), and Contesse (1980) on second-order optimality conditions for quadratic programs have been originated from that work of McCormick.

In mathematical programming theory, it is well known that the estimation like the one in (3.4) (resp., in (3.30)) is a consequence of a strict first-order sufficient optimality condition (resp., of a strong second-order sufficient optimality condition). In this chapter, the two estimations are obtained by simple direct proofs.

Chapter 4

Properties of the Solution Sets of Quadratic Programs

This chapter investigates the structure of the solution sets of quadratic programming problems. We consider the problem

$$(P) \quad \min\{f(x) = \frac{1}{2}x^T D x + c^T x \ : \ x \in R^n, \ Ax \geq b, \ Cx = d\},$$

where $D \in R_S^{n \times n}$, $A \in R^{m \times n}$, $C \in R^{s \times n}$, $c \in R^n$, $b \in R^m$, $d \in R^s$. Let

$$\Delta = \{x \in R^n \ : \ Ax \geq b, \ Cx = d\}, \ I = \{1, \ldots, m\}, \ J = \{1, \ldots, s\}.$$

Denote by $\mathrm{Sol}(P)$, $\mathrm{loc}(P)$ and $S(P)$, respectively, the solution set, the local-solution set and the KKT point set of (P). Our aim is to study such properties of the solution sets $\mathrm{Sol}(P)$, $\mathrm{loc}(P)$ and $S(P)$ as boundedness, closedness and finiteness. Note that sometimes the elements of $S(P)$ are called the *Karush-Kuhn-Tucker solutions* of (P). *The above notations will be kept throughout this chapter.*

4.1 Characterizations of the Unboundedness of the Solution Sets

Denote by $\mathrm{Sol}(P_0)$ the solution set of the following *homogeneous* quadratic program associated with (P):

$$(P_0) \qquad \min\left\{\frac{1}{2}v^T D v \ : \ v \in R^n, \ Av \geq 0, \ Cv = 0\right\}.$$

Definition 4.1. A half-line $\omega = \{\bar{x} + t\bar{v} : t \geq 0\}$, where $\bar{v} \in R^n \setminus \{0\}$, which is a subset of $\mathrm{Sol}(P)$ (resp., $\mathrm{loc}(P)$, $S(P)$), is called a *solution ray* (resp., a *local-solution ray*, a *KKT point ray*) of (P).

Theorem 4.1. *The set* $\mathrm{Sol}(P)$ *is unbounded if and only if* (P) *has a solution ray. A necessary and sufficient condition for* $\mathrm{Sol}(P)$ *to be unbounded is that there exist* $\bar{x} \in \mathrm{Sol}(P)$ *and* $\bar{v} \in \mathrm{Sol}(P_0) \setminus \{0\}$ *such that*

$$(D\bar{x} + c)^T \bar{v} = 0. \tag{4.1}$$

The following fact follows directly from the above theorem.

Corollary 4.1. *If the solution set* $\mathrm{Sol}(P_0)$ *is empty or it consists of just one element* 0 *then, for any* $(c, b, d) \in R^n \times R^m \times R^s$, *the solution set* $\mathrm{Sol}(P)$ *is bounded. In the case where* $\mathrm{Sol}(P_0)$ *contains a nonzero element, if*

$$(D\bar{x} + c)^T \bar{v} > 0 \quad \forall \bar{x} \in \mathrm{Sol}(P), \ \forall \bar{v} \in \mathrm{Sol}(P_0) \setminus \{0\},$$

then $\mathrm{Sol}(P)$ *is bounded.*

Proof of Theorem 4.1.

Suppose that $\mathrm{Sol}(P)$ is unbounded. Then there exists a sequence $\{x^k\}$ in $\mathrm{Sol}(P)$ such that $\|x^k\| \to +\infty$ as $k \to \infty$. Without loss of generality we can assume that $x^k \neq 0$ for all k and $\dfrac{x^k}{\|x^k\|} \to \bar{v}$ with $\|\bar{v}\| = 1$. We will show that $\bar{v} \in \mathrm{Sol}(P_0)$. Since $x^k \in \mathrm{Sol}(P)$, we have $Ax^k \geq b$ and $Cx^k = d$. This implies that $A\dfrac{x^k}{\|x^k\|} \geq \dfrac{b}{\|x^k\|}$ and $C\dfrac{x^k}{\|x^k\|} = \dfrac{d}{\|x^k\|}$. Letting $k \to \infty$ we obtain $A\bar{v} \geq 0$ and $C\bar{v} = 0$. Hence $\bar{v}$ is a feasible vector of (P_0). Since $\mathrm{Sol}(P) \neq \emptyset$, by the Eaves Theorem (see Corollary 2.6), $v^T Dv \geq 0$ for every $v \in R^n$ satisfying $Av \geq 0$, $Cv = 0$. In particular, $\bar{v}^T D\bar{v} \geq 0$. Fix a point $\hat{x} \in \Delta$. Since $x^k \in \mathrm{Sol}(P)$, we have

$$\frac{1}{2}(x^k)^T Dx^k + c^T x^k \leq f(\hat{x}) \quad (\forall k \in N).$$

Dividing this inequality by $\|x^k\|^2$ and letting $k \to \infty$, we obtain $\bar{v}^T D\bar{v} \leq 0$. Hence

$$\bar{v}^T D\bar{v} = 0. \tag{4.2}$$

Let $v \in R^n$ be any feasible vector of (P_0), that is $Av \geq 0$ and $Cv = 0$. On account of a preceding remark, we have $v^T Dv \geq 0$. Combining this with (4.2) we deduce that $\bar{v} \in \mathrm{Sol}(P_0) \setminus \{0\}$.

We now show that there exists $\bar{x} \in \mathrm{Sol}(P)$ satisfying (4.1). Since $Ax^k \geq b$ for every $k \in N$, arguing similarly as in the proof of Theorem 2.1 we can find a subsequence $\{k'\} \subset \{k\}$ such that for each $i \in I$ the limit $\lim\limits_{k' \to \infty} A_i x^{k'}$ exists (it may happen that $\lim\limits_{k' \to \infty} A_i x^{k'} = +\infty$). Obviously,

$$\lim_{k' \to \infty} A_i x^{k'} \geq b_i \quad (\forall i \in I)$$

and

$$\lim_{k' \to \infty} C_j x^{k'} = d_j \quad (\forall j \in J).$$

Without restriction of generality we can assume that $\{k'\} \equiv \{k\}$. Let

$$I_0 = \{i \in I : \lim_{k \to \infty} A_i x^k = b_i\}, \quad I_1 = \{i \in I : \lim_{k \to \infty} A_i x^k > b_i\}.$$

It is clear that there exist $\varepsilon > 0$ and $k_0 \in N$ such that

$$A_i x^k \geq b_i + \varepsilon \quad (\forall i \in I_1, \ \forall k \geq k_0).$$

We have

$$A_i \bar{v} = \lim_{k \to \infty} A_i \frac{x^k}{\|x^k\|} = \lim_{k \to \infty} \frac{b_i}{\|x^k\|} = 0 \quad (\forall i \in I_0),$$

and

$$A_i \bar{v} = \lim_{k \to \infty} A_i \frac{x^k}{\|x^k\|} \geq \lim_{k \to \infty} \frac{b_i + \varepsilon}{\|x^k\|} = 0 \quad (\forall i \in I_1).$$

Let $x^k(t) = x^k - t\bar{v}$, where $t > 0$ and $k \geq k_0$. We have

$$A_i x^k(t) = A_i x^k - t A_i \bar{v} = A_i x^k \geq b_i \quad (\forall i \in I_0),$$

$$A_i x^k(t) = A_i x^k - t A_i \bar{v} \geq b_i + \varepsilon - t A_i \bar{v} \quad (\forall i \in I_1).$$

Fix an index $k \geq k_0$. From what has been said it follows that there exists $\delta > 0$ such that, for every $t \in (0, \delta)$,

$$A_i x^k(t) \geq b_i \quad (\forall i \in I = I_0 \cup I_1).$$

It is obvious that $C_j x^k(t) = d_j$ for all $j \in J$. Hence $x^k(t) \in \Delta$ for every $t \in (0, \delta)$. Since

$$0 \leq f(x^k(t)) - f(x^k)$$
$$= \frac{1}{2}(x^k(t) - x^k)^T D(x^k(t) - x^k) + (Dx^k + c)^T (x^k(t) - x^k),$$

we have
$$-\frac{1}{2}t\bar{v}^T D\bar{v} + (Dx^k + c)^T\bar{v} \le 0.$$

Combining this with (4.2) we get

$$(Dx^k + c)^T\bar{v} \le 0.$$

On the other hand, applying Corollary 2.6 we can assert that $(Dx^k + c)^T\bar{v} \ge 0$. Hence $(Dx^k + c)^T\bar{v} = 0$. Taking $\bar{x} = x^k$ we see that (4.1) is satisfied.

Let us prove that if there exist $\bar{x} \in \text{Sol}(P)$ and $\bar{v} \in \text{Sol}(P_0)\setminus\{0\}$ such that (4.1) holds, then $\omega := \{\bar{x} + t\bar{v} : t \ge 0\}$ is a solution ray of (P). For each $t > 0$, since $\bar{x} \in \text{Sol}(P)$ and $\bar{v} \in \text{Sol}(P_0)$, we have

$$A(\bar{x} + t\bar{v}) = A\bar{x} + tA\bar{v} \ge b,$$
$$C(\bar{x} + t\bar{v}) = C\bar{x} + tC\bar{v} = d.$$

Hence $\bar{x} + t\bar{v} \in \Delta$. Since $\bar{v} \in \text{Sol}(P_0)$ and 0 is a feasible vector of (P_0), we have $\bar{v}^T D\bar{v} \le 0$. If $\bar{v}^T D\bar{v} < 0$ then we check at once that (P_0) have no solutions, which is impossible. Thus $\bar{v}^T D\bar{v} = 0$. Combining this with (4.1) we deduce that

$$\begin{aligned}
f(\bar{x} + t\bar{v}) &= \frac{1}{2}(\bar{x} + t\bar{v})^T D(\bar{x} + t\bar{v}) + c^T(\bar{x} + t\bar{v}) \\
&= (\frac{1}{2}\bar{x}^T D\bar{x} + c^T\bar{x}) + t(D\bar{x} + c)^T\bar{v} + \frac{1}{2}t^2\bar{v}^T D\bar{v} \\
&= f(\bar{x}).
\end{aligned}$$

Since $\bar{x} \in \text{Sol}(P)$, we conclude that $\bar{x} + t\bar{v} \in \text{Sol}(P)$ for all $t \ge 0$. We have thus shown that if $\text{Sol}(P)$ is unbounded then there exists $\bar{x} \in \text{Sol}(P)$ and $\bar{v} \in \text{Sol}(P_0)\setminus\{0\}$ satisfying (4.1) and $\omega = \{\bar{x} + t\bar{v} : t \ge 0\}$ is a solution ray of (P).

The claim that if (P) has a solution ray then $\text{Sol}(P)$ is unbounded is obvious. The proof is complete. $\quad\square$

Theorem 4.2. *The set* $\text{loc}(P)$ *is unbounded if and only if* (P) *has a local-solution ray.*

Proof. It suffices to prove that if $\text{loc}(P)$ is unbounded then (P) has a local-solution ray. Suppose that there is a sequence $\{x^k\}$ in $\text{loc}(P)$ satisfying the condition $\|x^k\| \to +\infty$. Let $\alpha \subset I$ be an index set. The set

$$F_\alpha = \{x \in R^n : A_\alpha x = b_\alpha, \ A_{I\setminus\alpha}x > b_{I\setminus\alpha}, \ Cx = d\}$$

(which may be empty) is called a *pseudo-face* (see, for instance, Bank et al. (1982), p. 102) of Δ corresponding to α. Recall (Rockafellar (1970), p. 162) that a *face* of a convex set $X \subset R^n$ is a convex subset F of X such that every closed line segment in X with a relative interior point in F has both endpoints in F. In agreement with this definition, the sets F of the form

$$F = \{x \in R^n : A_\alpha x = b_\alpha, \ A_{I \setminus \alpha} x \geq b_{I \setminus \alpha}, \ Cx = d\}$$

are the faces of the polyhedral convex set Δ under our consideration. Thus pseudo-faces are not faces in the sense of Rockafellar (1970). However, the closures of pseudo-faces are faces in that sense. It is clear that

$$\Delta = \bigcup \{F_\alpha : \alpha \subset I\}$$

and

$$F_\alpha \cap F_{\alpha'} = \emptyset \quad \text{whenever } \alpha \neq \alpha'.$$

It is a simple matter to show that for any $\alpha \subset I$ and for any $\bar{x} \in F_\alpha$ it holds

$$T_\Delta(\bar{x}) = \{v \in R^n : A_\alpha v \geq 0, \ Cv = 0\}.$$

Thus the tangent cone $T_\Delta(\bar{x})$ does not change when $\bar{x}$ varies inside a given pseudo-face F_α.

Since the number of pseudo-faces of Δ is finite, we conclude that there exist an index set $\alpha_* \subset I$ and a subsequence $\{k'\} \subset \{k\}$ such that $x^{k'} \in F_{\alpha_*}$ for every k'. There is no loss of generality in assuming that $\{k'\} \equiv \{k\}$.

We shall apply the construction due to Contesse which helped us to prove Theorem 3.4.

Since $x^k \in F_{\alpha_*}$ for all $k \in N$, we deduce that

$$T_\Delta(x^k) = \{v \in R^n : A_{\alpha_*} v \geq 0, \ Cv = 0\} \quad (\forall k \in N).$$

Let

$$M = \{v \in R^n : A_{\alpha_*} v = 0, \ Cv = 0\}.$$

Then M is a linear subspace and $M \subset T_\Delta(x^k)$. Let $M^\perp = \{v \in R^n : \langle v, u \rangle = 0 \text{ for every } u \in M\}$ and let

$$K = T_\Delta(x^k) \cap M^\perp = \Pr_{M^\perp}(T_\Delta(x^k)),$$

where $\Pr_{M^\perp}(\cdot)$ denotes the orthogonal projection of R^n onto the subspace $M^\perp$. We have

$$K = \{v \in R^n : A_{\alpha_*} v \geq 0, \ Cv = 0, \ v \in M^\perp\}$$

and $T_\Delta(x^k) = M + K$. Let

$$K_0^k = \{v \in K : \langle \nabla f(x^k), v \rangle = 0\}.$$

From the inclusion $x^k \in \text{loc}(P)$ and from Theorem 3.5 it follows that $\langle \nabla f(x^k), u \rangle = 0$ for every $u \in M$ and $\langle \nabla f(x^k), v \rangle \geq 0$ for every $v \in K$. This implies that K_0^k is a face of K.

Since the polyhedral convex cone K has only a finite number of faces, there must exist a face K_0 of K and a subsequence $\{k_l\} \subset \{k\}$ such that

$$K_0^{k_l} = K_0 \quad \forall l \in N.$$

Consider the sequence of unit vectors $\left\{ \dfrac{x^{k_l} - x^{k_1}}{\|x^{k_l} - x^{k_1}\|} \right\}$. Without loss of generality we can assume that

$$\frac{x^{k_l} - x^{k_1}}{\|x^{k_l} - x^{k_1}\|} \to \bar{z}, \quad \|\bar{z}\| = 1.$$

Set $\omega = \{x^{k_1} + t\bar{z} : t \geq 0\}$.

CLAIM 1. $\omega \subset S(P)$.

Let $x = x^{k_1} + t\bar{z}$, $t > 0$. For every $v \in M + K$ and $l \in N$, since $M + K = T_\Delta(x^{k_l})$ and $x^{k_l} \in \text{loc}(P)$, by Theorem 3.5 we have

$$(Dx^{k_l} + c)^T v = \langle \nabla f(x^{k_l}), v \rangle \geq 0.$$

Hence

$$\left(D\frac{x^{k_l} - x^{k_1}}{\|x^{k_l} - x^{k_1}\|} + \frac{c + Dx^{k_1}}{\|x^{k_l} - x^{k_1}\|} \right)^T v \geq 0.$$

Letting $k \to \infty$ we deduce that

$$(D\bar{z})^T v \geq 0 \quad (\forall v \in M + K). \tag{4.3}$$

Since $M + K = T_\Delta(x^{k_1})$ and $x^{k_1} \in \text{loc}(P)$, by Theorem 3.5 we have

$$(Dx^{k_1} + c)^T v = \langle \nabla f(x^{k_1}), v \rangle \geq 0 \quad (\forall v \in M + K). \tag{4.4}$$

Combining (4.4) with (4.3) gives

$$\langle \nabla f(x), v \rangle = (Dx + c)^T v = (Dx^{k_1} + c)^T v + t(D\bar{z})^T v \geq 0 \tag{4.5}$$

for all $v \in M + K$. We have $x \in F_{\alpha_*}$. Indeed, since $x^{k_l} \in F_{\alpha_*}$ for every $l \in N$, it follows that

$$A_i x = A_i(x^{k_1} + t\bar{z}) = A_i x^{k_1} + t \lim_{l \to \infty} \frac{A_i x^{k_l} - A_i x^{k_1}}{\|x^{k_l} - x^{k_1}\|} = b_i \quad (\forall i \in \alpha_*).$$

For every $i \in I$, we have

$$A_i \frac{x^{k_l} - x^{k_1}}{\|x^{k_l} - x^{k_1}\|} \geq \frac{b_i - A_i x^{k_1}}{\|x^{k_l} - x^{k_1}\|}.$$

Letting $l \to \infty$ yields

$$A_i \bar{z} \geq 0 \quad (\forall i \in I).$$

Consequently,

$$A_i x = A_i(x^{k_1} + t\bar{z}) = A_i x^{k_1} + t A_i \bar{z} > b_i \quad (\forall i \in I \setminus \alpha_*).$$

The equality $Cx = d$ can be established without any difficulty. From what has already been proved, we deduce that $x \in F_{\alpha_*}$. This implies that $T_\Delta(x) = M + K$. Hence from (4.5) it follows that

$$\langle \nabla f(x), v \rangle = (Dx + c)^T v \geq 0 \quad (\forall v \in T_\Delta(x)).$$

This shows that $x \in S(P)$. (Recall that property (i) in Theorem 3.5 is equivalent to the existence of a pair $(\bar{\lambda}, \bar{\mu}) \in R^m \times R^s$ satisfying system (3.11).)

CLAIM 2. $\omega \subset \text{loc}(P)$.

Let $x = x^{k_1} + t\bar{z}, t > 0$. By Claim 1, $x \in S(P)$, that is

$$\langle \nabla f(x), v \rangle = (Dx + c)^T v \geq 0 \quad (\forall v \in T_\Delta(x)). \qquad (4.6)$$

We want to show that

$$K \cap (\nabla f(x))^\perp = K_0. \qquad (4.7)$$

For each $l \in N$, we have

$$K_0 = \{ v \in K : \langle \nabla f(x^{k_l}), v \rangle = 0 \}.$$

So, for every $v \in K_0$, it holds $(Dx^{k_l} + c)^T v = \langle \nabla f(x^{k_l}), v \rangle = 0$ for all $l \in N$. Therefore

$$\left(D \frac{x^{k_l} - x^{k_1}}{\|x^{k_l} - x^{k_1}\|} + \frac{c + Dx^{k_1}}{\|x^{k_l} - x^{k_1}\|} \right)^T v = 0.$$

Letting $k \to \infty$ we deduce that

$$(D\bar{z})^T v = 0 \quad (\forall v \in K_0).$$

Hence

$$\langle \nabla f(x), v \rangle = \langle \nabla f(x^{k_1}), v \rangle + t(D\bar{z})^T v = 0 \quad (\forall v \in K_0).$$

This shows that $K_0 \subset K \cap (\nabla f(x))^\perp$. To prove the reverse inclusion, let us fix any $v \in K \cap (\nabla f(x))^\perp$. On account of (4.3) and (4.4), we have

$$0 = \langle \nabla f(x), v \rangle = \langle \nabla f(x^{k_1}), v \rangle + t(D\bar{z})^T v,$$

$$(D\bar{z})^T v \geq 0, \quad \langle \nabla f(x^{k_1}), v \rangle \geq 0.$$

This clearly forces $\langle \nabla f(x^{k_1}), v \rangle = 0$. So $v \in K_0$ whenever $v \in K \cap (\nabla f(x))^\perp$. The equality (4.7) has been established. We now show that

$$v^T D v \geq 0 \quad \forall v \in T_\Delta(x) \cap (\nabla f(x))^\perp. \tag{4.8}$$

In the proof of Claim 1, it has been shown that $T_\Delta(x) = M + K = T_\Delta(x^{k_1})$. Besides, from (4.6) we deduce that $M \subset (\nabla f(x))^\perp$. Hence, from (4.7) and the construction of the sequence $\{x^{k_l}\}$ it follows that

$$T_\Delta(x) \cap (\nabla f(x))^\perp = M + K_0 = T_\Delta(x^{k_1}) \cap (\nabla f(x^{k_1}))^\perp. \tag{4.9}$$

Since $x^{k_1} \in \text{loc}(P)$, Theorem 3.5 shows that

$$v^T D v \geq 0 \quad \forall v \in T_\Delta(x^{k_1}) \cap (\nabla f(x^{k_1})^\perp.$$

Combining this with (4.9) we get (4.8). From (4.6), (4.8) and Theorem 3.5, we deduce that $x \in \text{loc}(P)$. This completes the proof of Claim 2 and the proof of our theorem. $\quad\square$

Theorem 4.3. *The set $S(P)$ is unbounded if and only if (P) has a KKT point ray.*

Proof. By definition, $x \in S(P)$ if and only if there exists $(\lambda, \mu) \in R^m \times R^s$ such that

$$\begin{cases} Dx - A^T \lambda - C^T \mu + c = 0, \\ Ax \geq b, \quad Cx = d, \\ \lambda \geq 0, \quad \lambda^T (Ax - b) = 0. \end{cases} \tag{4.10}$$

Given a point $x \in S(P)$, we set $I_0 = \{i \in I : A_i x = b_i\}$, $I_1 = I \setminus I_0 = \{i \in I : A_i x > b_i\}$. From the last equality in (4.10) we get

$$\lambda_i = 0 \quad \forall i \in I_1.$$

Hence (x, λ, μ) satisfies the following system

$$
\begin{cases}
Dx - A^T\lambda - C^T\mu + c = 0, \\
A_{I_0}x = b_{I_0}, \quad \lambda_{I_0} \geq 0, \\
A_{I_1}x \geq b_{I_1}, \quad \lambda_{I_1} = 0, \\
Cx = d.
\end{cases}
\tag{4.11}
$$

Fix any $I_0 \subset I$ and denote by Q_{I_0} the set of all (x, λ, μ) satisfying (4.11). It is obvious that Q_{I_0} is a polyhedral convex set. From what has been said it follows that

$$
S(P) = \bigcup\{\mathrm{Pr}_{R^n}(Q_{I_0}) : I_0 \subset I\},
\tag{4.12}
$$

where $\mathrm{Pr}_{R^n}(x, \lambda, \mu) := x$. Since $\mathrm{Pr}_{R^n}(\cdot) : R^n \times R^m \times R^s \to R^n$ is a linear operator, $\mathrm{Pr}_{R^n}(Q_{I_0})$ is a polyhedral convex set for every $I_0 \subset I$. Indeed, as Q_{I_0} is a polyhedral convex set, it is *finitely generated*, i.e., there exist vectors $z^1, \ldots, z^k, w^1, \ldots, w^l$ in $R^n \times R^m \times R^s$ such that

$$
Q_{I_0} = \{z = \sum_{i=1}^k t_i z^i + \sum_{j=1}^l \theta_j w^j : t_i \geq 0 \text{ for all } i, \\
\theta_j \geq 0 \text{ for all } j, \text{ and } \sum_{i=1}^k t_i = 1\}
$$

(see Rockafellar (1970), Theorem 19.1). Then, by the linearity of the operator $\mathrm{Pr}_{R^n}(\cdot)$, we have

$$
\mathrm{Pr}_{R^n}(Q_{I_0}) = \{x = \sum_{i=1}^k t_i x^i + \sum_{j=1}^l \theta_j v^j : t_i \geq 0 \text{ for all } i, \\
\theta_j \geq 0 \text{ for all } j, \text{ and } \sum_{i=1}^k t_i = 1\},
$$

where $x^i = \mathrm{Pr}_{R^n}(z^i)$ for all i and $v^i = \mathrm{Pr}_{R^n}(w^j)$ for all j. This shows that the set $\mathrm{Pr}_{R^n}(Q_{I_0})$ is finitely generated, hence it is a polyhedral convex set (see Rockafellar (1970), Theorem 19.1).

If $S(P)$ is unbounded then from (4.12) it follows that there exists an index set $I_0 \subset I$ such that $\Omega_{I_0} := \mathrm{Pr}_{R^n}(Q_{I_0})$ is an unbounded set. Since Ω_{I_0} is a polyhedral convex set, it is an unbounded closed convex set. By Theorem 8.4 in Rockafellar (1970), Ω_{I_0} admits a direction of recession; that is there exists $\bar{v} \in R^n \setminus \{0\}$ such that

$$
x + t\bar{v} \in \Omega_{I_0} \quad \forall x \in \Omega_{I_0}, \forall t \geq 0.
\tag{4.13}
$$

Taking any $\bar{x} \in \Omega_{I_0}$ we deduce from (4.12) and (4.13) that $\bar{x} + t\bar{v} \in S(P)$ for all $t \geq 0$. Thus we have proved that (P) admits a KKT point ray. Conversely, it is obvious that if (P) admits a KKT point ray then $S(P)$ is unbounded. $\square$

Remark 4.1. Formula (4.12) shows that $S(P)$ is a union of finitely many polyhedral convex sets.

Let us derive another formula for the KKT point set of (P). For each index set $\alpha \subset I$, denote by F_α the pseudo-face of Δ corresponding to α; that is

$$F_\alpha = \{x \in R^n : A_\alpha x = b_\alpha, A_{I \setminus \alpha} x > b_{I \setminus \alpha}, Cx = d\}.$$

Since $\Delta = \cup\{F_\alpha : \alpha \subset I\}$, we deduce that

$$S(P) = \bigcup\{S(P) \cap F_\alpha : \alpha \subset I\}. \tag{4.14}$$

Lemma 4.1. For every $\alpha \subset I$, $S(P) \cap F_\alpha$ is a convex set.

Proof. Let $\alpha \subset I$ be any index set. From the definition of $S(P)$ it follows that $x \in S(P) \cap F_\alpha$ if and only if there exist $(\lambda, \mu) \in R^m \times R^s$ such that

$$\begin{cases} Dx - A^T\lambda - C^T\mu + c = 0, \\ A_\alpha x = b_\alpha, \quad \lambda_\alpha \geq 0, \\ A_{I\setminus\alpha}x > b_{I\setminus\alpha}, \quad \lambda_{I\setminus\alpha} = 0, \\ Cx = d. \end{cases} \tag{4.15}$$

Let Z_α denote the set of all the points $(x, \lambda, \mu) \in R^n \times R^m \times R^s$ satisfying the system (4.15). It is clear that Z_α is a convex set. From what has already been said it follows that $S(P) \cap F_\alpha = \mathrm{Pr}_{R^n}(Z_\alpha)$, where $\mathrm{Pr}_{R^n}(x, \lambda, \mu) := x$. Since $\mathrm{Pr}_{R^n}(\cdot)$ is a linear operator, we conclude that $S(P) \cap F_\alpha$ is a convex set. $\square$

Note that, in general, the convex sets $S(P) \cap F_\alpha$, $\alpha \subset I$, in the representation (4.14) may not be closed.

We know that $\mathrm{Sol}(D, A, c, b) \subset \mathrm{loc}(D, A, c, b) \subset S(D, A, c, b)$ (see (3.10)). We have characterized the unboundedness of these solution sets. The following questions arise:

QUESTION 1: Is it true that $\mathrm{Sol}(P)$ is unbounded whenever $\mathrm{loc}(P)$ is unbounded?

QUESTION 2: Is it true that $\mathrm{loc}(P)$ is unbounded whenever $S(P)$ is unbounded?

The following example gives a negative answer to Question 1.

Example 4.1. Consider the problem

$$(P_1) \quad \min\{f(x) = -x_2^2 + 2x_2 : x = (x_1, x_2), x_1 \geq 0, x_2 \geq 0\}.$$

Denote by Δ the feasible region of (P_1). We have

$$\mathrm{Sol}(P_1) = \emptyset, \quad \mathrm{loc}(P_1) = \{x \in R^2 : x_1 \geq 0, x_2 = 0\},$$
$$S(P_1) = \{x \in R^2 : x_1 \geq 0, x_2 = 0 \text{ or } x_2 = 1\}.$$

Thus $\mathrm{loc}(P_1)$ is unbounded, but $\mathrm{Sol}(P_1) = \emptyset$. In order to establish the above results, one can argue as follows. Since $I = \{1, 2\}$, the constraint set of (P_1) is composed by four pseudo-faces:

$$F_{\{1,2\}} = \{(0,0)\},$$
$$F_{\{1\}} = \{x \in R^2 : x_1 = 0, x_2 > 0\}$$
$$F_{\{2\}} = \{x \in R^2 : x_2 = 0, x_1 > 0\},$$
$$F_\emptyset = \{x \in R^2 : x_1 > 0, x_2 > 0\}.$$

Since $\nabla f(x) = (0, -2(x_2 - 1))$, by solving four KKT systems of the form (4.15) where C, d are vacuous,

$$D = \begin{pmatrix} 0 & 0 \\ 0 & -2 \end{pmatrix}, \quad A = \begin{pmatrix} 1 & 0 \\ 0 & 1 \end{pmatrix}, \quad b = \begin{pmatrix} 0 \\ 0 \end{pmatrix}, \quad c = \begin{pmatrix} 0 \\ 2 \end{pmatrix},$$

we obtain

$$S(P_1) \cap F_{\{1,2\}} = \{(0,0)\},$$
$$S(P_1) \cap F_{\{1\}} = \{(0,1)\},$$
$$S(P_1) \cap F_{\{2\}} = \{x \in R^2 : x_2 = 0, x_1 > 0\},$$
$$S(P_1) \cap F_\emptyset = \{x \in R^2 : x_1 > 0, x_2 = 1\}.$$

From formula (4.14) it follows that

$$S(P_1) = \{x \in R^2 : x_1 \geq 0, \ x_2 = 0 \text{ or } x_2 = 1\}.$$

Since $\lim_{x_2 \to +\infty} f(0, x_2) = -\infty$ and, for each $x_2 \geq 0$, $x = (0, x_2)$ is a feasible vector for (P_1), we conclude that $\mathrm{Sol}(P_1) = \emptyset$. For any $x = (x_1, 1) \in S(P_1) \cap F_\emptyset$, we have $T_\Delta(x) = R^2$, $(\nabla f(x))^\perp = R^2$. Then the condition

$$v^T D v \geq 0 \quad \forall v \in T_\Delta(x) \cap (\nabla f(x))^\perp, \tag{4.16}$$

which is equivalent to the condition

$$-2v_2^2 \geq 0 \quad \forall v_2 \in R,$$

cannot be satisfied. By Theorem 3.5, $x \notin \mathrm{loc}(P_1)$. Now, for any $x = (x_1, 0) \in S(P_1) \cap F_{\{2\}}$, we have

$$T_\Delta(x) = \{v = (v_1, v_2) \in R^2 : v_2 \geq 0\},$$

$$(\nabla f(x))^\perp = \{v = (v_1, v_2) \in R^2 : v_2 = 0\}.$$

Condition (4.16), which is now equivalent to the requirement

$$-2v_2^2 \geq 0 \quad \text{for } v_2 = 0,$$

is satisfied. Applying Theorem 3.5 we can assert that $x \in \mathrm{loc}(P_1)$. In the same manner we can see that the unique point $x = (0, 1)$ of the set $S(P_1) \cap F_{\{1\}}$ does not belong to $\mathrm{loc}(P_1)$, while the unique point $x = (0, 0)$ of the set $S(P_1) \cap F_{\{1,2\}}$ belongs to $\mathrm{loc}(P_1)$. Thus we have shown that $\mathrm{loc}(P_1) = \{x \in R^2 : x_1 \geq 0,\ x_2 = 0\}$.

The following example gives a negative answer to Question 2.

Example 4.2. Consider the problem

$$(P_2) \qquad \min\{f(x) = -x_2^2 : x = (x_1, x_2),\ x_1 \geq 0,\ x_2 \geq 0\}.$$

Analysis similar to that in the preceding example shows that

$$\mathrm{Sol}(P_2) = \mathrm{loc}(P_2) = \emptyset, \quad S(P_2) = \{x = (x_1, x_2) : x_1 \geq 0,\ x_2 = 0\}.$$

4.2 Closedness of the Solution Sets

Since $S(P)$ is a union of finitely many polyhedral convex set (see (4.12)), it is a closed set. The set $\mathrm{Sol}(P)$ is also closed. Indeed, we have

$$\mathrm{Sol}(P) = \{x \in \Delta : f(x) = v(P)\},$$

where $v(P) = \inf\{f(x) : x \in \Delta\}$. If $v(P)$ is finite then from the closedness of Δ, the continuity of f, and the above formula, it follows that $\mathrm{Sol}(P)$ is closed. If $v(P) = +\infty$ then $\Delta = \emptyset$, hence $\mathrm{Sol}(P) = \emptyset$. If $v(P) = -\infty$ then it is obvious that $\mathrm{Sol}(P) = \emptyset$. Thus we conclude that $\mathrm{Sol}(P)$ is always a closed set.

The following question arises:

QUESTION 3: Is it true that $\mathrm{loc}(P)$ is always a closed set?

The following example gives a negative answer to Question 3.

Example 4.3. Consider the problem

$$\min\{f(x) = -x_2^2 + x_1 x_2 : x = (x_1, x_2),\ x_1 \geq 0,\ x_2 \geq 0\}. \qquad (P_3)$$

Analysis similar to that in Example 4.1 shows that

$$\mathrm{Sol}(P_3) = \emptyset, \quad \mathrm{loc}(P_3) = \{x = (x_1, x_2) : x_1 > 0,\ x_2 = 0\},$$
$$S(P_3) = \{x = (x_1, x_2) : x_1 \geq 0,\ x_2 = 0\}.$$

Our next aim is to study the situation where $\mathrm{Sol}(P)$ (resp., $\mathrm{loc}(P)$, $S(P)$) is a bounded set having infinitely many elements.

4.3 A Property of the Bounded Infinite Solution Sets

Definition 4.2. A line segment $\omega_\delta = \{\bar{x} + t\bar{v} : t \in [0, \delta)\}$, where $\bar{v} \in R^n \setminus \{0\}$ and $\delta > 0$, which is a subset of $\mathrm{Sol}(P)$ (resp., $\mathrm{loc}(P)$, $S(P)$), is called a *solution interval* (resp., a *local-solution interval*, a *KKT point interval*) of (P).

Proposition 4.1. *If the set* $\mathrm{Sol}(P)$ *is bounded and infinite, then* (P) *has a solution interval.*

Proof. For each index set $\alpha \subset I$, denote by F_α the pseudo-face of Δ corresponding to α. As $\mathrm{Sol}(P) \subset \Delta$ has infinitely many elements and $\Delta = \cup\{F_\alpha : \alpha \subset I\}$, there must exists some $\alpha_* \subset I$ such that the intersection $\mathrm{Sol}(P) \cap F_{\alpha_*}$ has infinitely many elements. For each $x \in \mathrm{Sol}(P) \cap F_{\alpha_*}$ we have $T_\Delta(x) = \{v \in R^n : A_{\alpha_*}v \geq 0, \ Cv = 0\}$ and, by Theorem 3.5, $\langle \nabla f(x), v \rangle \geq 0$ for every $v \in T_\Delta(x)$. Hence $T := T_\Delta(x)$ is a constant polyhedral convex cone which does not depend on the position of x in the pseudo-face F_{α_*} of Δ, and

$$T_0^x := \{v \in T : \langle \nabla f(x), v \rangle = 0\}$$

is a face of T. Since the number of faces of T is finite, from what has already been said it follows that there must exist two different points x and y of $\mathrm{Sol}(P) \cap F_{\alpha_*}$ such that $T_0^x = T_0^y$. Set $T_0 = T_0^x$. By Lemma 4.1, the intersection $S(P) \cap F_{\alpha_*}$ is convex. Since $x \in \mathrm{Sol}(P)$, $y \in \mathrm{Sol}(P)$ and $\mathrm{Sol}(P) \subset S(P)$, it follows that $z_t := (1 - t)x + ty$ belongs to $S(P) \cap F_{\alpha_*}$ for every $t \in [0, 1]$. By the remark following Theorem 3.5, for every $t \in (0, 1)$, we have

$$\langle \nabla f(z_t), v \rangle \geq 0 \quad \forall v \in T.$$

Therefore

$$
\begin{aligned}
&T_\Delta(z_t) \cap (\nabla f(z_t))^\perp \\
&= \{v \in T : \langle \nabla f(z_t), v \rangle = 0\} \\
&= \{v \in T : (1 - t)\underbrace{\langle \nabla f(x), v \rangle}_{\geq 0} + t\underbrace{\langle \nabla f(y), v \rangle}_{\geq 0} = 0\} \\
&= \{v \in T : \langle \nabla f(x), v \rangle = 0, \ \langle \nabla f(y), v \rangle = 0\} \\
&= T_0.
\end{aligned}
$$

Since $x \in \mathrm{Sol}(P)$ and $T_\Delta(x) \cap (\nabla f(x))^\perp = T_0$, from Theorem 3.5 it follows that

$$v^T D v \geq 0 \quad \forall v \in T_0.$$

We have shown that, for every $t \in (0,1)$, $z_t \in S(P)$ and

$$T_\Delta(z_t) \cap (\nabla f(z_t))^\perp = T_0.$$

Combining these facts and applying Theorem 3.5 we deduce that $z_t \in \mathrm{loc}(P)$. Consider the function $\varphi : [0,1] \to R$ defined by setting $\varphi(t) = f(z_t)$ for all $t \in [0,1]$. It is clear that φ is a continuous function which is differentiable at each $t \in (0,1)$. Since z_t is a local solution of (P), φ attains a local minimum at every $t \in (0,1)$. Hence $\varphi'(t) = 0$ for every $t \in (0,1)$. Consequently, $\varphi(t)$ is a constant function. Since $x \in \mathrm{Sol}(P)$, we see that $z_t \in \mathrm{Sol}(P)$ for all $t \in [0,1]$. Thus $[x,y] := \{(1-t)x + ty : t \in [0,1)\}$ is a solution interval of (P). $\square$

Proposition 4.2. (See Phu and Yen (2001), Theorem 3) *If the set $\mathrm{loc}(P)$ is bounded and infinite, then (P) has a local-solution interval.*

Proof. For each index set $\alpha \subset I$, denote by F_α the pseudo-face of Δ corresponding to α. As $\mathrm{loc}(P) \subset \Delta$ is an infinite set and $\Delta = \cup\{F_\alpha : \alpha \subset I\}$, there must exists some $\alpha_* \subset I$ such that the intersection $\mathrm{loc}(P) \cap F_{\alpha_*}$ has infinitely many elements. For each $x \in \mathrm{loc}(P) \cap F_{\alpha_*}$ we have $T_\Delta(x) = \{v \in R^n : A_{\alpha_*}v \geq 0, \, Cv = 0\}$ and, by Theorem 3.5, $\langle \nabla f(x), v \rangle \geq 0$ for every $v \in T_\Delta(x)$. Hence $T := T_\Delta(x)$ is a constant polyhedral convex cone which does not depend on the position of x in the pseudo-face F_{α_*} of Δ, and

$$T_0^x := \{v \in T : \langle \nabla f(x), v \rangle = 0\}$$

is a face of T. Since the number of faces of T is finite, it follows that there must exist two different point x and y of $\mathrm{loc}(P) \cap F_{\alpha_*}$ such that $T_0^x = T_0^y$. Set $T_0 = T_0^x$. By Lemma 4.1, the intersection $S(P) \cap F_{\alpha_*}$ is convex. Since $x \in \mathrm{loc}(P)$, $y \in \mathrm{loc}(P)$ and $\mathrm{loc}(P) \subset S(P)$, it follows that $z_t := (1-t)x + ty$ belongs to $S(P) \cap F_{\alpha_*}$ for every $t \in [0,1]$. According to the remark following Theorem 3.5, for every $t \in (0,1)$ we have

$$\langle \nabla f(z_t), v \rangle \geq 0 \quad \forall v \in T.$$

As in the proof of Proposition 4.1, we have

$$T_\Delta(z_t) \cap (\nabla f(z_t))^\perp = T_0.$$

Since $x \in \mathrm{loc}(P)$ and $T_\Delta(x) \cap (\nabla f(x))^\perp = T_0$, from Theorem 3.5 it follows that

$$v^T Dv \geq 0 \quad \forall v \in T_0.$$

Therefore, for every $t \in (0,1)$, $z_t \in S(P)$ and

$$T_\Delta(z_t) \cap (\nabla f(z_t))^\perp = T_0.$$

On account of these facts and of Theorem 3.5, we conclude that z_t is a local solution of (P). Thus $[x, y)$ is a local-solution interval of (P). $\quad\square$

Proposition 4.3. *If the set $S(P)$ is bounded and infinite, then (P) has a KKT point interval.*

Proof. For each index set $\alpha \subset I$, denote by F_α the pseudo-face of Δ corresponding to α. As $S(P) \subset \Delta$ is an infinite set and $\Delta = \cup\{F_\alpha : \alpha \subset I\}$, there must exists some $\alpha_* \subset I$ such that the intersection $S(P) \cap F_{\alpha_*}$ has infinitely many elements. Hence there must exist two different point x and y of $S(P) \cap F_{\alpha_*}$. By Lemma 4.1, the intersection $S(P) \cap F_{\alpha_*}$ is convex. This implies that $[x, y)$ is a KKT point interval of (P). $\quad\square$

4.4 Finiteness of the Solution Sets

Theorem 4.4. *The following assertions are valid:*

(i) *If D is a positive definite matrix and Δ is nonempty, then (P) has a unique solution and it holds $\mathrm{Sol}(P) = \mathrm{loc}(P) = S(P)$.*

(ii) *If D is a negative definite matrix then each local solution of (P) is an extreme point of Δ. In this case, $\mathrm{Sol}(P) \subset \mathrm{loc}(P) \subset \mathrm{extr}\Delta$. Hence, if D is a negative definite matrix then the number of solutions of (P) (resp., the number of local solutions of (P)) is always less than or equal to the number of extreme points of Δ. Besides, if $\mathrm{Sol}(P)$ is nonempty then Δ is a compact polyhedral convex set.*

(iii) *If D is a positive semidefinite matrix then $\mathrm{Sol}(P)$ is a closed convex set and it holds $\mathrm{Sol}(P) = \mathrm{loc}(P) = S(P)$. Hence, if D is a positive semidefinite matrix then $\mathrm{Sol}(P)$ is finite if and only if it is a singleton or it is empty.*

Proof. (i) Suppose that the symmetric matrix D is positive definite and the set $\Delta := \{x \in R^n : Ax \geq b, Cx = d\}$ is nonempty. Setting

$$\varrho := \inf\{v^T D v : v \in R^n, \|v\| = 1\} > 0$$

we deduce that $x^T D x \geq \varrho \|x\|^2$ for every $x \in R^n$. Fix any $x^0 \in \Delta$. Note that

$$
\begin{aligned}
f(x) - f(x_0) &= \frac{1}{2}(x - x^0)^T D(x - x^0) + (Dx^0 + c)^T (x - x^0) \\
&\geq \frac{1}{2}\varrho \|x - x^0\|^2 - \|Dx^0 + c\| \|x - x^0\|.
\end{aligned}
$$

The last expression tends to $+\infty$ as $\|x - x^0\| \to +\infty$. Hence there exists $\gamma > 0$ such that

$$
f(x) - f(x^0) \geq 1 \quad \forall x \in \Delta \setminus \bar{B}(x^0, \gamma). \tag{4.17}
$$

From (4.17) it follows that (P) cannot have solutions in $\Delta \setminus \bar{B}(x^0, \gamma)$. Since $\Delta \cap \bar{B}(x^0, \gamma) \neq \emptyset$ the problem $\min\{f(x) : x \in \Delta \cap \bar{B}(x^0, \gamma)\}$ possesses a solution $\bar{x}$. By (4.17), $\bar{x} \in \mathrm{Sol}(P)$. Assume, contrary to our claim, that there are two different solutions $\bar{x}$ and $\bar{y}$ of (P). Since $\bar{y} - \bar{x} \in T_\Delta(\bar{x})$ and $\bar{x} \in \mathrm{Sol}(P)$, by Theorem 3.1 we have $(D\bar{x} + c)^T(\bar{y} - \bar{x}) \geq 0$. As $\bar{y} \neq \bar{x}$ and D is positive definite, we have $(\bar{y} - \bar{x})^T D(\bar{y} - \bar{x}) > 0$. It follows that

$$
0 = f(\bar{y}) - f(\bar{x}) = \frac{1}{2}(\bar{y} - \bar{x})^T D(\bar{y} - \bar{x}) + (D\bar{x} + c)^T(\bar{y} - \bar{x}) > 0,
$$

a contradiction. The equalities $\mathrm{Sol}(P) = \mathrm{loc}(P) = S(P)$ follow from the fact that, under our assumptions, f is a convex function (see Proposition 1.2).

(ii) Let D be a negative definite matrix and $\bar{x} \in \mathrm{loc}(P)$. If $\bar{x} \notin \mathrm{extr}\Delta$ then there exist $x \in \Delta$, $y \in \Delta$, $x \neq y$, and $t \in (0, 1)$ such that $\bar{x} = (1 - t)x + ty$. Since $\bar{x} \in \Delta$, $x - \bar{x} \in T_\Delta(\bar{x})$, $y - \bar{x} \in T_\Delta(\bar{x})$, and $y - \bar{x} = -\dfrac{1 - t}{t}(x - \bar{x})$, applying Theorem 3.5 we get $\langle \nabla f(\bar{x}), x - \bar{x} \rangle \geq 0$ and

$$
-\frac{1 - t}{t} \langle \nabla f(\bar{x}), x - \bar{x} \rangle = \langle \nabla f(\bar{x}), y - \bar{x} \rangle \geq 0.
$$

Therefore $\langle \nabla f(\bar{x}), x - \bar{x} \rangle = 0$. This equality and the assumption $\bar{x} \in \mathrm{loc}(P)$ allows us to apply Theorem 3.5 to obtain $(x - \bar{x})^T D(x - \bar{x}) \geq 0$. This contradicts the fact that matrix D is negative definite. We have thus proved that $\mathrm{loc}(P) \subset \mathrm{extr}\Delta$. Consequently, $\mathrm{Sol}(P) \subset \mathrm{loc}(P) \subset \mathrm{extr}\Delta$. We now suppose that $\mathrm{Sol}(P) \neq \emptyset$. By Corollary 2.6,

$$
(v \in R^n, \ Av \geq 0, \ Cv = 0) \implies v^T D v \geq 0.
$$

Combining this with the negative definiteness of D we conclude that

$$\{v \in R^n : Av \geq 0, \ Cv = 0\} = \{0\}.$$

Hence Δ has no directions of recession. By Theorem 8.4 in Rock-afellar (1970), Δ is a compact set.

(iii) Let D be a positive semidefinite matrix. By Proposition 1.2, f is a convex function. Hence $\mathrm{Sol}(P)$ is a closed convex set and it holds $\mathrm{Sol}(P) = \mathrm{loc}(P) = S(P)$. $\square$

Example 4.4. Consider problem (P) of the following form

$$\min\{f(x) = -x_1^2 - x_2^2 + 1 : x = (x_1, x_2), \ x_1 \geq -1,$$
$$-x_1 \geq -1, \ x_2 \geq -1, \ -x_2 \geq -1\}.$$

The matrix $D = \begin{pmatrix} -2 & 0 \\ 0 & -2 \end{pmatrix}$ corresponding to this problem is negative definite. It is easily seen that

$$\mathrm{Sol}(P) = \mathrm{loc}(P) = \{(1, 1), (1, -1), (-1, 1), (-1, -1)\},$$

$$S(P) = \mathrm{Sol}(P) \cup \{(0, 0), (1, 0), (0, 1), (-1, 0), (0, -1)\}.$$

Theorem 4.5. *If* $\mathrm{Sol}(P)$ *(resp.,* $S(P)$, $\mathrm{loc}(P)$*) is a finite set, then each pseudo-face of* Δ *cannot contain more than one element of* $\mathrm{Sol}(P)$ *(resp.,* $S(P)$, $\mathrm{loc}(P)$*). Hence, if* $\mathrm{Sol}(P)$ *(resp.,* $S(P)$, $\mathrm{loc}(P)$*) is a finite set, then* $\mathrm{Sol}(P)$ *(resp.,* $S(P)$, $\mathrm{loc}(P)$*) cannot have more than* 2^m *elements, where m is the number of inequality constraints of* (P).

Before proving this theorem let us establish the following two auxiliary results.

Proposition 4.4. *Assume that* $x \in \mathrm{loc}(P) \cap F_\alpha$, $y \in S(P) \cap F_\alpha$, $y \neq x$, *where* F_α *is a pseudo-face of* Δ. *Then there exists* $\delta > 0$ *such that, for every* $t \in (0, \delta)$, $a(t) := (1 - t)x + ty$ *is a local solution of* (P).

Proof. Let $x \in \mathrm{loc}(P) \cap F_\alpha$, $y \in S(P) \cap F_\alpha$, $y \neq x$, where $\alpha \subset I = \{1, \ldots, m\}$ and $F_\alpha = \{x \in R^n : A_\alpha x = b_\alpha, A_{I \setminus \alpha} x > b_{I \setminus \alpha}, Cx = d\}$. Since $x \in F_\alpha$, we have $T_\Delta(x) = \{v \in R^n : A_\alpha v \geq 0, Cv = 0\}$. Let

$$M = \{v \in R^n : A_\alpha v = 0, \ Cv = 0\}.$$

Then M is a linear subspace and $M \subset T_\Delta(x)$. Let $M^\perp = \{v \in R^n : \langle v, u \rangle = 0 \text{ for every } u \in M\}$ and let

$$K = T_\Delta(x) \cap M^\perp = \mathrm{Pr}_{M^\perp}(T_\Delta(x)),$$

where $\mathrm{Pr}_{M^\perp}(\cdot)$ denotes the orthogonal projection of R^n onto the subspace $M^\perp$. We have

$$K = \{v \in R^n : A_\alpha v \geq 0, \ Cv = 0, \ v \in M^\perp\}$$

and $T_\Delta(x) = M + K$. Since K is a pointed polyhedral convex cone (see the proof of Theorem 3.4), according to Theorem 19.1 in Rockafellar (1970), there exists a finite system of generators $\{z^1, \ldots, z^q\}$ of K. By convention, if $K = \{0\}$ then the system is vacuous. In the case where that system is not vacuous, we have

$$K = \{v = \sum_{j=1}^{q} t_j z^j : t_j \geq 0 \text{ for all } j = 1, \ldots, q\}.$$

Let $Q = \{1, \ldots, q\}$,

$$Q_0 = \{j \in Q : \langle \nabla f(x), z^j \rangle = 0\}, \ Q_1 = \{j \in Q : \langle \nabla f(x), z^j \rangle > 0\}. \tag{4.18}$$

Since $x \in \mathrm{loc}(P)$, we must have

$$\langle \nabla f(x), v \rangle \geq 0 \quad \forall v \in T_\Delta(x).$$

From this we deduce that $Q = Q_0 \cup Q_1$. For every $a \in F_\alpha$, let $K_0^a = \{v \in K : \langle \nabla f(a), v \rangle = 0\}$. For every $t \in [0,1]$, we set $a(t) = (1-t)x + ty$. Since $x \in F_\alpha$ and $y \in F_\alpha$, it follows that $a(t) \in F_\alpha$ for every $t \in [0,1]$. Consequently,

$$T_\Delta(a(t)) = T_\Delta(x) = \{v \in R^n : A_\alpha v \geq 0, \ Cv = 0\} = M + K. \tag{4.19}$$

It follows from (4.18) that there exists $\delta \in (0,1)$ such that

$$\langle \nabla f(a(t)), z^j \rangle > 0 \quad \forall t \in (0, \delta), \ \forall j \in Q_1. \tag{4.20}$$

For any $t \in (0, \delta)$, by Lemma 4.1 we have $a(t) \in S(P)$. Therefore $\langle \nabla f(a(t)), v \rangle \geq 0$ for every $v \in T_\Delta(a(t))$. Combining this with (4.20) we deduce that

$$K_0^{a(t)} \subset K_0^x = \{v = \sum_{j \in Q_0} t_j z^j : t_j \geq 0 \text{ for all } j \in Q_0\}. \tag{4.21}$$

We claim that $a(t) \in \mathrm{loc}(P)$ for every $t \in (0, \delta)$. Indeed, since $\langle \nabla f(a(t)), v \rangle \geq 0$ for every $v \in T_\Delta(a(t))$, we get

$$\langle \nabla f(a(t)), v \rangle = 0 \quad \forall v \in M, \ \forall t \in (0, \delta).$$

By (4.19) and (4.21),

$$T_\Delta(a(t)) \cap (\nabla f(a(t))^\perp = M + K_0^{a(t)} \subset M + K_0^x. \qquad (4.22)$$

As $x \in \mathrm{loc}(P)$, by Theorem 3.5 we have

$$v^T D v \geq 0 \quad \forall v \in T_\Delta(x) \cap (\nabla f(x))^\perp = M + K_0^x.$$

Combining this with (4.22) yields

$$v^T D v \geq 0 \quad \forall v \in T_\Delta(a(t)) \cap (\nabla f(a(t)))^\perp = M + K_0^{a(t)}.$$

Since $a(t) \in S(P)$, from the last fact and Theorem 3.5 we conclude that $a(t) \in \mathrm{loc}(P)$. □

Proposition 4.5. *Assume that x and y are two different Karush-Kuhn-Tucker points of (P) belonging to the same pseudo-face of Δ. Then the function $\varphi(t) := f((1 - t)x + ty)$ is constant on $[0,1]$.*

Proof. Let $\alpha \subset I$, $x \in S(P) \cap F_\alpha$, $y \in S(P) \cap F_\alpha$, $x \neq y$. For every $t \in [0,1]$, we define $a(t) = (1 - t)x + ty$. Since $a(t) \in F_\alpha$, it follows that $T_\Delta(a(t)) = \{v \in R^n : A_\alpha v \geq 0, \, Cv = 0\}$. By Lemma 4.1, $a(t) \in S(P)$. Hence $\langle \nabla f(a(t)), v \rangle \geq 0$ for every $v \in T_\Delta(a(t))$. Combining these facts we see that $\langle \nabla f(a(t)), v \rangle = 0$ for every $v \in M := \{v \in R^n : A_\alpha v = 0, \, Cv = 0\}$. It is easy to check that $y - x \in M$. So we have $\langle \nabla f(a(t)), y - x \rangle = 0$. From this and the obvious relation

$$\nabla \varphi(t) = \langle \nabla f(a(t)), y - x \rangle$$

we deduce that the function φ is constant on $[0,1]$, as desired. □

Proof of Theorem 4.5.

We first consider the case where $\mathrm{Sol}(P)$ is a finite set. Suppose, contrary to our claim, that there exists a pseudo-face F_α of Δ containing two different elements x, y of $\mathrm{Sol}(P)$. By Proposition 4.5, the function $\varphi(t) := f((1 - t)x + ty)$ is constant on $[0,1]$. From this and the inclusion $x \in \mathrm{Sol}(P)$ we conclude that whole the segment $[x, y]$ is contained in $\mathrm{Sol}(P)$. This contradicts the finiteness of $\mathrm{Sol}(P)$.

The fact that if $S(P)$ is finite then each pseudo-face of Δ cannot contain more than one element of $S(P)$ follows immediately from Lemma 4.1.

The fact that if loc(P) is finite then each pseudo-face of Δ cannot contain more than one element of loc(P) is a direct consequence of Proposition 4.4.   □

Actually, in the course of the preceding proof we have established the following useful fact.

Proposition 4.6. *If the intersection* Sol(P) $\cap F_\alpha$ *of the solution set of* (P) *with a pseudo-face of Δ is nonempty, then* $S(P) \cap F_\alpha =$ Sol(P) $\cap F_\alpha$.

Combining the last proposition with Lemma 4.1 we obtain the following statement.

Proposition 4.7. *The intersection* Sol(P)$\cap F_\alpha$ *of the solution set of* (P) *with a pseudo-face of Δ is always a convex set (may be empty).*

In connection with Propositions 4.4 and 4.7, the following two open questions seem to be interesting:

QUESTION 4: Let $x \in$ loc(P) $\cap F_\alpha$, $y \in S(P) \cap \overline{F}_\alpha$, $y \neq x$, where F_α is a pseudo-face of Δ and $\overline{F}_\alpha$ denotes the topological closure of F_α. Is it true that there must exist some $\delta > 0$ such that, for every $t \in (0, \delta)$, $a(t) := (1 - t)x + ty$ is a local solution of (P)?

QUESTION 5: Is it true that the intersection loc(P) $\cap F_\alpha$ of the local-solution set of (P) with a pseudo-face of Δ is always a convex set?

4.5   Commentaries

The notion of solution ray has proved to be very efficient for studying the structure of the solution set of linear complementarity problems (see, for instance, Cottle et al. (1992)) and affine variational inequalities (see, for instance, Gowda and Pang (1994a)).

This chapter shows that the notions of solution ray and solution interval interval are also useful for studying the structure of the solution sets of (nonconvex) quadratic programs.

Lemma 4.1, Propositions 4.2 and 4.3, and Theorems 4.3 and 4.4 are well known facts. Other results might be new.

Chapter 5

Affine Variational Inequalities

In this chapter, the notions of affine variational inequality and linear complementarity problem are discussed in a broader context of variational inequalities and complementarity problems. Besides, a characterization of the solutions of affine variational inequalities via Lagrange multipliers and a basic formula for representing the solution sets will be given.

5.1 Variational Inequalities

Variational inequality problems arise in a natural way in the framework of optimization problems.

Let $f : R^n \to R$ be a C^1-function and $\Delta \subset R^n$ a nonempty, closed, convex set.

Proposition 5.1. *If $\bar{x}$ is a local solution of the optimization problem*

$$\min\{f(x) : x \in \Delta\} \tag{5.1}$$

then

$$\langle \nabla f(\bar{x}), y - \bar{x} \rangle \geq 0 \quad \forall y \in \Delta. \tag{5.2}$$

Proof. Similar to the proof of Theorem 3.1 (i). □

Setting

$$\phi(x) = \nabla f(x) = \begin{pmatrix} \dfrac{\partial f(x)}{\partial x_1} \\ \vdots \\ \dfrac{\partial f(x)}{\partial x_n} \end{pmatrix} \quad \forall x \in R^n, \tag{5.3}$$

we see that (5.2) can be rewritten as

$$\langle \phi(\bar{x}), y - \bar{x} \rangle \geq 0 \quad \forall y \in \Delta. \tag{5.4}$$

Definition 5.1. If $\Delta \subset R^n$ is a nonempty, closed, convex subset and $\phi : \Delta \to R^n$ is a given operator (mapping) then the problem of finding some $\bar{x} \in \Delta$ satisfying (5.4) is called a *variational inequality problem* or, simply, a *variational inequality* (VI, for brevity). It is denoted by $VI(\phi, \Delta)$. The *solution set* $\text{Sol}(VI(\phi, \Delta))$ of $VI(\phi, \Delta)$ is the set of all $\bar{x} \in \Delta$ satisfying (5.4).

It is easy to check that $\bar{x} \in \text{Sol}(VI(\phi, \Delta))$ if and only if the inclusion

$$0 \in \phi(\bar{x}) + N_\Delta(\bar{x}),$$

where $N_\Delta(\bar{x})$ denotes the normal cone to Δ at $\bar{x}$ (see Definition 1.9), is satisfied.

Proposition 5.1 shows how smooth optimization problems can lead to variational inequalities. A natural question arises: *Given a variational inequality* $VI(\phi, \Delta)$ *with a continuous operator* $\phi : R^n \to R^n$, *can one find a* C^1-*function* $f : R^n \to R$ *such that* $VI(\phi, \Delta)$ *can be obtained from optimization problem* (5.1) *by the above-described procedure or not?* If such a function f exists, we must have

$$\phi(x) = \nabla f(x) \quad \forall x \in \Delta. \tag{5.5}$$

One can observe that if f is a C^2-function then the operator $\phi : R^n \to R^n$ defined by (5.3) has a *symmetric* Jacobian matrix. Recall that if a vector-valued function $\phi : R^n \to R^n$ has smooth components $\phi_1, \ldots, \phi_n$ then the *Jacobian matrix* of ϕ at x is defined by the formula

$$J\phi(x) = \begin{pmatrix} \dfrac{\partial \phi_1(x)}{\partial x_1} & \dfrac{\partial \phi_1(x)}{\partial x_2} & \cdots & \dfrac{\partial \phi_1(x)}{\partial x_n} \\ \vdots & \vdots & & \vdots \\ \dfrac{\partial \phi_n(x)}{\partial x_1} & \dfrac{\partial \phi_n(x)}{\partial x_2} & \cdots & \dfrac{\partial \phi_n(x)}{\partial x_n} \end{pmatrix}.$$

Since f is assumed to be a C^2-function, from (5.3) we deduce that

$$\frac{\partial \phi_i(x)}{\partial x_j} = \frac{\partial^2 f(x)}{\partial x_j \partial x_i} = \frac{\partial^2 f(x)}{\partial x_i \partial x_j} = \frac{\partial \phi_j(x)}{\partial x_i}$$

for all i, j. This shows that $J\phi(x)$ is a symmetric matrix.

Proposition 5.2. (See, for instance, Nagurney (1993)) *Let $\Delta \subset R^n$ be a nonempty, closed, convex set. If $\phi : R^n \to R^n$ is such a vector-valued function with smooth components that $\dfrac{\partial \phi_i(x)}{\partial x_j} = \dfrac{\partial \phi_j(x)}{\partial x_i}$ for all i and j (a smooth symmetric operator), then there exists a C^2-function $f : R^n \to R$ such that the relation (5.5) is satisfied. This means that the variational inequality problem $\mathrm{VI}(\phi, \Delta)$ can be regarded as the first-order necessary optimality condition of the optimization problem (5.1).*

So, we have seen that *C^2-smooth optimization problems correspond to variational inequalities with smooth symmetric operators.* However, when one studies the VI model, one can consider also VI problems with asymmetric discontinuous operators. Thus the VI model is a mathematical subject which is treated independently from its original interpretation as the first-order necessary optimality condition of a smooth optimization problem.

The following simple statement shows that, unlike the solutions of mathematical programming problems, solutions of VI problems have a local character. From this point of view, VI problems should be regarded as *generalized equations* (see, for instance, Robinson (1979, 1981)), but not as something similar to optimization problems.

Proposition 5.3. *Let $\bar{x} \in \Delta$. If there exists $\varepsilon > 0$ such that*

$$\langle \phi(\bar{x}), y - \bar{x} \rangle \geq 0 \quad \forall y \in \Delta \cap \bar{B}(\bar{x}, \varepsilon), \tag{5.6}$$

then $\bar{x} \in \mathrm{Sol}(\mathrm{VI}(\phi, \Delta))$.

Proof. Suppose that $\varepsilon > 0$ satisfies (5.6). Obviously, for each $y \in \Delta$ there exists $t = t(y) \in (0, 1)$ such that $y(t) := \bar{x} + t(y - \bar{x})$ belongs to $\Delta \cap \bar{B}(\bar{x}, \varepsilon)$. By (5.6), $0 \leq \langle \phi(\bar{x}), y(t) - \bar{x} \rangle = t\langle \phi(\bar{x}), y - \bar{x} \rangle$. This implies that $\langle \phi(\bar{x}), y - \bar{x} \rangle \geq 0$ for every $y \in \Delta$. Hence $\bar{x} \in \mathrm{Sol}(\mathrm{VI}(\phi, \Delta))$. $\square$

Problem $\mathrm{VI}(\phi, \Delta)$ depends on two data: the set Δ and the operator ϕ. Structure of the solution set $\mathrm{Sol}(\mathrm{VI}(\phi, \Delta))$ is decided by the properties of the set and the operator. In variational inequality theory, the following topics are fundamental: solution existence and uniqueness, stability and sensitivity of the solution sets with respect to perturbations of the problem data, algorithms for finding all the solutions or one part of the solution set.

The following Hartman-Stampacchia Theorem is a fundamental existence theorem for VI problems. It is proved by using the Brouwer fixed point theorem.

Theorem 5.1. (See Hartman and Stampacchia (1966), Kinderlehrer and Stampacchia (1980), Theorem 3.1 in Chapter 1) *If $\Delta \subset R^n$ is nonempty, compact, convex and $\phi : \Delta \to R^n$ is continuous, then problem* $\mathrm{VI}(\phi, \Delta)$ *has a solution.*

Under suitable *coercivity conditions*, one can have existence theorems for problems on noncompact convex sets. For example, the following result is valid.

Theorem 5.2. (See Kinderlehrer and Stampacchia (1980), p. 14) *Let $\Delta \subset R^n$ be a nonempty, closed, convex set and $\phi : \Delta \to R^n$ a continuous operator. If there exists $x^0 \in \Delta$ such that*

$$\langle \phi(y) - \phi(x^0), y - x^0 \rangle \|y - x^0\| \to +\infty \quad as \quad \|y\| \to +\infty, \ y \in \Delta, \tag{5.7}$$

then problem $\mathrm{VI}(\phi, \Delta)$ *has a solution.*

The exact meaning of (5.7) is as follows: Given any $\gamma > 0$ one can find $\rho > 0$ such that

$$\frac{\langle \phi(y) - \phi(x^0), y - x^0 \rangle}{\|y - x^0\|} \geq \gamma \quad \text{for every } y \in \Delta \quad \text{satisfying } \|y\| > \rho.$$

It is obvious that if Δ is compact then, for any $x^0 \in \Delta$, (5.7) is valid. If there exists $x^0 \in \Delta$ such that (5.7) holds then one says that the *coercivity condition* is satisfied. Coercivity conditions play an important role in the study of variational inequalities on noncompact constraint sets. Note that (5.7) is only one of the most well-known forms of coercivity conditions.

If there exists $x^0 \in \Delta$ and $\alpha > 0$ such that

$$\langle \phi(y) - \phi(x^0), y - x^0 \rangle \geq \alpha \|y - x^0\|^2 \quad \forall y \in \Delta \tag{5.8}$$

then, surely, (5.7) holds. It is clear that there exists $\alpha > 0$ such that

$$\langle \phi(y) - \phi(x), y - x \rangle \geq \alpha \|y - x\|^2 \quad \forall x \in \Delta, \ \forall y \in \Delta, \tag{5.9}$$

then (5.8) is satisfied.

Definition 5.2. If there exists $\alpha > 0$ such that (5.9) holds then ϕ is said to be *strongly monotone* on Δ. If the following weaker conditions

$$\langle \phi(y) - \phi(x), y - x \rangle > 0 \quad \forall x \in \Delta, \ \forall y \in \Delta, \ x \neq y, \qquad (5.10)$$

and

$$\langle \phi(y) - \phi(x), y - x \rangle \geq 0 \quad \forall x \in \Delta, \ \forall y \in \Delta, \qquad (5.11)$$

hold, then ϕ is said to be *strictly monotone* on Δ and *monotone* on Δ, respectively.

Example 5.1. Let $\Delta \subset R^n$ is a nonempty, closed, convex set. Let $D \in R^{n \times n}$ and $c \in R^n$. If matrix D is positive definite then the operator $\phi : \Delta \to R^n$ defined by $\phi(x) = Dx + c$, $x \in \Delta$, is strongly monotone on Δ. In this case, it is easily verified that $\alpha > 0$ required for the fulfilment of (5.9) can be defined by setting

$$\alpha = \inf\{v^T D v : v \in R^n, \ \|v\| = 1\}.$$

Likewise, if D is positive semidefinite then the formula $\phi(x) = Dx + c$, $x \in \Delta$, defines a monotone operator.

Proposition 5.4. *The following statements are valid:*

(i) *If ϕ is strictly monotone on Δ then problem VI(ϕ, Δ) cannot have more than one solution;*

(ii) *If ϕ is continuous and monotone on Δ then the solution set of problem VI(ϕ, Δ) is closed and convex (possibly empty).*

For proving the second statement in the preceding proposition we shall need the following useful fact about monotone VI problems.

Lemma 5.1. (The Minty Lemma; Kinderlehrer and Stampacchia (1980), Lemma 1.5 in Chapter 3) *If $\Delta \subset R^n$ is a closed, convex set and $\phi : \Delta \to R^n$ is a continuous, monotone operator, then $\bar{x} \in \mathrm{Sol}(\mathrm{VI}((\phi, \Delta))$ if and only if $\bar{x} \in \Delta$ and*

$$\langle \phi(y), y - \bar{x} \rangle \geq 0 \quad \forall y \in \Delta. \qquad (5.12)$$

Proof. *Necessity:* Let $\bar{x} \in \mathrm{Sol}(\mathrm{VI}((\phi, \Delta))$. By the monotonicity of ϕ, we have

$$\langle \phi(y) - \phi(\bar{x}), y - \bar{x} \rangle \geq 0 \quad \forall y \in \Delta.$$

Combining this with (5.4) yields

$$\langle \phi(y), y - \bar{x} \rangle \geq \langle \phi(\bar{x}), y - \bar{x} \rangle \geq 0 \quad \forall y \in \Delta.$$

Property (5.12) has been established.

Sufficiency: Suppose that $\bar{x} \in \Delta$ and (5.12) is satisfied. Fix any $y \in \Delta$. By the convexity of Δ, $y(t) := \bar{x} + t(y - \bar{x})$ belongs to Δ for every $t \in (0, 1)$. Substituting $y = y(t)$ into (5.12) gives

$$0 \leq \langle \phi(y(t)), y(t) - \bar{x} \rangle = \langle \phi(\bar{x} + t(y - \bar{x}), t(y - \bar{x}) \rangle.$$

This implies that

$$\langle \phi(\bar{x} + t(y - \bar{x}), y - \bar{x} \rangle \geq 0 \quad \forall t \in (0, 1).$$

Letting $t \to 0$, by the continuity of ϕ we obtain $\langle \phi(\bar{x}), y - \bar{x} \rangle \geq 0$. Since the last inequality holds for every $y \in \Delta$, we conclude that $\bar{x} \in \mathrm{Sol}(\mathrm{VI}((\phi, \Delta))$. $\quad \square$

Proof of Proposition 5.4.

(i) Suppose, contrary to our claim, that ϕ is strictly monotone on Δ but problem $\mathrm{VI}(\phi, \Delta)$ has two different solutions $\bar{x}$ and $\bar{y}$. Then $\langle \phi(\bar{x}), \bar{y} - \bar{x} \rangle \geq 0$ and $\langle \phi(\bar{y}), \bar{x} - \bar{y} \rangle \geq 0$. Combining these inequalities we get $\langle \phi(\bar{x}) - \phi(\bar{y}), \bar{y} - \bar{x} \rangle \geq 0$. The last inequality contradicts the fact that $\langle \phi(\bar{y}) - \phi(\bar{x}), \bar{y} - \bar{x} \rangle > 0$.

(ii) Assume that ϕ is continuous and monotone on Δ. For every $y \in \Delta$, denote by $\Omega(y)$ the set of all $\bar{x} \in \Delta$ satisfying the inequality $\langle \phi(y), y - \bar{x} \rangle \geq 0$. It is clear that $\Omega(y)$ is closed and convex. From Lemma 5.1 it follows that

$$\mathrm{Sol}(\mathrm{VI}((\phi, \Delta)) = \bigcap_{y \in \Delta} \Omega(y).$$

Hence $\mathrm{Sol}(\mathrm{VI}((\phi, \Delta))$ is closed and convex (possibly empty). $\quad \square$

From Theorem 5.2 and Proposition 5.4(i) it follows that if Δ is nonempty and $\phi : \Delta \to R^n$ is a continuous, strongly monotone operator then problem $\mathrm{VI}(\phi, \Delta)$ has a unique solution.

In the next section, we will consider variational inequality problems in the case where the constraint set Δ is a cone.

5.2 Complementarity Problems

The following fact paves a way to the notion of (nonlinear) complementarity problem.

Proposition 5.5. *If Δ is a closed convex cone, then problem $VI(\phi, \Delta)$ can be rewritten equivalently as follows*

$$\bar{x} \in \Delta, \quad \phi(\bar{x}) \in \Delta^+, \quad \langle \phi(\bar{x}), \bar{x} \rangle = 0, \tag{5.13}$$

where $\Delta^+ = \{\xi \in R^n : \langle \xi, v \rangle \geq 0 \ \forall v \in \Delta\}$ denotes the positive dual cone of Δ.

Proof. Let $\bar{x}$ be a solution of (5.4). For any $v \in \Delta$, since Δ is a convex cone, we have $\bar{x} + v \in \Delta$. From (5.4) we deduce that

$$0 \leq \langle \phi(\bar{x}), (\bar{x} + v) - \bar{x} \rangle = \langle \phi(\bar{x}), v \rangle.$$

So $\phi(\bar{x}) \in \Delta^+$. Furthermore, since $\frac{1}{2}\bar{x} \in \Delta$ and $2\bar{x} \in \Delta$, by (5.4) we have

$$0 \leq \langle \phi(\bar{x}), \frac{1}{2}\bar{x} - \bar{x} \rangle = -\frac{1}{2}\langle \phi(\bar{x}), \bar{x} \rangle$$

and

$$0 \leq \langle \phi(\bar{x}), 2\bar{x} - \bar{x} \rangle = \langle \phi(\bar{x}), \bar{x} \rangle.$$

Hence $\langle \phi(\bar{x}), \bar{x} \rangle = 0$. We have proved that (5.13) is satisfied.

Now, let $\bar{x}$ be such that (5.13) holds. For every $y \in \Delta$, since $\langle \phi(\bar{x}), \bar{x} \rangle = 0$ and $\phi(\bar{x}) \in \Delta^+$, we have

$$\langle \phi(\bar{x}), y - \bar{x} \rangle = \langle \phi(\bar{x}), y \rangle \geq 0.$$

This shows that $\bar{x} \in \text{Sol}(VI((\phi, \Delta))$. $\square$

Definition 5.3. Problem (5.13) where $\Delta \subset R^n$ is a closed convex cone and $\phi : R^n \to R^n$, is denoted by $NCP(\phi, \Delta)$ and is called the *(nonlinear) complementarity problem* defined by ϕ and Δ.

Since complementarity problems are variational inequality problems of a special type, existence theorems for VI problems can be applied to them.

5.3 Affine Variational Inequalities

By Theorem 3.1, if $\bar{x}$ is a local solution of the quadratic program

$$\min \left\{ f(x) = \frac{1}{2}x^T M x + q^T x : x \in \Delta \right\}, \tag{5.14}$$

where $M \in R_S^{n \times n}$, $q \in R^n$, and $\Delta \subset R^n$ is a polyhedral convex set, then $\langle M\bar{x} + q, y - \bar{x} \rangle \geq 0$ for every $y \in \Delta$. This implies that $\bar{x}$ is a solution of the problem VI(ϕ, Δ) where $\phi(x) = Mx + q$ is an *affine operator* having the constant symmetric Jacobian matrix M.

Definition 5.4. Let $M \in R^{n \times n}$, $q \in R^n$. Let $\Delta \subset R^n$ be a polyhedral convex set. The variational inequality problem

$$\text{Find } \bar{x} \in \Delta \text{ such that } \langle M\bar{x} + q, y - \bar{x} \rangle \geq 0 \quad \forall y \in \Delta \qquad (5.15)$$

is called the *affine variational inequality* (AVI, for brevity) problem defined by the data set $\{M, q, \Delta\}$ and is denoted by AVI(M, q, Δ). The solution set of this problem is abbreviated to Sol(AVI(M, q, Δ)).

The remarks given at the beginning of this section show that quadratic programs lead to *symmetric* AVI problems. Later on, in the study of AVI problems we will not restrict ourselves only to the case of the symmetric problems.

The following theorem shows that solutions of an AVI problem can be characterized by using some Lagrange multipliers.

Theorem 5.3. (See, for instance, Gowda and Pang (1994b), p. 834) *Vector $\bar{x} \in R^n$ is a solution of* (5.15) *where Δ is given by the formula*

$$\Delta = \{x \in R^n : Ax \geq b\} \qquad (5.16)$$

with $A \in R^{m \times n}$, $b \in R^m$, if and only if there exists $\bar{\lambda} = (\bar{\lambda}_1, \ldots, \bar{\lambda}_m) \in R^m$ such that

$$\begin{cases} M\bar{x} - A^T\bar{\lambda} + q = 0, \\ A\bar{x} \geq b, \quad \bar{\lambda} \geq 0, \\ \bar{\lambda}^T(A\bar{x} - b) = 0. \end{cases} \qquad (5.17)$$

Proof. The necessity part of this proof is very similar to the proof of Theorem 3.3. As in the preceding chapters, we denote by A_i the i-th row of A and by b_i the i-th component of vector b. We set $a_i = A_i^T$ for every $i = 1, \ldots, m$. Let $\bar{x} \in$ Sol(AVI(M, q, Δ)). Define $I = \{1, \ldots, m\}$, $I_0 = \{i \in I : \langle a_i, \bar{x} \rangle = b_i\}$ and $I_1 = \{i \in I : \langle a_i, \bar{x} \rangle > b_i\}$. For any $v \in R^n$ satisfying

$$\langle a_i, v \rangle \geq 0 \quad \text{for every } i \in I_0,$$

it is easily seen that there exists $\delta_1 > 0$ such that $\langle a_i, \bar{x} + tv \rangle \geq b_i$ for every $i \in I$ and $t \in (0, \delta_1)$. Substituting $y = \bar{x} + tv$, where $t \in (0, \delta_1)$, into (5.15) gives $\langle M\bar{x} + q, v \rangle \geq 0$. Thus

$$\langle -M\bar{x} - q, v \rangle \leq 0$$

for any $v \in R^n$ satisfying

$$\langle -a_i, v \rangle \leq 0 \quad \text{for every } i \in I_0.$$

By the Farkas Lemma (see Theorem 3.2), there exist non-negative real numbers $\bar{\lambda}_i$ ($i \in I_0$) such that

$$\sum_{i \in I_0} \bar{\lambda}_i(-a_i) = -M\bar{x} - q. \tag{5.18}$$

Put $\bar{\lambda}_i = 0$ for all $i \in I_1$ and $\bar{\lambda} = (\bar{\lambda}_1, \ldots, \bar{\lambda}_m)$. Since $a_i = A_i^T$ for every $i \in I$, from (5.18) we obtain the first equality in (5.17). Since $\bar{x} \in \Delta(A, b)$ and $\bar{\lambda}_i(A_i\bar{x} - b_i) = 0$ for each $i \in I$, the other conditions in (5.17) are also satisfied.

In order to prove the sufficiency part, suppose that there exists $\bar{\lambda} = (\bar{\lambda}_1, \ldots, \bar{\lambda}_m) \in R^m$ such that (5.17) holds. Then, for any $y \in \Delta$, one has

$$\begin{aligned}
\langle M\bar{x} + q, y - \bar{x} \rangle = \langle A^T\bar{\lambda}, y - \bar{x} \rangle &= \langle \bar{\lambda}, (Ay - b) - (A\bar{x} - b) \rangle \\
&= \bar{\lambda}^T(Ay - b) + \bar{\lambda}^T(A\bar{x} - b) \\
&= \bar{\lambda}^T(Ay - b) \geq 0.
\end{aligned}$$

This shows that $\bar{x}$ is a solution of (5.15). The proof is complete.
$\square$

One can derive from Theorem 5.3 the following two corollaries, one of which is applicable to the situation where Δ has the representation

$$\Delta = \{x \in R^n : Ax \geq b, \ x \geq 0\} \tag{5.19}$$

and the other is applicable to the situation where Δ has the representation

$$\Delta = \{x \in R^n : Ax \geq b, \ Cx = d\}. \tag{5.20}$$

Here $A \in R^{m \times n}$, $b \in R^m$, $C \in R^{s \times n}$, and $d \in R^s$.

Corollary 5.1. *Vector $\bar{x} \in R^n$ is a solution of (5.15) where Δ is given by (5.19) if and only if there exists $\bar{\lambda} = (\bar{\lambda}_1, \ldots, \bar{\lambda}_m) \in R^m$ such that*

$$\begin{cases}
M\bar{x} - A^T\bar{\lambda} + q \geq 0, \\
A\bar{x} \geq b, \quad \bar{x} \geq 0, \quad \bar{\lambda} \geq 0, \\
\bar{x}^T(M\bar{x} - A^T\bar{\lambda} + q) + \bar{\lambda}^T(A\bar{x} - b) = 0.
\end{cases} \tag{5.21}$$

Proof. Define matrix $\tilde{A} \in R^{(m+n) \times n}$ and vector $\tilde{b} \in R^{m+n}$ as in the proof of Corollary 2.5. Then problem (5.15), where Δ is given by (5.19), is equivalent to the problem

$$\begin{cases} \text{Find } \bar{x} \in \tilde{\Delta} := \{x \in R^n : \tilde{A}x \geq \tilde{b}\} \text{ such that} \\ \langle M\bar{x} + q, y - \bar{x} \rangle \geq 0 \quad \forall y \in \tilde{\Delta}. \end{cases}$$

Applying Theorem 5.3 to this AVI problem we deduce that $\bar{x}$ is a solution of the latter if and only if there exists $\tilde{\lambda} = (\lambda_1, \ldots, \lambda_{m+n}) \in R^{m+n}$ such that

$$M\bar{x} - \tilde{A}^T \tilde{\lambda} + q = 0, \quad \tilde{A}\bar{x} \geq \tilde{b}, \quad \tilde{\lambda} \geq 0, \quad \tilde{\lambda}^T(\tilde{A}\bar{x} - \tilde{b}) = 0.$$

Taking $\bar{\lambda} = (\bar{\lambda}_1, \ldots, \bar{\lambda}_m)$ where $\bar{\lambda}_i = \lambda_i$ for every $i \in \{1, \ldots, m\}$, we can obtain the desired properties in (5.21) from the last ones. $\square$

Corollary 5.2. *Vector $\bar{x} \in R^n$ is a solution of (5.15) where Δ is given by (5.20) if and only if there exist $\bar{\lambda} = (\bar{\lambda}_1, \ldots, \bar{\lambda}_m) \in R^m$ and $\bar{\mu} = (\bar{\mu}_1, \ldots, \bar{\mu}_s) \in R^s$ such that*

$$\begin{cases} M\bar{x} - A^T\bar{\lambda} - C^T\bar{\mu} + q = 0, \\ A\bar{x} \geq b, \quad C\bar{x} = d, \quad \bar{\lambda} \geq 0, \\ \bar{\lambda}^T(A\bar{x} - b) = 0. \end{cases} \tag{5.22}$$

Proof. Define $\tilde{A} \in R^{(m+2s) \times n}$ and $\tilde{b} \in R^{m+2s}$ as in the proof of Corollary 2.6. Then problem (5.15), where Δ is given by (5.20), is equivalent to the problem

$$\begin{cases} \text{Find } \bar{x} \in \tilde{\Delta} := \{x \in R^n : \tilde{A}x \geq \tilde{b}\} \text{ such that} \\ \langle M\bar{x} + q, y - \bar{x} \rangle \geq 0 \quad \forall y \in \tilde{\Delta}. \end{cases}$$

Applying Theorem 5.3 to this AVI problem we deduce that $\bar{x}$ is a solution of the latter if and only if there exists $\tilde{\lambda} = (\lambda_1, \ldots, \lambda_{m+2s}) \in R^{m+2s}$ such that

$$M\bar{x} - \tilde{A}^T \tilde{\lambda} + q = 0, \quad \tilde{A}\bar{x} \geq \tilde{b}, \quad \tilde{\lambda} \geq 0, \quad \tilde{\lambda}^T(\tilde{A}\bar{x} - \tilde{b}) = 0.$$

Taking $\bar{\lambda} = (\bar{\lambda}_1, \ldots, \bar{\lambda}_m)$ and $\bar{\mu} = (\bar{\mu}_1, \ldots, \bar{\mu}_s)$ where $\bar{\lambda}_i = \lambda_i$ for every $i \in \{1, \ldots, m\}$ and $\bar{\mu}_j = \lambda_{m+j} - \lambda_{m+s+j}$ for every $j \in \{1, \ldots, s\}$, we can obtain the properties stated in (5.22) from the last ones. $\square$

Unlike the solution set and the local solution set of a nonconvex quadratic program, the solution set of an AVI problem has a rather simple structure.

Theorem 5.4. *The solution set of any affine variational inequality problem is the union of finitely many polyhedral convex sets.*

Proof. This proof follows the idea of the proof of formula (4.12). Consider a general AVI problem in the form (5.15). Since Δ is a polyhedral convex set, there exists $m \in N$, $A \in R^{m \times n}$, $b \in R^m$ such that $\Delta = \{x \in R^n : Ax \geq b\}$. According to Theorem 5.3, $x \in \text{Sol}(\text{AVI}(M, q, \Delta))$ if and only if there exists $\lambda = (\lambda_1, \ldots, \lambda_m) \in R^m$ such that

$$\begin{cases} Mx - A^T\lambda + q = 0, \\ Ax \geq b, \quad \lambda \geq 0, \\ \lambda^T(Ax - b) = 0. \end{cases} \tag{5.23}$$

Let $I = \{1, \ldots, m\}$. Given a point $x \in \text{Sol}(\text{AVI}(M, q, \Delta))$, we set $I_0 = \{i \in I : A_ix = b_i\}$, $I_1 = I \setminus I_0 = \{i \in I : A_ix > b_i\}$. From the last equality in (5.23) we get

$$\lambda_i = 0 \quad \forall i \in I_1.$$

Hence (x, λ) satisfies the system

$$\begin{cases} Mx - A^T\lambda + q = 0, \\ A_{I_0}x = b_{I_0}, \quad \lambda_{I_0} \geq 0, \\ A_{I_1}x \geq b_{I_1}, \quad \lambda_{I_1} = 0. \end{cases} \tag{5.24}$$

Fix any subset $I_0 \subset I$ and denote by Q_{I_0} the set of all (x, λ) satisfying (5.24). It is obvious that Q_{I_0} is a polyhedral convex set. From what has been said it follows that

$$\text{Sol}(\text{AVI}(M, q, \Delta)) = \bigcup \{\text{Pr}_{R^n}(Q_{I_0}) : I_0 \subset I\}, \tag{5.25}$$

where $\text{Pr}_{R^n}(x, \lambda) := x$. Since $\text{Pr}_{R^n}(\cdot) : R^n \times R^m \to R^n$ is a linear operator, for every $I_0 \subset I$, $\text{Pr}_{R^n}(Q_{I_0})$ is a polyhedral convex set. From (5.25) it follows that $\text{Sol}(\text{AVI}(M, q, \Delta))$ is the union of finitely many polyhedral convex sets. $\square$

Definition 5.5. A half-line $\omega = \{\bar{x} + t\bar{v} : t \geq 0\}$, where $\bar{v} \in R^n \setminus \{0\}$, which is a subset of $\text{Sol}(\text{AVI}(M, q, \Delta))$, is called a *solution ray* of problem (5.15).

Definition 5.6. A line segment $\omega_\delta = \{\bar{x} + t\bar{v} : t \in [0, \delta)\}$, where $\bar{v} \in R^n \setminus \{0\}$ and $\delta > 0$, which is a subset of $\text{Sol}(\text{AVI}(M, q, \Delta))$, is called a *solution interval* of problem (5.15).

Corollary 5.3. *The following statements hold:*

(i) *The solution set of any affine variational inequality is a closed set (possibly empty);*

(ii) *If the solution set of an affine variational inequality is unbounded, then the problem has a solution ray;*

(iii) *If the solution set of an affine variational inequality is infinite, then the problem has a solution interval.*

Proof. Statement (i) follows directly from formula (5.25) because, for any $I_0 \subset I$, the set $\mathrm{Pr}_{R^n}(Q_{I_0})$, being polyhedral convex, is closed. If $\mathrm{Sol}(\mathrm{AVI}(M, q, \Delta))$ is unbounded then from (5.25) it follows that there exists an index set $I_0 \subset I$ such that

$$\Omega_{I_0} := \mathrm{Pr}_{R^n}(Q_{I_0}) \tag{5.26}$$

is an unbounded set. Since Ω_{I_0} is a polyhedral convex set, it is an unbounded closed convex set. By Theorem 8.4 in Rockafellar (1970), Ω_{I_0} admits a direction of recession; that is there exists $\bar{v} \in R^n \setminus \{0\}$ such that

$$x + t\bar{v} \in \Omega_{I_0} \forall x \in \Omega_{I_0}, \ \forall t \geq 0. \tag{5.27}$$

Taking any $\bar{x} \in \Omega_{I_0}$ we deduce from (5.25) and (5.27) that $\bar{x} + t\bar{v} \in \mathrm{Sol}(\mathrm{AVI}(M, q, \Delta))$ for all $t \geq 0$. Thus we have proved that problem (5.15) has a solution ray. If $\mathrm{Sol}(\mathrm{AVI}(M, q, \Delta))$ is infinite then from (5.25) we deduce that there is an index set $I_0 \subset I$ such that the polyhedral convex set Ω_{I_0} defined by (5.26) is infinite. Then there must exist two different points $x \in \Omega_{I_0}$ and $y \in \Omega_{I_0}$. It is clear that the set $[x, y] := \{x + t(y - x) : t \in [0, 1)\}$ is a solution interval of (5.15). $\square$

Using Theorem 5.4 one can obtain a complete characterization for the unboundedness property of the solution set of an AVI problem. Let us consider problem (5.15) where Δ is given by (5.16) and introduce the following notations:

$$\delta(A) = \{v \in R^n : Av \geq 0\},$$
$$\delta(A)^+ = \{z \in R^n : z^T v \geq 0 \ \ \forall v \in \delta(A)\},$$
$$\ell(M) = \{z \in R^n : z^T M z = 0\}.$$

Note that $\delta(A)$ and $\{v \in R^n : Av \in \delta(A)^+\}$ are polyhedral convex cones, while $\ell(M)$ is, in general, a nonconvex closed cone. Note also that $\delta(A) = 0^+\Delta$ and $\delta(A)^+ = (0^+\Delta)^+$.

Theorem 5.5 (cf. Gowda and Pang (1994a)). *The solution set of* (5.15) *is unbounded if and only if there exists a pair* $(v, u^0) \in R^n \times R^n$, $v \neq 0$, $u^0 \in \mathrm{Sol}(\mathrm{AVI}(M, q, \Delta))$, *such that*

(i) $v \in \delta(A), \quad Mv \in \delta(A)^+, \quad v \in \ell(M)$;

(ii) $(Mu^0 + q)^T v = 0$;

(iii) $\langle Mv, y - u^0 \rangle \geq 0 \quad \forall y \in \Delta$.

Proof. *Sufficiency:* Suppose that there is a pair $(v, u^0) \in R^n \times R^n$, $v \neq 0$, $u^0 \in \mathrm{Sol}(\mathrm{AVI}(M, q, \Delta))$, such that (i)-(iii) are fulfilled. Let $x_t = u^0 + tv$, $t > 0$. Given any $y \in \Delta$, we deduce from (i)-(iii) that

$$
\begin{aligned}
\langle Mx_t + q, y - x_t \rangle &= \langle M(u^0 + tv) + q, y - (u^0 + tv) \rangle \\
&= \underbrace{\langle Mu^0 + q, y - u^0 \rangle}_{\geq 0} - t \underbrace{\langle Mu^0 + q, v \rangle}_{=0} \\
&\quad + t \underbrace{\langle Mv, y - u^0 \rangle}_{\geq 0} - t^2 \underbrace{\langle Mv, v \rangle}_{=0} \\
&\geq 0.
\end{aligned}
$$

This implies that $x_t \in \mathrm{Sol}(\mathrm{AVI}(M, q, \Delta))$ for every $t > 0$. Hence the solution set is unbounded.

Necessity: Suppose that the set $\mathrm{Sol}(\mathrm{AVI}(M, q, \Delta))$ is unbounded. By (5.25), there exists $I_0 \subset I$ such that the set Ω_{I_0} defined by (5.26) is unbounded. Applying Theorem 8.4 from Rockafellar (1970), we can assert that there exist $v \in R^n$, $v \neq 0$, and $u^0 \in \Omega_{I_0}$ such that

$$
u^0 + tv \in \Omega_{I_0} \subset \mathrm{Sol}(\mathrm{AVI}(M, q, \Delta)) \quad \forall t \geq 0. \tag{5.28}
$$

Since $A(u^0 + tv) \geq b$ for every $t > 0$, we can deduce that $Av \geq 0$. This means that $v \in \delta(A)$. By (5.28), we have

$$
\langle M(u^0 + tv) + q, y - (u^0 + tv) \rangle \geq 0 \quad \forall y \in \Delta. \tag{5.29}
$$

Fixing any $y \in \Delta$, we deduce from (5.29) that

$$
\left\langle \frac{1}{t} Mu^0 + Mv + \frac{1}{t} q, \frac{1}{t} y - \frac{1}{t} u_0 - v \right\rangle \geq 0 \quad \forall t > 0.
$$

Therefore

$$
\langle Mv, -v \rangle \geq 0. \tag{5.30}
$$

Substituting $y = u^0 + t^2 v$, where $t > 1$, into (5.29) and dividing the inequality by $t(t^2 - t)$, we obtain

$$\left\langle \frac{1}{t} M u^0 + M v + \frac{1}{t} q, v \right\rangle \geq 0 \quad \forall t > 1.$$

Letting $t \to +\infty$ yields $\langle M v, v \rangle \geq 0$. Combining this with (5.30) we get

$$\langle M v, v \rangle = 0. \tag{5.31}$$

This shows that $v \in \ell(M)$. Substituting $y = u^0$ into (5.29) and taking account of (5.31) we have $\langle M u^0 + q, v \rangle \leq 0$. Substituting $y = u^0 + t^2 v$, where $t > 1$, into (5.29) and using (5.31) we can deduce that $\langle M u^0 + q, v \rangle \geq 0$. This and the preceding inequality shows that (ii) is satisfied. By (5.29), (5.31) and (ii), for every $y \in \Delta$ we have

$$\begin{aligned} 0 &\leq \langle M u^0 + q + t M v, y - u^0 - t v \rangle \\ &= \langle M u^0 + q, y - u^0 \rangle + t \langle M v, y - u^0 \rangle \end{aligned}$$

for all $t > 0$. This implies that the inequality $\langle M v, y - u^0 \rangle < 0$ must be false. So we have

$$\langle M v, y - u^0 \rangle \geq 0 \quad \forall y \in \Delta. \tag{5.32}$$

Substituting $y = u^0 + w$, where $w \in \delta(A)$, into the inequality in (5.32) we deduce that $\langle M v, w \rangle \geq 0$ for every $w \in \delta(A)$. This means that $M v \in \delta(A)^+$. We have thus shown that all the three inclusions in (i) are valid. The proof is complete. $\square$

Several simple sufficient conditions for (5.15) to have a compact solution set can be obtained directly from the preceding theorem.

Corollary 5.4. *Problem (5.15) has a compact solution set (possibly empty) if one of the following conditions is satisfied:*

(γ_1) *the cone $\ell(M)$ consists of only one element 0;*

(γ_2) *the intersection of the cones $\ell(M)$ and $\{v \in R^n : M v \in \delta(A)^+\}$ consists of only one element 0;*

(γ_3) *the intersection of the cones $\ell(M)$, $\{v \in R^n : M v \in \delta(A)^+\}$ and $\delta(A)$, consists of only one element 0.*

Examples given in the next section will show how the above sufficient conditions can be used in practice.

5.4 Linear Complementarity Problems

We now consider a special case of the model (5.13) which plays a very important role in theory of finite-dimensional variational inequalities and complementarity problems (see, for instance, Harker and Pang (1990) and Cottle et al. (1992)).

Definition 5.7. Problem (5.13) with $\Delta = R_+^n$ and $\phi(x) = Mx + q$ where $M \in R^{n \times n}$ and $q \in R^n$, is denoted by LCP(M, q) and is called the *linear complementarity problem* defined by M and q. The solution set of this problem is denoted by Sol(M, q).

We can write LCP(M, q) as follows

$$\bar{x} \geq 0, \quad M\bar{x} + q \geq 0, \quad \bar{x}^T (M\bar{x} + q) = 0. \tag{5.33}$$

Thus LCP problem is a special case of the NCP problem where $\Delta = R_+^n$ and ϕ is an affine operator.

If $\bar{x}$ is a local solution of quadratic program (3.1) where $\Delta = R_+^n$, then $\bar{x} \in R_+^n$ and, by Theorem 3.1,

$$\langle D\bar{x} + c, y - \bar{x} \rangle \geq 0 \quad \forall y \in R_+^n.$$

This amounts to saying that $\bar{x}$ is a solution of the linear complementarity problem LCP(D, c) defined D and c.

By Corollary 3.1, if $\bar{x}$ is a local solution of the quadratic program (2.26) then there exists $\lambda = (\lambda_1, \ldots, \lambda_m) \in R^m$ such that (3.8) holds. Setting

$$M = \begin{bmatrix} D & -A^T \\ A & 0 \end{bmatrix}, \quad q = \begin{pmatrix} c \\ -b \end{pmatrix}, \quad \bar{z} = \begin{pmatrix} \bar{x} \\ \lambda \end{pmatrix}, \tag{5.34}$$

we have $M \in R^{(n+m) \times (n+m)}$, $q \in R^{n+m}$, $\bar{z} \in R^{n+m}$. It is easily verified that (3.8) is equivalent to the system

$$\bar{z} \geq 0, \quad M\bar{z} + q \geq 0, \quad \bar{z}^T (M\bar{z} + q) = 0.$$

Thus (3.8) can be interpreted as a LCP problem.

Definition 5.8. If Δ is a polyhedral convex cone and there exist $M \in R^{n \times n}$, $q \in R^n$, such that $\phi(x) = Mx + q$ for every $x \in \Delta$, then (5.13) is said to be a *generalized linear complementarity problem*. It is denoted by GLCP(M, q, Δ).

From the above definition we see that generalized linear complementarity problems are the AVI problems of a special type. Comparing Definition 5.8 with Definition 5.7 we see at once that the structure of GLCP problems is very similar to that of LCP problems. This explains why many results concerning LCP problems can be extended to GLCP problems.

It is easily seen that if in (5.15) one chooses $\Delta = R_+^n$ then one obtains the linear complementarity problem $LCP(M, q)$. Hence linear complementarity problems are the AVI problems of a special type. In this book, as a rule, we try first to prove theorems (on the solution existence, on the solution stability, etc.) for AVI problems then apply them to LCP problems.

Theorem 5.5 can be specialized for LCP problems as follows.

Proposition 5.6. (See Yen and Hung (2001), Theorem 2) *The solution set of (5.33) is unbounded if and only if there exists a pair* $(v, u^0) \in R^n \times R^n$, $v \neq 0$, $u^0 \in Sol(M, q)$, *such that*

(i) $v \geq 0$, $Mv \geq 0$, $v \in \ell(M)$;

(ii) $(Mu^0 + q)^T v = 0$;

(iii) $\langle Mv, u^0 \rangle = 0$.

Corollary 5.4 is specialized for LCP problems as follows.

Corollary 5.5. *Problem (5.33) has a compact solution set (possibly empty) if one of the following conditions is satisfied:*

(γ_1) *the cone* $\ell(M)$ *consists of only one element 0;*

(γ_2) *the intersection of the cones* $\ell(M)$ *and* $\{v \in R^n : Mv \geq 0\}$ *consists of only one element 0;*

(γ_3) *the intersection of the cones* $\ell(M)$, $\{v \in R^n : Mv \geq 0\}$ *and* R_+^n, *consists of only one element 0.*

Example 5.2. (See Yen and Hung (2001)) Consider problem (5.33) with

$$M = \begin{pmatrix} 1 & 0 \\ 0 & -1 \end{pmatrix}, \quad q = \begin{pmatrix} q_1 \\ q_2 \end{pmatrix} \in R^2, \quad n = 2.$$

A direct computation shows that the intersection of the cones $\ell(M)$, $\{v : Mv \geq 0\}$ and $\{v : v \geq 0\}$, consists of only 0. By Corollary 5.5, $Sol(M, q)$ is a compact set.

Observe that M in the above example is a nondegenerate matrix, so from the theory in Chapter 3 of Cottle et al. (1982), it follows that $Sol(M, q)$ is a finite set. By definition, $M = (a_{ij})$ is said to be a *nondegenerate matrix* if, for any nonempty subset $\alpha \subset \{1, \ldots, n\}$, the determinant of the principal submatrix $M_{\alpha\alpha}$ consisting of the elements a_{ij} $(i \in \alpha,\ j \in \alpha)$ of M is nonzero.

Example 5.3. (See Yen and Hung (2001)) Consider problem (5.13) with

$$M = \begin{pmatrix} 1 & 0 \\ 0 & 0 \end{pmatrix}, \quad q = \begin{pmatrix} -1 \\ \mu \end{pmatrix} \in R^2, \quad \mu \neq 0, \quad n = 2.$$

A direct computation shows that the intersection of the cones $\ell(M)$, $\{v : Mv \geq 0\}$ and $\{v : v \geq 0\}$, is the set $\{v = (0, v_2) \in R^2 : v_2 \geq 0\}$. For verifying condition (ii) in Proposition 5.6, there is no loss of generality in assuming that $v = (0, 1)$. It is easy to show that there is no $u^0 \geq 0$ such that $(Mu^0 + q)^T v = 0$ and $\langle Mv, u^0 \rangle = 0$. By Proposition 5.6, $\mathrm{Sol}(M, q)$ is a compact set.

Example 5.4. (See Yen and Hung (2001)) Consider problem (5.13) with

$$M = \begin{pmatrix} 1 & 0 \\ 0 & 0 \end{pmatrix}, \quad q = \begin{pmatrix} -1 \\ 0 \end{pmatrix} \in R^2, \quad n = 2.$$

The intersection of the cones $\ell(M)$, $\{v : Mv \geq 0\}$ and $\{v : v \geq 0\}$, is the set $\{v = (0, v_2) \in R^2 : v_2 \geq 0\}$. It is easy to show that conditions (i)-(iii) in Proposition 5.6 are satisfied if we choose $v = (0, 1)$ and $u^0 = (1, 0)$. By Proposition 5.6, $\mathrm{Sol}(M, q)$ is an unbounded set.

5.5 Commentaries

Problem (5.4) is finite-dimensional. Infinite-dimensional VI problems are not studied in this book. Systematic studies on infinite-dimensional VI problems with applications to mathematical physics (obstacle problems, etc.) can be found, for example, in Kinderlehrer and Stampacchia (1980), Rodrigues (1987).

The important role of finite-dimensional VI problems and complementarity problems in mathematics and in mathematical applications is well known (see, for instance, Harker and Pang (1990), Nagurney (1993), and Patriksson (1999)).

A comprehensive theory on LCP problems was given by Cottle, Pang and Stone (1992). Several key results on LCP problems have been extended to the case of AVI problems.

The first volume of the book by Facchinei and Pang (2003) describes the basic theory on finite-dimensional VI problems and complementarity problems, while its second volume concentrates on iterative algorithms for solving these problems. The book aims at being an enduring reference on the subject and at providing the foundation for its continued growth.

Robinson (see Robinson (1979), Theorem 2, and Robinson (1981), Proposition 1) obtained two fundamental theorems on Lipschitz continuity of the solution map in general AVI problems, which he called the *linear generalized equations*. In Chapter 7 we will study these theorems.

Chapter 6

Solution Existence for Affine Variational Inequalities

In this chapter, some basic theorems on the solution existence of affine variational inequalities will be proved. Different conditions on monotonicity of the linear operator represented by matrix M and the relative position of vector q with respect to the constraint set Δ and the recession cone $0^+\Delta$ will be used in these theorems. As in the preceding chapter, we denote the problem

$$\text{Find } \bar{x} \in \Delta \text{ such that } \langle M\bar{x} + q, y - \bar{x}\rangle \geq 0 \ \forall y \in \Delta \qquad (6.1)$$

by $\text{AVI}(M, q, \Delta)$. Here $M \in R^{n \times n}$, $q \in R^n$, and Δ is a nonempty polyhedral convex set in R^n.

6.1 Solution Existence under Monotonicity

Consider problem (6.1). Since Δ is a polyhedral convex set, there exist $m \in N$, $A \in R^{m \times n}$ and $b \in R^m$ such that

$$\Delta = \{x \in R^n \ : \ Ax \geq b\}. \qquad (6.2)$$

Theorem 6.1. (See Gowda and Pang (1994a), p. 432) *If the following two conditions are satisfied*

(i) *there exists $\bar{x} \in \Delta$ such that $(M\bar{x}+q)^T v \geq 0$ for every $v \in 0^+\Delta$;*

(ii) $(y - x)^T M(y - x) \geq 0$ for all $x \in \Delta$ and $y \in \Delta$;

then the solution set $\text{Sol}(\text{AVI}(M, q, \Delta))$ is nonempty.

Let

$$\tilde{\Delta} = \left\{ z = \begin{pmatrix} x \\ \lambda \end{pmatrix} \in R^{n+m} : (M \quad -A^T)z = -q, \right.$$
$$\left. \begin{pmatrix} A & 0 \\ 0 & E \end{pmatrix} z \geq \begin{pmatrix} b \\ 0 \end{pmatrix} \right\}$$
$$= \left\{ z = \begin{pmatrix} x \\ \lambda \end{pmatrix} \in R^{n+m} : Mx - A^T\lambda + q = 0, \ Ax \geq b, \ \lambda \geq 0 \right\},$$

where I denotes the unit matrix in $R^{m \times m}$. Let

$$f(z) = \frac{1}{2}z^T(\widetilde{M} + \widetilde{M}^T)z + \begin{pmatrix} q \\ -b \end{pmatrix}^T z,$$

where

$$\widetilde{M} = \begin{pmatrix} M & -A^T \\ A & 0 \end{pmatrix} \in R^{(n+m) \times (n+m)}, \quad z = \begin{pmatrix} x \\ \lambda \end{pmatrix} \in R^{n+m}.$$

Consider the following auxiliary quadratic program

$$\min\{f(z) : z \in \tilde{\Delta}\}. \tag{6.3}$$

Lemma 6.1. The set $\tilde{\Delta}$ is nonempty if and only if there exists $\bar{x} \in \Delta$ such that $(M\bar{x} + q)^T v \geq 0$ for every $v \in 0^+\Delta$.

Proof. Necessity: If $\tilde{\Delta} \neq \emptyset$ then there exists $\bar{z} = \begin{pmatrix} \bar{x} \\ \bar{\lambda} \end{pmatrix} \in R^{n+m}$
such that
$$M\bar{x} - A^T\bar{\lambda} + q = 0, \ A\bar{x} \geq b, \ \bar{\lambda} \geq 0. \tag{6.4}$$
Let $v \in 0^+\Delta$. By (6.2), we have $Av \geq 0$. From (6.4) we deduce that

$$0 = (M\bar{x} - A^T\bar{\lambda} + q)^T v = (M\bar{x} + q)^T v - \bar{\lambda}^T Av.$$

Hence $(M\bar{x} + q)^T v = \bar{\lambda}^T Av \geq 0$.

Sufficiency: Suppose that there exists $\bar{x} \in \Delta$ such that $(M\bar{x} + q)^T v \geq 0$ for every $v \in 0^+\Delta = \delta(A)$. Consider the following linear program
$$\min\{c^T y : y \in \Delta\}, \tag{6.5}$$
where $c := M\bar{x} + q$. From our assumption it follows that $\Delta \neq \emptyset$ and $(M\bar{x} + q)^T v \geq 0$ whenever $v \in R^n$, $Av \geq 0$. By Theorem 2.2, (6.5)

has a solution. According to Theorem 3.3, there exists $\bar{\lambda} \in R^m$ such that

$$-A^T\bar{\lambda} + c = 0, \quad \bar{\lambda} \geq 0. \tag{6.6}$$

Since $\bar{x} \in \Delta$, we have $A\bar{x} \geq b$. Combining this with (6.6) we deduce that $\bar{z} := \begin{pmatrix} \bar{x} \\ \bar{\lambda} \end{pmatrix}$ belongs to $\tilde{\Delta}$. So $\tilde{\Delta} \neq \emptyset$. $\quad\square$

Lemma 6.2. *If there exists $\bar{x} \in \Delta$ such that $(M\bar{x} + q)^T v \geq 0$ for every $v \in 0^+\Delta$ then the auxiliary quadratic program (6.3) has a solution.*

Proof. By Lemma 6.1, from the assumption it follows that $\tilde{\Delta}$ is nonempty. Let $z = \begin{pmatrix} x \\ \lambda \end{pmatrix} \in \tilde{\Delta}$. We have

$$
\begin{aligned}
f(z) &= \frac{1}{2}z^T(\widetilde{M} + \widetilde{M}^T)z + \begin{pmatrix} q \\ -b \end{pmatrix}^T z \\
&= \frac{1}{2}\begin{pmatrix} x \\ \lambda \end{pmatrix}^T \left[\begin{pmatrix} M & -A^T \\ A & 0 \end{pmatrix} + \begin{pmatrix} M & -A^T \\ A & 0 \end{pmatrix}^T \right] \begin{pmatrix} x \\ \lambda \end{pmatrix} \\
&\quad + \begin{pmatrix} q \\ -b \end{pmatrix}^T \begin{pmatrix} x \\ \lambda \end{pmatrix} \\
&= \frac{1}{2}\begin{pmatrix} x \\ \lambda \end{pmatrix}^T \begin{pmatrix} M + M^T & 0 \\ 0 & 0 \end{pmatrix} \begin{pmatrix} x \\ \lambda \end{pmatrix} + q^T x - b^T \lambda \\
&= \frac{1}{2}x^T(M + M^T)x + q^T x - b^T \lambda \\
&= x^T M x + q^T x - b^T \lambda \\
&= x^T(Mx + q) - b^T \lambda \\
&= x^T(A^T \lambda) - b^T \lambda \\
&= \lambda^T(Ax - b) \geq 0.
\end{aligned}
$$

So $f(z)$ is bounded from below on $\tilde{\Delta}$. By the Frank-Wolfe Theorem (see Theorem 2.1), (6.3) has a solution. $\quad\square$

Proof of Theorem 6.1.

By assumption (i) and by Lemma 6.2, the auxiliary quadratic problem (6.3) has a solution $\bar{z} = \begin{pmatrix} \bar{x} \\ \bar{\lambda} \end{pmatrix}$. Hence, by Corollary 3.2 there exist Lagrange multipliers $\theta = \begin{pmatrix} \theta^1 \\ \theta^2 \end{pmatrix} \in R^{2m}$ and $\mu \in R^n$ such that

$$\begin{cases} (\widetilde{M} + \widetilde{M}^T) \begin{pmatrix} \bar{x} \\ \bar{\lambda} \end{pmatrix} - \begin{pmatrix} A & 0 \\ 0 & I \end{pmatrix}^T \begin{pmatrix} \theta^1 \\ \theta^2 \end{pmatrix} \\ \qquad - (M - A^T)^T \mu + \begin{pmatrix} q \\ -b \end{pmatrix} = 0, \\ \begin{pmatrix} A & 0 \\ 0 & I \end{pmatrix} \begin{pmatrix} \bar{x} \\ \bar{\lambda} \end{pmatrix} \geq \begin{pmatrix} b \\ 0 \end{pmatrix}, \quad \theta \geq 0, \\ \theta^T \left[\begin{pmatrix} A & 0 \\ 0 & I \end{pmatrix} \begin{pmatrix} \bar{x} \\ \bar{\lambda} \end{pmatrix} - \begin{pmatrix} b \\ 0 \end{pmatrix} \right] = 0. \end{cases}$$

This system can be written as the following one

$$\begin{cases} \begin{pmatrix} M & -A^T \\ A & 0 \end{pmatrix} \begin{pmatrix} \bar{x} \\ \bar{\lambda} \end{pmatrix} + \begin{pmatrix} M^T & A^T \\ -A & 0 \end{pmatrix} \begin{pmatrix} \bar{x} \\ \bar{\lambda} \end{pmatrix} - \begin{pmatrix} A^T & 0 \\ 0 & I \end{pmatrix} \begin{pmatrix} \theta^1 \\ \theta^2 \end{pmatrix} \\ \qquad - \begin{pmatrix} M^T \\ -A \end{pmatrix} \mu + \begin{pmatrix} q \\ -b \end{pmatrix} = 0 \\ A\bar{x} \geq b, \ \bar{\lambda} \geq 0, \ \theta^1 \geq 0, \ \theta^2 \geq 0, \\ (\theta^1)^T (A\bar{x} - b) = 0, \ (\theta^2)^T \bar{\lambda} = 0. \end{cases}$$

In its turn, the latter is equivalent to the system (6.7)–(6.10) below:

$$(M\bar{x} - A^T\bar{\lambda} + q) + M^T\bar{x} + A^T\bar{\lambda} - A^T\theta^1 - M^T\mu = 0, \qquad (6.7)$$

$$A\bar{x} - A\bar{x} - \theta^2 + A\mu - b = 0, \qquad (6.8)$$

$$A\bar{x} \geq b, \ \bar{\lambda} \geq 0, \ \theta^1 \geq 0, \ \theta^2 \geq 0, \qquad (6.9)$$

$$(\theta^1)^T (A\bar{x} - b) = 0, \ (\theta^2)^T \bar{\lambda} = 0. \qquad (6.10)$$

From (6.7) and the inclusion $\bar{z} = \begin{pmatrix} \bar{x} \\ \bar{\lambda} \end{pmatrix} \in \tilde{\Delta}$ it follows that

$$M^T(\bar{x} - \mu) = A^T(\theta^1 - \bar{\lambda}). \qquad (6.11)$$

From (6.8) it follows that

$$A(\bar{x} - \mu) = A\bar{x} - \theta^2 - b. \qquad (6.12)$$

By (6.10)–(6.12),

$$\begin{aligned} (\bar{x} - \mu)^T M^T (\bar{x} - \mu) &= (\bar{x} - \mu)^T A^T (\theta^1 - \bar{\lambda}) \\ &= (\theta^1 - \bar{\lambda})^T A (\bar{x} - \mu) \\ &= (\theta^1 - \bar{\lambda})^T (A\bar{x} - \theta^2 - b) \\ &= \underbrace{(\theta^1)^T (A\bar{x} - b)}_{=0} - (\theta^1)^T \theta^2 \\ &\qquad + \underbrace{(\theta^2)^T \bar{\lambda}}_{=0} - \bar{\lambda}^T (A\bar{x} - b) \\ &= -(\theta^1)^T \theta^2 - \bar{\lambda}^T (A\bar{x} - b). \end{aligned}$$

Hence, by virtue of (6.9), we have

$$(\bar{x} - \mu)^T M^T (\bar{x} - \mu) = -(\theta^1)^T \theta^2 - \bar{\lambda}^T (A\bar{x} - b) \leq 0. \qquad (6.13)$$

From (6.8) it follows that

$$A\mu - b = \theta^2 \geq 0.$$

So $\mu \in \Delta$. Since $\bar{x} \in \Delta$, we can deduce from assumption (ii) that $(\bar{x} - \mu)^T M^T (\bar{x} - \mu) \geq 0$. Combining this with (6.9) and (6.13) gives $\bar{\lambda}^T (A\bar{x} - b) = 0$. Since $\left(\begin{smallmatrix} \bar{x} \\ \bar{\lambda} \end{smallmatrix}\right) \in \tilde{\Delta}$, $M\bar{x} - A^T \bar{\lambda} + q = 0$. Thus we have shown that

$$\begin{cases} M\bar{x} - A^T\bar{\lambda} + q = 0, \\ A\bar{x} \geq b, \ \bar{\lambda} \geq 0, \\ \bar{\lambda}^T (A\bar{x} - b) = 0. \end{cases}$$

Then, according to Theorem 5.3, $\bar{x} \in \mathrm{Sol}(\mathrm{AVI}(M, q, \Delta))$. $\square$

Assumption (ii) is crucial for the validity of the conclusion of the above theorem. It is easily seen that (ii) is equivalent to the requirement that the operator $\phi : \Delta \to R^n$ defined by setting $\phi(x) = Mx + q$ is monotone on Δ (see Definition 5.2).

Definition 6.1 (cf. Cottle et al. (1992), p. 176). By abuse of terminology, we say that matrix $M \in R^{n \times n}$ is *monotone* on a closed convex set $\Delta \subset R^n$ if the linear operator corresponding to M is monotone on Δ, that is

$$(y - x)^T M(y - x) \geq 0 \quad \forall x \in \Delta, \ \forall y \in \Delta. \qquad (6.14)$$

Matrix M is said to be *copositive* on Δ if

$$v^T M v \geq 0 \quad \forall v \in 0^+ \Delta. \qquad (6.15)$$

If M is copositive on R^n_+ then one simply says that M is a *copositive matrix*. Matrix M is said to be *strictly copositive* on Δ if

$$v^T M v > 0 \quad \forall v \in 0^+ \Delta \setminus \{0\}. \qquad (6.16)$$

Remark 6.1. Monotonicity implies copositivity. But the reverse implication, in general, is false. Indeed, if (6.14) holds and if Δ is nonempty then, for any $\bar{x} \in \Delta$ and $v \in 0^+ \Delta$, we have

$$v^T M v = ((\bar{x} + v) - \bar{x})^T M((\bar{x} + v) - \bar{x}) \geq 0.$$

Hence M is copositive on Δ. To show that in general copositivity does not imply monotonicity, we consider the following example. Let

$$M = \begin{pmatrix} 1 & 2 \\ 2 & 1 \end{pmatrix} \in R^{2\times 2}, \quad \Delta = R^2_+.$$

For every $v \in 0^+\Delta = R^2_+$ we have $v^T M v \geq 0$. So M is copositive on Δ. But M is not monotone on Δ. Indeed, choosing $x = (0,1)$ and $y = (1,0)$, we see that

$$(y-x)^T M(y-x) = (1 \ -1) \begin{pmatrix} 1 & 2 \\ 2 & 1 \end{pmatrix} \begin{pmatrix} 1 \\ -1 \end{pmatrix} = -2 < 0.$$

Remark 6.2. If $\mathrm{int}\Delta \neq \emptyset$ then matrix M is monotone on Δ if and only if M is positive semidefinite. Indeed, it is clear that if $M \in R^{n\times n}$ is a positive semidefinite matrix then, for any nonempty closed convex set $\Delta \subset R^n$, M is copositive on Δ. On the other hand, if $\mathrm{int}\Delta \neq \emptyset$ then there exists $\bar{x} \in \Delta$ and $\varepsilon > 0$ such that $B(\bar{x}, \varepsilon) \subset \Delta$. For every $z \in R^n$ there exists $t > 0$ such that $y := \bar{x} + tz \in B(\bar{x}, \varepsilon) \subset \Delta$. Then we have

$$0 \leq (y - \bar{x})^T M(y - \bar{x}) = t^2 z^T M z.$$

Hence $z^T M z \geq 0$ for every $z \in R^n$.

Remark 6.3. It is clear that if M is strictly copositive on Δ then it is copositive on Δ. The converse is not true in general. For example, if $\Delta = R^2_+$ and $M = \begin{pmatrix} 1 & 0 \\ 0 & 0 \end{pmatrix}$, then M is copositive but not strictly copositive on Δ. Indeed, choosing $\bar{v} = (0,1)$ we see that $\bar{v} \in 0^+\Delta \setminus \{0\} = R^2_+ \setminus \{0\}$ but $\bar{v}^T M \bar{v} = 0$.

We now consider a simple example to see how Theorem 6.1 can be used.

Example 6.1. Let

$$M = \begin{pmatrix} 1 & 0 \\ 0 & -1 \end{pmatrix} \in R^{2\times 2}, \quad q = (q_1, q_2) \in R^2,$$
$$\Delta = \{x = (x_1, x_2) \in R^2_+ : x_1 \geq 0, \ x_2 = 0\}.$$

Theorem 6.1 can be applied to this problem. Indeed, since M is monotone on Δ, it suffices to show that there exists $\bar{x} \in \Delta$ such that $(M\bar{x} + q)^T v \geq 0$ for every $v \in 0^+\Delta$. If $q_1 < 0$ then $\bar{x} = (-q_1, 0)$ satisfies the last condition. If $q_1 \geq 0$ then $\bar{x} = (0,0)$ satisfies that condition. Further investigation on the problem shows that $\mathrm{Sol}(\mathrm{AVI}(M, q, \Delta)) = \{(-q_1, 0)\}$ if $q_1 < 0$ and $\mathrm{Sol}(\mathrm{AVI}(M, q, \Delta)) = \{(0, 0)\}$ if $q_1 \geq 0$.

From Theorem 6.1 it is easy to deduce the following result.

Theorem 6.2. (See Gowda and Pang (1994a), Theorem 1) *If M is a positive semidefinite matrix and there exists $\bar{x} \in \Delta$ such that $(M\bar{x} + q)^T v \geq 0$ for every $v \in 0^+\Delta$, then problem (6.1) has a solution.*

Corollary 6.1. (See Cottle et al. (1982), Theorem 3.1.2) *Suppose that M is a positive semidefinite matrix. Then the linear complementarity problem $\mathrm{LCP}(M, q)$ has a solution if and only if there exists $\bar{x}$ such that*

$$\bar{x} \geq 0, \quad M\bar{x} + q \geq 0. \tag{6.17}$$

Proof. Put $\Delta = R_+^n$. Note that $(M\bar{x} + q)^T v \geq 0$ for every $v \in 0^+\Delta = R_+^n$ for some $\bar{x} \in \Delta$ if and only if there exists $\bar{x}$ satisfying (6.17). Applying Theorem 6.2 we obtain the desired conclusion. $\square$

In the terminology of Cottle et al. (1992), if there exists $\bar{x} \in R^n$ satisfying (6.17) then problem $\mathrm{LCP}(M, q)$ is said to be *feasible*. The set of all $\bar{x} \in R^n$ satisfying (6.17) is called *the feasible region* of that problem. Corollary 6.1 asserts that *a linear complementarity problem with a positive semidefinite matrix M is solvable if and only it is feasible.*

6.2 Solution Existence under Copositivity

In this section we obtain some existence theorems for the AVI problem (6.1) where M is not assumed to be monotone on Δ. It is assumed only that M is copositive on Δ.

We first establish an existence theorem under strict copositivity.

Theorem 6.3. *If matrix M is strictly copositive on a nonempty polyhedral convex set Δ then, for any $q \in R^n$, problem $\mathrm{AVI}(M, q, \Delta)$ has a solution.*

The following auxiliary fact shows that the strict copositivity assumption in the above theorem is, in fact, equivalent to a coercivity condition of the form (5.7).

Lemma 6.3. *Matrix $M \in R^{n \times n}$ is strictly copositive on a nonempty polyhedral convex set $\Delta \subset R^n$ if and only if there exists $x^0 \in \Delta$ such that*

$$\frac{\langle My - Mx^0, y - x^0 \rangle}{\|y - x^0\|} \to +\infty \text{ as } \|y\| \to +\infty, \ y \in \Delta. \qquad (6.18)$$

Proof. *Necessity:* Suppose that Δ is nonempty and M is strictly copositive on Δ. If $0^+\Delta = \{0\}$ then, according to Theorem 8.4 in Rockafellar (1970), Δ is compact. So, for an arbitrarily chosen $x^0 \in \Delta$, condition (6.18) is satisfied. Now consider the case where $0^+\Delta \neq \{0\}$. select any $x^0 \in \Delta$. We claim that (6.18) is valid. On the contrary, suppose that (6.18) is false. Then there must exist $\gamma > 0$ and a sequence $\{y^k\} \subset \Delta$ such that $\|y^k\| \to +\infty$ and

$$\frac{\langle My^k - Mx^0, y^k - x^0 \rangle}{\|y^k - x^0\|} \leq \gamma \quad \forall k \in N. \qquad (6.19)$$

Since Δ is a nonempty polyhedral convex set, by Theorems 19.1 and 19.5 from Rockafellar (1970) one can find a compact set $K \subset \Delta$ such that

$$\Delta = K + 0^+\Delta.$$

Hence, for each $k \in N$ there exist $u^k \in K$ and $v^k \in 0^+\Delta$ such that $y^k = u^k + v^k$. It is easily seen that $\|v^k\| \to +\infty$. Therefore, without loss of generality we can assume that

$$u^k \to \bar{u}, \quad v^k \neq 0 \text{ for every } k \in N, \quad \frac{v^k}{\|v^k\|} \to \bar{v},$$

for some $\bar{u} \in \Delta$ and $\bar{v} \in 0^+\Delta$ with $\|\bar{v}\| = 1$. From (6.19) it follows that

$$\frac{\|v^k\|^{-2} \langle Mv^k + M(u^k - x^0), v^k + (u^k - x^0) \rangle}{\|v^k\|^{-1}\|v^k + u^k - x^0\|} \leq \frac{1}{\|v^k\|}\gamma$$

for all $k \in N$. Letting $k \to \infty$, from the above inequality we obtain

$$\frac{\langle M\bar{v}, \bar{v} \rangle}{\|\bar{v}\|} \leq 0,$$

which contradicts the assumed strict copositivity of M on Δ. We have thus proved that (6.18) is valid.

Sufficiency: Suppose that there exists $x^0 \in \Delta$ such that (6.18) is fulfilled. Let $v \in 0^+\Delta \setminus \{0\}$ be given arbitrarily. Since $y(t) :=$

$x^0 + tv \in \Delta$ for every $t > 0$ and $\|y(t)\| \to +\infty$ as $t \to +\infty$, substituting $y = y(t)$ into (6.18) gives

$$t\frac{\langle Mv, v \rangle}{\|v\|} \to +\infty \quad \text{as } t \to +\infty.$$

This implies that $v^T Mv = \langle Mv, v \rangle > 0$. We have thus shown that M is strictly copositive on Δ. $\square$.

Proof of Theorem 6.3.

Suppose that M is strictly copositive on Δ and $q \in R^n$ is an arbitrarily given vector. Consider the affine operator $\phi(x) = Mx + q$. Applying Lemma 6.3 we can assert that there exists $x^0 \in \Delta$ such that the coercivity condition (5.17) is satisfied. According to Theorem 5.2, problem VI(ϕ, Δ) has solutions. Since the latter is exactly the problem AVI(M, q, Δ), the desired conclusion follows. $\square$

One can derive Theorem 6.3 directly from Theorem 5.1 without appealing to Theorem 5.2 and Lemma 6.3.

Another proof of Theorem 6.3.

Suppose that $\Delta \neq \emptyset$, M is strictly copositive on Δ, and $q \in R^n$ is given arbitrarily. Let $m \in N$, $A \in R^{m \times n}$ and $b \in R^m$ be such that Δ has the representation (6.2). Then $0^+\Delta = \{v \in R^n : Av \geq 0\}$ (see Rockafellar (1970), p. 62). Select a point $x^0 \in \Delta$. For each $k \in N$, we set

$$\Delta_k = \Delta \cap \{x = (x_1, \ldots, x_n) \in R^n : x_i^0 - k \leq x_i \leq x_i^0 + k \\ \text{for every } i = 1, 2, \ldots, n\}.$$

(6.20)

It is clear that, for every $k \in N$, Δ_k is a nonempty, compact, polyhedral convex set. Given any $k \in N$, we consider the problem AVI(M, q, Δ_k). According to the Hartman-Stampacchia Theorem (see Theorem 5.1), Sol(AVI(M, q, Δ_k)) $\neq \emptyset$. For each $k \in N$, select a point $x^k \in$ Sol(AVI(M, q, Δ_k)). We claim that the sequence $\{x^k\}$ is bounded. To obtain a contradiction, suppose that $\{x^k\}$ is unbounded. Without restriction of generality we can assume that $x^k \neq 0$ for all k, $\|x^k\| \to +\infty$ as $k \to \infty$, and there exists $\bar{v} \in R^n$ such that

$$\frac{x^k}{\|x^k\|} \to \bar{v}, \quad \|\bar{v}\| = 1.$$

Since $x^0 \in \Delta_k$ for every $k \in N$, we have

$$\langle Mx^k + q, x^0 - x^k \rangle \geq 0 \quad \forall k \in N$$

or, equivalently,

$$\langle Mx^k + q, x^0 \rangle \geq \langle Mx^k + q, x^k \rangle \quad \forall k \in N. \tag{6.21}$$

Dividing the inequality in (6.21) by $\|x^k\|^2$ and letting $k \to \infty$ we get

$$0 \geq \langle M\bar{v}, \bar{v} \rangle. \tag{6.22}$$

Since $x^k \in \Delta$, we have $Ax^k \geq b$. Dividing the last inequality by $\|x^k\|$ and taking the limit as $k \to \infty$ we obtain $A\bar{v} \geq 0$. This shows that $\bar{v} \in 0^+\Delta$. Since $\|\bar{v}\| = 1$, (6.22) contradicts the assumed strict copositivity of M on Δ. We have thus proved that the sequence $\{x^k\}$ is bounded. There is no loss of generality in assuming that $x^k \to \bar{x}$ for some $\bar{x} \in \Delta$. For each $x \in \Delta$ one can find and index $k_x \in N$ such that $x \in \Delta_k$ for all $k \geq k_x$. Consequently, for every $k \geq k_x$, it holds

$$\langle Mx^k + q, x - x^k \rangle \geq 0.$$

Letting $k \to \infty$ we obtain

$$\langle M\bar{x} + q, x - \bar{x} \rangle \geq 0.$$

Since the last inequality is valid for any $x \in \Delta$, we conclude that $\bar{x} \in \text{Sol}(\text{AVI}(M, q, \Delta))$. The proof is complete. $\quad\square$

Example 6.2. Let

$$M = \begin{pmatrix} 1 & 2 \\ 2 & 1 \end{pmatrix} \in R^{2 \times 2}, \quad \Delta = R_+^2.$$

In Remark 6.1 we have observed that M is not monotone on Δ. However, M is strictly copositive on Δ. Indeed, since Δ is a cone, we have $0^+\Delta = \Delta$. For any nonzero vector $v = (v_1, v_2) \in 0^+\Delta = R_+^2$ it holds

$$v^T M v = v_1^2 + 4v_1v_2 + v_2^2 > 0.$$

This shows that M is strictly copositive on Δ. According to Theorem 6.3, for any $q \in R^2$, problem $\text{AVI}(M, q, \Delta)$ is solvable. Note that Theorem 6.1 cannot be applied to this problem because M is not monotone on Δ.

In the following existence theorem we need not to assume that the matrix M strictly copositive on Δ. Instead of the strict copositivity, a weaker assumption is employed.

Theorem 6.4. *If matrix M is copositive on a nonempty polyhedral convex set Δ and there exists no $\bar{v} \in R^n \setminus \{0\}$ such that*

$$\bar{v} \in 0^+\Delta, \quad M\bar{v} \in (0^+\Delta)^+, \quad \bar{v}^T M\bar{v} = 0, \tag{6.23}$$

where

$$(0^+\Delta)^+ = \{\xi \in R^n : \xi^T v \geq 0 \ \forall v \in 0^+\Delta\}$$

then, for any $q \in R^n$, problem $\mathrm{AVI}(M, q, \Delta)$ has a solution.

Proof. Suppose that $\Delta \neq \emptyset$, M is copositive on Δ and there exists no $\bar{v} \in R^n \setminus \{0\}$ satisfying (6.23). Suppose that $q \in R^n$ is given arbitrarily. Let $m \in N$, $A \in R^{m \times n}$ and $b \in R^m$ be such that Δ has the representation (6.2). Then $0^+\Delta = \{v \in R^n : Av \geq 0\}$. Let $x^0 \in \Delta$. For each $k \in N$, we define

$$\Delta_k = \Delta \cap \bar{B}(x^0, k). \tag{6.24}$$

Note that, for every $k \in N$, Δ_k is a nonempty, compact, convex set. Given any $k \in N$, we consider the VI problem

Find $x \in \Delta_k$ such that $\langle Mx + q, y - x \rangle \geq 0 \ \forall y \in \Delta_k$

and denote its solution set by $\mathrm{Sol}(\mathrm{VI}((M, q, \Delta_k))$. By Theorem 5.1, $\mathrm{Sol}(\mathrm{VI}((M, q, \Delta_k)) \neq \emptyset$. For each $k \in N$, select a point $x^k \in \mathrm{Sol}(\mathrm{VI}((M, q, \Delta_k))$. We claim that the sequence $\{x^k\}$ is bounded. Suppose, contrary to our claim, that $\{x^k\}$ is unbounded. There is no loss of generality in assuming that $x^k \neq 0$ for all k, $\|x^k\| \to +\infty$ as $k \to \infty$, and there exists $\bar{v} \in R^n$ such that

$$\frac{x^k}{\|x^k\|} \to \bar{v}, \quad \|\bar{v}\| = 1.$$

Since $x^0 \in \Delta_k$ for every $k \in N$, we have

$$\langle Mx^k + q, x^0 - x^k \rangle \geq 0 \quad \forall k \in N.$$

As in the second proof of Theorem 6.3, from the last property we deduce that

$$0 \geq \langle M\bar{v}, \bar{v} \rangle.$$

Since $x^k \in \Delta$, we have $Ax^k \geq b$ for every $k \in N$. This implies $A\bar{v} \geq 0$. So $\bar{v} \in 0^+\Delta$. Since M is copositive on Δ, from what has already been said it follows that

$$\bar{v}^T M \bar{v} = 0. \tag{6.25}$$

For any $w \in 0^+\Delta \setminus \{0\}$, from (6.24) and the fact that $x^0 + tw \in \Delta$ for every $t \geq 0$ we deduce that

$$x^0 + \|x^k - x^0\| \frac{w}{\|w\|} \in \Delta_k.$$

Since $x^k \in \mathrm{Sol}(\mathrm{VI}((M, q, \Delta_k)))$, we have

$$\left\langle Mx^k + q, x^0 + \|x^k - x^0\| \frac{w}{\|w\|} - x^k \right\rangle \geq 0.$$

Dividing this inequality by $\|x^k\|^2$, letting $k \to \infty$ and noting that $\lim_{k \to \infty} \dfrac{\|x^k - x^0\|}{\|x^k\|} \to 1$, by virtue of (6.25) we obtain $\left\langle M\bar{v}, \dfrac{w}{\|w\|} \right\rangle \geq 0$. Hence $\langle M\bar{v}, w \rangle \geq 0$ for every $w \in 0^+\Delta$. This means that $M\bar{v} \in (0^+\Delta)^+$. We see that vector $\bar{v} \in R^n \setminus \{0\}$ satisfies all the three conditions described in (6.23). This contradicts our assumption. We have thus proved that the sequence $\{x^k\}$ is bounded. Without loss of generality we can assume that $x^k \to \bar{x}$ for some $\bar{x} \in \Delta$. For each $x \in \Delta$ there exists $k_x \in N$ such that $x \in \Delta_k$ for all $k \geq k_x$. Consequently, for every $k \geq k_x$, we have $\langle Mx^k + q, x - x^k \rangle \geq 0$. Letting $k \to \infty$ we obtain $\langle M\bar{x} + q, x - \bar{x} \rangle \geq 0$. Since this inequality holds for any $x \in \Delta$, we can assert that $\bar{x} \in \mathrm{Sol}(\mathrm{AVI}(M, q, \Delta))$. The proof is complete. $\square$

Example 6.3. Let M and Δ be the same as in Example 6.1. It is a simple matter to verify that there exists no $\bar{v} \in R^n \setminus \{0\}$ satisfying the three conditions in (6.23). Since M is copositive on Δ, Theorem 6.4 asserts that, for any $q = (q_1, q_2) \in R^2$, problem $\mathrm{AVI}(M, q, \Delta)$ has a solution.

In the sequel, sometimes we shall use the following simple fact.

Lemma 6.4. Let $K \subset R^n$ be a nonempty closed cone. Let $q \in R^n$. Then $q \in \mathrm{int}K^+$, where $\mathrm{int}K^+$ denotes the interior of the positive dual cone K^+ of K, if and only if

$$v^T q > 0 \quad \forall v \in K \setminus \{0\}. \tag{6.26}$$

Proof. Suppose that $q \in \text{int} K^+$. If there exists $\bar{v} \in K \setminus \{0\}$ such that $\bar{v}^T q \leq 0$ then $\bar{v}^T q = 0$ because the condition $q \in K^+$ implies that $v^T q \geq 0$ for every $v \in K$. From this we see that the linear functional $\xi \rightarrow \bar{v}^T \xi$ achieves its global minimum on K^+ at q. As $q \in \text{int} K^+$, there exists $\varepsilon > 0$ such that $\bar{B}(q, \varepsilon) \subset K^+$. Then

$$\bar{v}^T \xi \geq 0 \quad \forall \xi \in \bar{B}(q, \varepsilon).$$

This implies that $\bar{v} = 0$, a contradiction. We have thus proved that if $q \in \text{int} K^+$ then (6.26) is valid.

Conversely, assume that (6.26) holds. To obtain a contradiction, suppose that $q \notin \text{int} K^+$. Then there exists a sequence $\{q^k\}$ in $R^n \setminus K^+$ such that $q^k \rightarrow q$. Consequently, for each $k \in N$ there exists $v^k \in K$ such that $(v^k)^T q^k < 0$. Without loss of generality we can assume that $\dfrac{v^k}{\|v^k\|} \rightarrow \bar{v}$ with $\|\bar{v}\| = 1$. We have

$$\frac{(v^k)^T q^k}{\|v^k\|} < 0, \quad \frac{v^k}{\|v^k\|} \in K \quad \forall k \in N.$$

Taking the limits as $k \rightarrow \infty$ we obtain $\bar{v}^T q \leq 0$ and $\bar{v} \in K$, contrary to (6.26). $\square$

In the case where Δ is a cone, we have the following existence theorem.

Theorem 6.5. *Assume that Δ is a polyhedral convex cone. If matrix M is copositive on Δ and*

$$q \in \text{int}([\text{Sol}(\text{AVI}(M, 0, \Delta))]^+), \tag{6.27}$$

then problem $\text{AVI}(M, q, \Delta)$ *has a solution.*

Note that $\text{AVI}(M, q, \Delta)$ is a generalized linear complementarity problem (see Definition 5.8). From the definition it follows that $v \in \text{Sol}(\text{AVI}(M, 0, \Delta))$ if and only if

$$v \in \Delta, \quad Mv \in \Delta^+, \quad v^T M v = 0.$$

Hence, applying Lemma 6.4 to the cone $K := \text{Sol}(\text{AVI}(M, 0, \Delta))$ we see that condition (6.27) is equivalent to the requirement that there exists no $\bar{v} \in R^n \setminus \{0\}$ such that

$$\bar{v} \in \Delta, \quad M\bar{v} \in \Delta^+, \quad \bar{v}^T M \bar{v} = 0, \quad q^T \bar{v} \leq 0. \tag{6.28}$$

Proof of Theorem 6.5.

Suppose that Δ is a polyhedral convex cone, M is copositive on Δ, and q is such that (6.27) holds. For each $k \in N$, we set

$$\Delta_k = \Delta \cap \{x \in R^n : -k \le x_i \le k \text{ for every } i = 1, 2, \dots, n\}.$$

It is clear that, for each $k \in N$, $0 \in \Delta_k$ and Δ_k is a compact, polyhedral convex set. Consider the problem $\text{AVI}(M, q, \Delta_k)$. By Theorem 5.1, we can find a point $x^k \in \text{Sol}(\text{AVI}((M, q, \Delta_k))$. If the sequence $\{x^k\}$ is unbounded then without loss of generality we can assume that $x^k \ne 0$ for all k, $\|x^k\| \to +\infty$ as $k \to \infty$, and there exists $\bar{v} \in R^n$ such that

$$\frac{x^k}{\|x^k\|} \to \bar{v}, \ \|\bar{v}\| = 1.$$

Since $0 \in \Delta_k$, we have

$$\langle Mx^k + q, 0 - x^k \rangle \ge 0.$$

Hence

$$-q^T x^k \ge (x^k)^T M x^k \quad (\forall k \in N). \tag{6.29}$$

Dividing the inequality in (6.29) by $\|x^k\|^2$ and taking limits as $k \to \infty$ we get

$$0 \ge \bar{v}^T M \bar{v}. \tag{6.30}$$

It is clear that $\bar{v} \in \Delta$. Since M is copositive on Δ, we have $v^T M v \ge 0$ for every $v \in \Delta$. Combining this fact with (6.30) yields

$$\bar{v}^T M \bar{v} = 0. \tag{6.31}$$

From (6.29) and the copositivity of M on Δ it follows that $-q^T x^k \ge 0$ for every $k \in N$. This implies that

$$-q^T \bar{v} \ge 0. \tag{6.32}$$

Fix any $w \in \Delta \setminus \{0\}$. It is evident that

$$\|x^k\| \frac{w}{\|w\|} \in \Delta.$$

Since $x^k \in \text{Sol}(\text{AVI}(M, q, \Delta_k))$, we have

$$\left\langle Mx^k + q, \|x^k\| \frac{w}{\|w\|} - x^k \right\rangle \ge 0 \quad (\forall k \in N).$$

From this and (6.31) we deduce that $\langle M\bar{v}, w \rangle \geq 0$. Since the last inequality is valid for every $w \in \Delta \setminus \{0\}$, we see that $M\bar{v} \in \Delta^+$. Combining this with (6.31) and (6.32) we can assert that (6.28) is satisfied. Then (6.27) is false. We have arrived at a contradiction. Thus the sequence $\{x^k\}$ must be bounded. Analysis similar to that in the final part of the proof of Theorem 6.4 shows that problem AVI(M, q, Δ) has a solution. $\quad\square$

Example 6.4. Let

$$M = \begin{pmatrix} 1 & 0 \\ 0 & -1 \end{pmatrix} \in R^{2\times 2}, \quad q = (1,1) \in R^2,$$
$$\Delta = \{x = (x_1, x_2) \in R^2 : x_1 \geq 0, \ x_2 \geq 0, \ x_1 = x_2\}.$$

Theorem 6.5 can be applied to the problem AVI(M, q, Δ). Indeed, we have

$$v^T M v = v_1^2 - v_2^2 = 0 \quad \forall v \in 0^+\Delta = \Delta.$$

This shows that M is copositive on Δ. Furthermore, we have

$$\text{Sol(AVI}(M, 0, 0^+\Delta)) = \{v \in R^2 : v \in 0^+\Delta, \ Mv \in (0^+\Delta)^+, \\ v^T M v = 0\} = \Delta.$$

Therefore

$$q^T v = v_1 + v_2 > 0 \quad \forall v = (v_1, v_2) \in \text{Sol(AVI}(M, 0, \Delta)) \setminus \{0\}.$$

So (6.27) is satisfied. By Theorem 6.5, problem AVI(M, q, Δ) is solvable. In fact, we have

$$\text{Sol(AVI}(M, q, \Delta)) = \{(0, 0)\}.$$

It is worth pointing that, since M is not strictly copositive on Δ, Theorem 6.3 cannot be applied to this problem. Since all the three conditions described in (6.23) are satisfied if one chooses $\bar{v} = (1,1) \in R^2 \setminus \{0\}$, Theorem 6.4 also cannot be applied to this problem.

Remark 6.4. In the case where Δ is a polyhedral convex cone, the conclusion of Theorem 6.4 follows from Theorem 6.5. Indeed, in this case, under the assumption of Theorem 6.4 we have

$$\text{Sol(AVI}(M, 0, \Delta)) = \{0\}.$$

Hence $[\text{Sol(AVI}(M, 0, \Delta))]^+ = R^n$. So (6.27) is satisfied for any $q \in R^n$. By Theorem 6.5, problem AVI(M, q, Δ) is solvable.

Applying Theorem 6.5 to LCP problems we obtain the following corollary.

Corollary 6.2. *If M is a copositive matrix and*

$$q \in \text{int}([\text{Sol}(M,0)]^+), \tag{6.33}$$

then the problem LCP(M,q) *has a solution.*

Note that condition (6.33) is stronger than condition (6.34) in the following existence theorem for LCP problems.

Theorem 6.6. (See Cottle et al. (1992), Theorem 3.8.6) *If M is a copositive matrix and*

$$q \in [\text{Sol}(M,0)]^+, \tag{6.34}$$

then problem LCP(M,q) *has a solution.*

It is clear that (6.34) can be rewritten in the following form:

$$[v \in R^n, \ v \geq 0, \ Mv \geq 0, \ v^T Mv = 0] \implies [q^T v \geq 0].$$

Meanwhile, by Lemma 6.4, condition (6.33) is equivalent to the following one:

$$[v \in R^n \setminus \{0\}, \ v \geq 0, \ Mv \geq 0, \ v^T Mv = 0] \implies [q^T v > 0].$$

In connection with Theorems 6.5, the following open question seems to be interesting.

QUESTION: Whether the conclusion of Theorem 6.5 is still valid if in the place of (6.27) one uses the following weaker condition

$$q \in [\text{Sol}(\text{AVI}(M,0,\Delta))]^+?$$

Note that the last inclusion can be rewritten in the form:

$$[v \in R^n, \ v \in 0^+\Delta, \ Mv \in \Delta^+, \ v^T Mv = 0] \implies [q^T v \geq 0].$$

6.3   Commentaries

In this chapter, we have considered a variety of solution existence theorems for affine variational inequalities. Here the compactness of the constraint set Δ is not assumed. But we have to employ a monotonicity property of the matrix M with respect to Δ. Namely, we have had deal with the monotonicity, the strict monotonicity, and the copositivity of M w.r.t. Δ.

The interested reader is referred to Gowda and Pang (1994a) for an insightful study on existence theorems for AVI problems.

Chapter 7

Upper-Lipschitz Continuity of the Solution Map in Affine Variational Inequalities

In this chapter we shall discuss two fundamental theorems due to Robinson (1979, 1981) on the upper-Lipschitz continuity of the solution map in affine variational inequality problems. The theorem on the upper-Lipschitz continuity of the solution map in linear complementarity problems due to Cottle et al. (1992) is also studied in this chapter. The Walkup-Wets Theorem (see Walkup and Wets (1969)), which we analyze in Section 7.1, is the basis for obtaining these results.

7.1   The Walkup-Wets Theorem

Let $\Delta \subset R^n$ be a nonempty subset. Let $\tau : R^n \to R^m$ be an affine operator; that is there exist a linear operator $A : R^n \to R^m$ and a vector $b \in R^m$ such that $\tau(x) = Ax + b$ for every $x \in R^n$. Define

$$
\Delta(y) = \tau^{-1}(y) \cap \Delta \;\; = \{x \in \Delta : \tau(x) = y\} \\
= \{x \in \Delta : Ax + b = y\}. \tag{7.1}
$$

Definition 7.1. (See Walkup and Wets (1969), Definition 1) A subset $\Delta \subset R^n$ is said to have *property* $\mathcal{L}_j$ if for every affine operator $\tau : R^n \to R^m$, $m \in N$, with $\dim(\ker(\tau)) = j$, the inverse mapping

$y \to \Delta(y)$ is Lipschitz on its effective domain. This means that there exists a constant $\ell > 0$ such that

$$\Delta(y') \subset \Delta(y) + \ell\|y' - y\|\bar{B}_{R^n} \quad \text{whenever } \Delta(y) \neq \emptyset, \ \Delta(y') \neq \emptyset. \tag{7.2}$$

In the above definition, $\dim(\ker(\tau))$ denotes the dimension of the affine set

$$\ker(\tau) = \{x \in R^n : \tau(x) = 0\}.$$

The following theorem is a key tool for proving other results in this chapter.

Theorem 7.1 (The Walkup-Wets Theorem; see Walkup and Wets (1969), Theorem 1). *Let $\Delta \subset R^n$ be a nonempty closed convex set and let $j \in N$, $1 \leq j \leq n - 1$. Then Δ is a polyhedral convex set if and only if it has property $\mathcal{L}_j$.*

In the sequel, we will use only one assertion of this theorem: *If Δ is a polyhedral convex set, then it has property $\mathcal{L}_j$.* A detailed proof of this assertion can be found in Mangasarian and Shiau (1987).

Corollary 7.1. *If $\Delta \subset R^n$ is a polyhedral convex set and if $\tau : R^n \to R^m$ is an affine operator, then there exists a constant $\ell > 0$ such that (7.2), where $\Delta(y)$ is defined by (7.1) for all $y \in R^n$, holds.*

Proof. If $j := \dim(\ker(\tau))$ satisfies the condition $1 \leq j \leq n - 1$, then the conclusion is immediate from Theorem 7.1. If $\dim(\ker(\tau)) = n$ then $\ker(\tau) = R^n$, and we have

$$\Delta(y) = \tau^{-1}(y) \cap \Delta = \begin{cases} \Delta & \text{if } y = 0, \\ \emptyset & \text{if } y \neq 0. \end{cases}$$

This shows that (7.2) is fulfilled with any $\ell > 0$. We now suppose that $\dim(\ker(\tau)) = 0$. Let $\tau(x) = Ax + b$, where $A : R^n \to R^m$ is a linear operator and $b \in R^m$. Since τ is an injective mapping, $Y := \tau(R^n)$ is an affine set in R^m with $\dim Y = n$, and that $n \leq m$. Likewise, the set $Y_0 := A(R^n)$ is a linear subspace of R^m with $\dim Y_0 = n$. Let $\tilde{A} : R^n \to Y_0$ be the linear operator defined by setting $\tilde{A}x = Ax$ for every $x \in R^n$. It is easily shown that

$$\|\tau^{-1}(y') - \tau^{-1}(y)\| \leq \|\tilde{A}^{-1}\|\|y' - y\|$$

for every $y \in Y$ and $y' \in Y$. From this we deduce that (7.2) is satisfied with $\ell := \|\tilde{A}^{-1}\|$. $\square$

Remark 7.1. Under the assumptions of Corollary 7.1, for every $y \in R^m$, $\Delta(y)$ is a polyhedral convex set (possibly empty).

Remark 7.2. The conclusion of Theorem 7.1 is not true if one chooses $j = 0$. Namely, the arguments described in the final part of the proof of Corollary 7.1 show that any nonempty set $\Delta \subset R^n$ has property $\mathcal{L}_0$. Similarly, the conclusion of Theorem 7.1 is not valid if $j = n$.

Corollary 7.2. *For any nonempty polyhedral convex set $\Delta \subset R^n$ and any matrix $C \in R^{s \times n}$ there exists a constant $\ell > 0$ such that*

$$\Delta(C, d'') \subset \Delta(C, d') + \ell \|d'' - d'\| \bar{B}_{R^n} \qquad (7.3)$$

whenever $\Delta(C, d')$ and $\Delta(C, d'')$ are nonempty; where

$$\Delta(C, d) := \{x \in \Delta : Cx = d\}$$

for every $d \in R^s$.

Proof. Set $\tau(x) = Cx$. Since

$$\Delta(C, y) = \tau^{-1}(y) \cap \Delta = \Delta(y)$$

where $\Delta(y)$ is defined by (7.1), applying Corollary 7.1 we can find $\ell > 0$ such that the Lipschitz continuity property stated in (7.3) is satisfied. $\square$

Corollary 7.3. *For any nonempty polyhedral convex set $\Delta \subset R^n$, any matrix $A \in R^{m \times n}$ and matrix $C \in R^{s \times n}$ there exists a constant $\ell > 0$ such that*

$$\Delta(A, C, b'', d'') \subset \Delta(A, C, b', d') + \ell(\|b'' - b'\| + \|d'' - d'\|) \bar{B}_{R^n} \quad (7.4)$$

whenever $\Delta(A, C, b', d')$ and $\Delta(A, C, b'', d'')$ are nonempty; where

$$\Delta(A, C, b, d) := \{x \in \Delta : Ax \geq b, \ Cx = d\}$$

for every $b \in R^m$ and $d \in R^s$.

Proof. Define

$$\tilde{C} = \begin{pmatrix} A & -E \\ C & 0 \end{pmatrix} \in R^{(m+s) \times (n+m)},$$

where E denotes the unit matrix in $R^{m \times m}$ and 0 denotes the null in $R^{s \times m}$. Let

$$\tilde{\Delta} = \{(x, w) \in R^n \times R^m : x \in \Delta, \ w \geq 0\}.$$

By Corollary 7.2, there exists $\ell > 0$ such that

$$\widetilde{\Delta}(\widetilde{C}, b'', d'') \subset \widetilde{\Delta}(\widetilde{C}, b', d') + \ell(\|b'' - b'\| + \|d'' - d'\|)\bar{B}_{R^{n+m}} \quad (7.5)$$

whenever $\widetilde{\Delta}(\widetilde{C}, b', d') \neq \emptyset$ and $\widetilde{\Delta}(\widetilde{C}, b'', d'') \neq \emptyset$, where

$$\widetilde{\Delta}(\widetilde{C}, b, d) := \left\{ (x, w) \in \widetilde{\Delta} \ : \ \widetilde{C}\begin{pmatrix} x \\ w \end{pmatrix} = \begin{pmatrix} b \\ d \end{pmatrix} \right\}.$$

Since

$$\begin{aligned} \Delta(A, C, b, d) &= \{x \in \Delta \ : \ \exists w \in R^m, \ w \geq 0, \ Ax - w = b, \ Cx = d\} \\ &= \mathrm{Pr}_{R^n}(\widetilde{\Delta}(\widetilde{C}, b, d)) \end{aligned}$$

where $\mathrm{Pr}_{R^n}(x, w) = x$ for every $(x, w) \in R^n \times R^m$, we see at once that (7.5) implies (7.4). □

7.2　Upper-Lipschitz Continuity with respect to Linear Variables

The notion of polyhedral multifunction was proposed by Robinson (see Robinson (1979, 1981)). We now study several basic facts concerning polyhedral multifunctions.

Definition 7.2. If $\Phi : R^n \to 2^{R^m}$ is a multifunction then its *graph* and *effective domain* are defined, respectively, by setting

$$\begin{aligned} \mathrm{graph}\Phi &= \{(x, y) \in R^n \times R^m \ : \ y \in \Phi(x)\}, \\ \mathrm{dom}\Phi &= \{x \in R^n \ : \ \Phi(x) \neq \emptyset\}. \end{aligned}$$

Definition 7.3. A set-valued mapping $\Phi : R^n \to 2^{R^m}$ is called a *polyhedral multifunction* if its graph can be represented as the union of finitely many polyhedral convex sets in $R^n \times R^m$.

The following statement shows that the *normal-cone operator* corresponding to a polyhedral convex set is a polyhedral multifunction.

Proposition 7.1. (See Robinson (1981)) *Suppose that* $\Delta \subset R^n$ *is a nonempty polyhedral convex set. Then the formula*

$$\Phi(x) = N_\Delta(x) \quad (x \in R^n)$$

defines a polyhedral multifunction $\Phi : R^n \to 2^{R^n}$.

Proof. Let $m \in N$, $A \in R^{n \times n}$ and $b \in R^m$ be such that $\Delta = \{x \in R^n : Ax \geq b\}$. Set $I = \{1, \ldots, m\}$. Let

$$F_\alpha = \{x \in R^n : A_\alpha x = b_\alpha, \ A_{I\backslash\alpha} x > b_{I\backslash\alpha}\}$$

be the pseudo-face of Δ corresponding to an index set $\alpha \subset I$. For every $x \in F_\alpha$ we have

$$T_\Delta(x) = \{v \in R^n : A_\alpha v \geq 0\}.$$

(See the proof of Theorem 4.2.) Since

$$N_\Delta(x) = \{\xi \in R^n : \langle \xi, v \rangle \leq 0 \ \forall v \in T_\Delta(x)\},$$

we have $\xi \in N_\Delta(x)$ if and only if the inequality $\langle \xi, v \rangle \leq 0$ is a consequence of the inequality system $A_\alpha v \geq 0$. Consequently, applying Farkas' Lemma (see Theorem 3.2) we deduce that $\xi \in N_\Delta(x)$ if and only if there exist $\lambda_1 \geq 0, \ldots, \lambda_m \geq 0$ such that

$$\xi = \sum_{i \in \alpha} \lambda_i(-A_i^T),$$

where A_i denotes the i-th row of matrix A. (Note that if $\alpha = \emptyset$ and $x \in F_\alpha$, then $x \in \text{int}\Delta$; hence $\xi = 0$ for every $\xi \in N_\Delta(x)$.) Define

$$\Omega_\alpha = \{(x, \xi) \in R^n \times R^n : x \in F_\alpha, \ \xi \in N_\Delta(x)\}.$$

Obviously, $\Omega_\alpha \subset \text{graph}\Phi$. Note that

$$\Omega_\alpha = \{(x, \xi) \in R^n \times R^n : A_\alpha x = b_\alpha, \ A_{I\backslash\alpha} x > b_{I\backslash\alpha},$$
$$\xi = \textstyle\sum_{i \in \alpha} \lambda_i(-A_i^T) \text{ for some } \lambda_\alpha \in R_+^{|\alpha|}\}.$$

is a convex set. Here $|\alpha|$ denotes the number of elements in α. It is easily seen that the topological closure $\overline{\Omega}_\alpha$ of Ω_α is given by the formula

$$\overline{\Omega}_\alpha = \{(x, \xi) \in R^n \times R^n : A_\alpha x = b_\alpha, \ A_{I\backslash\alpha} x \geq b_{I\backslash\alpha},$$
$$\xi = \textstyle\sum_{i \in \alpha} \lambda_i(-A_i) \text{ for some } \lambda_\alpha \in R_+^{|\alpha|}\}$$
$$= \text{Pr}_{R^n \times R^n}\{(x, \xi, \lambda_\alpha) \in R^n \times R^n \times R_+^{|\alpha|} : A_\alpha x = b_\alpha,$$
$$A_{I\backslash\alpha} x \geq b_{I\backslash\alpha}, \ \textstyle\sum_{i \in \alpha} \lambda_i A_i^T + \xi = 0\},$$

where $\text{Pr}_{R^n \times R^n}(x, \xi, \lambda_\alpha) = (x, \xi)$. It is clear that the set in the last curly brackets is a polyhedral convex set. From this fact, the above formula for $\overline{\Omega}_\alpha$ and Theorem 19.1 in Rockafellar (1970) we deduce that $\overline{\Omega}_\alpha$ is a polyhedral convex set (see the proof of Theorem 4.3). Since $\Delta = \bigcup_{\alpha \subset I} F_\alpha$, we have

$$\text{graph}\Phi = \bigcup_{\alpha \subset I} \Omega_\alpha. \tag{7.6}$$

Observe that graphΦ is a closed set. Indeed, suppose that $\{(x^k, \xi^k)\}$ is a sequence satisfying $(x^k, \xi^k) \to (\bar{x}, \bar{\xi}) \in R^n \times R^n$, and $(x^k, \xi^k) \in$ graphΦ for every $k \in N$. On account of formula (1.12), we have

$$\langle \xi^k, y - x^k \rangle \leq 0 \quad \forall y \in \Delta, \ \forall k \in N.$$

Fixing any $y \in \Delta$ and taking limit as $k \to \infty$, from the last inequality we obtain $\langle \bar{\xi}, y - \bar{x} \rangle \leq 0$. Since this inequality holds for each $y \in \Delta$, we see that $\bar{\xi} \in N_\Delta(\bar{x})$. Hence $(\bar{x}, \bar{\xi}) \in$ graphΦ. We have thus proved that the set graphΦ is closed. On account of this fact, from (7.6) we deduce that

$$\text{graph}\Phi = \bigcup_{\alpha \subset I} \overline{\Omega}_\alpha.$$

This shows that graphΦ can be represented as the union of finitely many polyhedral convex sets. The proof is complete. $\quad\square$

The following statement shows that the solution map of a parametric affine variational inequality problem is a polyhedral multifunction (on the linear variables of the problem).

Proposition 7.2. *Suppose that $M \in R^{n \times n}$, $A \in R^{m \times n}$ and $C \in R^{s \times n}$ are given matrices. Then the formula*

$$\Phi(q, b, d) = \text{Sol}(\text{AVI}(M, q, \Delta(b, d))),$$

where $(q, b, d) \in R^n \times R^m \times R^s$, $\Delta(b, d) := \{x \in R^n : Ax \geq b, Cx = d\}$ and $\text{Sol}(\text{AVI}(M, q, \Delta(b, d)))$ denotes the solution set of problem (6.1) with $\Delta = \Delta(b, d)$, defines a polyhedral multifunction

$$\Phi : R^n \times R^m \times R^s \to 2^{R^n}.$$

Proof. According to Corollary 5.2, $x \in \text{Sol}(\text{AVI}(M, q, \Delta(b, d)))$ if and only if there exist $\lambda = (\lambda_1, \ldots, \lambda_m) \in R^m$ and $\mu = (\mu_1, \ldots, \mu_s) \in R^s$ such that

$$\begin{cases} Mx - A^T\lambda - C^T\mu + q = 0, \\ Ax \geq b, \ Cx = d, \ \lambda \geq 0, \\ \lambda^T(Ax - b) = 0. \end{cases} \tag{7.7}$$

Let $I = \{1, \ldots, m\}$. For each index set $\alpha \subset I$, we define

$$
\begin{aligned}
Q_\alpha = \mathrm{Pr}_1\Big(\{(x, q, b, d, \lambda, \mu) \ : \ & Mx - A^T\lambda - C^T\mu + q = 0, \\
& A_\alpha x = b_\alpha, \ A_{I\setminus\alpha}x \geq b_{I\setminus\alpha}, \\
& Cx = d, \ \lambda_\alpha \geq 0, \ \lambda_{I\setminus\alpha} = 0\}\Big),
\end{aligned}
\tag{7.8}
$$

where

$$
\mathrm{Pr}_1(x, q, b, d, \lambda, \mu) = (x, q, b, d)
$$

for all $(x, q, b, d, \lambda, \mu) \in R^n \times R^n \times R^m \times R^s \times R^m \times R^s$. Hence Q_α is a polyhedral convex set. Note that

$$
\mathrm{graph}\Phi = \bigcup_{\alpha \subset I} Q_\alpha.
\tag{7.9}
$$

Indeed, for each $(x, q, b, d) \in \mathrm{graph}\Phi$ we have

$$
x \in \mathrm{Sol}(\mathrm{AVI}(M, q, \Delta(b, d))).
$$

So there exist $\lambda = (\lambda_1, \ldots, \lambda_m) \in R^m$ and $\mu = (\mu_1, \ldots, \mu_s) \in R^s$ satisfying (7.7). Let $\alpha = \{i \in I : A_ix = b_i\}$. For every $i \in I \setminus \alpha$, we have $A_ix > b_i$. Then from the equality $\lambda_i(A_ix - b_i) = 0$ we deduce that $\lambda_i = 0$ for every $i \in I \setminus \alpha$. On account of this remark, we see that $(x, q, b, d, \lambda, \mu)$ satisfies all the conditions described in the curly braces in formula (7.8). This implies that $(x, q, b, d) \in Q_\alpha$. We thus get

$$
\mathrm{graph}\Phi \subset \bigcup_{\alpha \subset I} Q_\alpha.
$$

Since the reverse inclusion is obvious, we obtain formula (7.9), which shows that $\mathrm{graph}\Phi$ can be represented as the union of finitely many polyhedral convex sets. $\square$

Theorem 7.2. (See Robinson (1981), Proposition 1) *If* $\Phi : R^n \rightarrow 2^{R^m}$ *is a polyhedral multifunction, then there exists a constant* $\ell > 0$ *such that for every* $\bar{x} \in R^n$ *there is a neighborhood* $U_{\bar{x}}$ *of* $\bar{x}$ *satisfying*

$$
\Phi(x) \subset \Phi(\bar{x}) + \ell\|x - \bar{x}\|\bar{B}_{R^m} \quad \forall x \in U_{\bar{x}}.
\tag{7.10}
$$

Definition 7.4. (See Robinson (1981)) Suppose that $\Phi : R^n \rightarrow 2^{R^m}$ is a multifunction and $\bar{x} \in R^n$ is a given point. If there exist $\ell > 0$ and a neighborhood $U_{\bar{x}}$ of $\bar{x}$ such that property (7.10) is valid, then Φ is said to be *locally upper-Lipschitz* at $\bar{x}$ with the Lipschitz constant ℓ.

The locally upper-Lipschitz property is weaker than the locally Lipschitz property which is described as follows.

Definition 7.5. A multifunction $\Phi : R^n \to 2^{R^m}$ is said to be *locally Lipschitz* at $\bar{x} \in R^n$ if there exist a constant $\ell > 0$ and a neighborhood $U_{\bar{x}}$ of $\bar{x}$ such that

$$\Phi(x) \subset \Phi(u) + \ell\|x - u\|\bar{B}_{R^m} \quad \forall x \in U_{\bar{x}}, \forall u \in U_{\bar{x}}.$$

If there exists a constant $\ell > 0$ such that

$$\Phi(x) \subset \Phi(u) + \ell\|x - u\|\bar{B}_{R^m}$$

for all x and u from a subset $\Omega \subset R^n$, then Φ is said to be *Lipschitz* on Ω.

From Theorem 7.2 it follows that if Φ is a polyhedral multifunction then it is locally upper-Lipschitz at any point in R^n with the same Lipschitz constant. Note that the *diameter* diam$U_{\bar{x}} :=$ $\sup\{\|y - x\| : x \in U_{\bar{x}}, y \in U_{\bar{x}}\}$ of neighborhood $U_{\bar{x}}$ depends on $\bar{x}$ and it can change greatly from one point to another.

Proof of Theorem 7.2.

Since Φ is a polyhedral multifunction, there exist nonempty polyhedral convex sets $Q_j \subset R^n \times R^m$ $(j = 1, \ldots, k)$ such that

$$\text{graph}\Phi = \bigcup_{j \in J} Q_j, \tag{7.11}$$

where $J = \{1, \ldots, k\}$. For each $j \in J$ we consider the multifunction $\Phi_j : R^n \to 2^{R^m}$ defined by setting

$$\Phi_j(x) = \{y \in R^m : (x, y) \in Q_j\}. \tag{7.12}$$

Obviously, graph$\Phi_j = Q_j$. From (7.11) and (7.12) we deduce that

$$\text{graph}\Phi = \bigcup_{j \in J} \text{graph}\Phi_j, \quad \Phi(x) = \bigcup_{j \in J} \Phi_j(x).$$

CLAIM 1. *For each $j \in J$ there exists a constant $\ell_j > 0$ such that*

$$\Phi_j(x) \subset \Phi_j(u) + \ell_j\|x - u\|\bar{B}_{R^m} \tag{7.13}$$

whenever $\Phi_j(x) \neq \emptyset$ *and* $\Phi_j(u) \neq \emptyset$. *(This means that* Φ_j *is Lipschitz on its effective domain.)*

For proving the claim, consider the linear operator $\tau : R^n \times R^m \to R^n$ defined by setting $\tau(x, y) = x$ for every $(x, y) \in R^n \times R^m$. Let

$$Q_j(x) = \{z \in Q_j : \tau(z) = x\}. \tag{7.14}$$

By Corollary 7.1, there exists $\ell_j > 0$ such that

$$Q_j(x) \subset Q_j(u) + \ell_j \|x - u\| \bar{B}_{R^{n+m}} \tag{7.15}$$

whenever $Q_j(x) \neq \emptyset$ and $Q_j(u) \neq \emptyset$. From (7.12) and (7.14) it follows that

$$Q_j(x) = \{x\} \times \Phi_j(x) \quad \forall x \in R^n. \tag{7.16}$$

In particular, $Q_j(x) \neq \emptyset$ if and only if $\Phi_j(x) \neq \emptyset$. Given any $x \in R^n$, $u \in R^n$ and $y \in \Phi_j(x)$, from (7.15) and (7.16) we see that there exist $v \in \Phi(u)$ such that

$$\|(x, y) - (u, v)\| \leq \ell_j \|x - u\|.$$

Since $\|(x, y) - (u, v)\| = (\|x - u\|^2 + \|y - v\|^2)^{1/2}$, the last inequality implies that $\|y - v\| \leq \ell_j \|x - u\|$. From what has already been proved, it may be concluded that (7.13) holds whenever $\Phi_j(x) \neq \emptyset$ and $\Phi_j(u) \neq \emptyset$.

We set $\ell = \max\{l_j : j \in J\}$. The proof will be completed if we can establish the following fact.

CLAIM 2. *For each* $\bar{x} \in R^n$ *there exists a neighborhood* $U_{\bar{x}}$ *of* $\bar{x}$ *such that (7.10) holds.*

Let $\bar{x} \in R^n$ be given arbitrarily. Define

$$J_0 = \{j \in J : \bar{x} \in \mathrm{dom}\Phi_j\}, \quad J_1 = J \setminus J_0.$$

Since $\mathrm{dom}\Phi_j = \tau(Q_j)$, where τ is the linear operator defined above, we see that $\mathrm{dom}\Phi_j$ is a polyhedral convex set. This implies that the set $\bigcup_{j \in J_1} \mathrm{dom}\Phi_j$ is closed. (Note that if $J_1 = \emptyset$ then this set is empty.) As $\bar{x} \notin \bigcup_{j \in J_1} \mathrm{dom}\Phi_j$, there must exist $\varepsilon > 0$ such that the neighborhood $U_{\bar{x}} := B(\bar{x}, \varepsilon)$ of $\bar{x}$ does not intersect the set

$$\bigcup_{j \in J_1} \mathrm{dom}\Phi_j.$$

Let $x \in U_{\bar{x}}$. If $x \notin \bigcup_{j \in J_0} \mathrm{dom}\Phi_j$, then

$$\Phi(x) = \left(\bigcup_{j \in J_0} \Phi_j(x)\right) \bigcup \left(\bigcup_{j \in J_1} \Phi_j(x)\right) = \emptyset.$$

So the inclusion (7.10) is valid. If $x \in \bigcup_{j \in J_0} \mathrm{dom}\Phi_j$, then we have

$$\Phi(x) = \bigcup_{j \in J_0} \Phi_j(x) = \bigcup_{j \in J_0'} \Phi_j(x),$$

where $J_0' = \{j \in J_0 : x \in \mathrm{dom}\Phi_j\}$. For each $j \in J_0'$, according to Claim 1, we have

$$\Phi_j(x) \subset \Phi_j(\bar{x}) + \ell_j \|x - \bar{x}\| \bar{B}_{R^m} \subset \Phi(\bar{x}) + \ell\|x - \bar{x}\| \bar{B}_{R^m}.$$

Therefore

$$\Phi(x) = \bigcup_{j \in J_0'} \Phi_j(x) \subset \Phi(\bar{x}) + \ell\|x - \bar{x}\| \bar{B}_{R^m}.$$

Claim 2 has been proved.   □

Remark 7.3. From the proof of Theorem 7.2 it is easily seen that Φ is Lipschitz on the set $\bigcap_{j \in J} \mathrm{dom}\Phi_j$ with the Lipschitz constant ℓ.

Combining Theorem 7.2 with Proposition 7.2 we obtain the next result on upper-Lipschitz continuity of the solution map in a general AVI problem where the linear variables are subject to perturbation.

Theorem 7.3. *Suppose that $M \in R^{n \times n}$, $A \in R^{m \times n}$ and $C \in R^{s \times n}$ are given matrices. Then there exists a constant $\ell > 0$ such that the multifunction $\Phi : R^n \times R^m \times R^s \to 2^{R^n}$ defined by the formula*

$$\Phi(q, b, d) = \mathrm{Sol}(\mathrm{AVI}(M, q, \Delta(b, d))),$$

where $(q, b, d) \in R^n \times R^m \times R^s$ and $\Delta(b, d) := \{x \in R^n : Ax \geq b, \; Cx = d\}$, is locally upper-Lipschitz at any point $(\bar{q}, \bar{b}, \bar{d}) \in R^n \times R^m \times R^s$ with the Lipschitz constant ℓ.

Applying Theorem 7.3 to the case where the constraint set $\Delta(b, d)$ of the problem $\mathrm{AVI}(M, q, \Delta(b, d))$ is fixed (i.e., the pair (b, d) is not subject to perturbations), we have the following result.

Corollary 7.4. *Suppose that $M \in R^{n \times n}$ is a given matrix and $\Delta \subset R^n$ is a nonempty polyhedral convex set. Then there exists a constant $\ell > 0$ such that the multifunction $\Phi : R^n \to 2^{R^n}$ defined by the formula*

$$\Phi(q) = \mathrm{Sol}(\mathrm{AVI}(M, q, \Delta)),$$

where $q \in R^n$, is locally upper-Lipschitz at any point $\bar{q} \in R^n$ with the Lipschitz constant ℓ.

7.3   Upper-Lipschitz Continuity with respect to all Variables

Our aim in this section is to study some results on locally upper-Lipschitz continuity of the multifunction $\Phi : R^{n \times n} \times R^n \to 2^{R^n}$ defined by the formula

$$\Phi(M, q) = \text{Sol}(\text{AVI}(M, q, \Delta)),$$

where $\text{Sol}(\text{AVI}(M, q, \Delta))$ denotes the solution set of the problem (6.1). First we consider the case where Δ is a polyhedral convex cone. Then we consider the case where Δ is an arbitrary nonempty polyhedral convex set.

The following theorem specializes to Theorem 7.5.1 in Cottle et al. (1992) about the solution map in parametric linear complementarity problems if $\Delta = R^n_+$.

Theorem 7.4. *Suppose that $\Delta \subset R^n$ is a polyhedral convex cone. Suppose that $M \in R^{n \times n}$ is a given matrix and $q \in R^n$ is a given vector. If M is copositive on Δ and*

$$q \in \text{int}\left([\text{Sol}(\text{AVI}(M, 0, \Delta))]^+ \right), \tag{7.17}$$

then there exist constants $\varepsilon > 0$, $\delta > 0$ and $\ell > 0$ such that if $(\widetilde{M}, \widetilde{q}) \in R^{n \times n} \times R^n$, $\widetilde{M}$ is copositive on Δ, and if

$$\max\{\|\widetilde{M} - M\|, \|\widetilde{q} - q\|\} < \varepsilon, \tag{7.18}$$

then the set $\text{Sol}(\text{AVI}(\widetilde{M}, \widetilde{q}, \Delta))$ is nonempty,

$$\text{Sol}(\text{AVI}(\widetilde{M}, \widetilde{q}, \Delta)) \subset \delta \bar{B}_{R^n}, \tag{7.19}$$

and

$$\text{Sol}(\text{AVI}(\widetilde{M}, \widetilde{q}, \Delta)) \subset \text{Sol}(\text{AVI}(M, q, \Delta)) + \ell(\|\widetilde{M} - M\| + \|\widetilde{q} - q\|)\bar{B}_{R^n}. \tag{7.20}$$

Proof. Suppose that M is copositive on Δ and (7.17) is satisfied. Since Δ is a polyhedral convex cone, we see that for every $(\widetilde{M}, \widetilde{q}) \in R^{n \times n} \times R^n$ the problem $\text{AVI}(\widetilde{M}, \widetilde{q}, \Delta)$ is an GLCP. In particular, $\text{AVI}(M, 0, \Delta)$ is a GLCP problem and we have

$$\text{Sol}(\text{AVI}(M, 0, \Delta)) = \{v \in \Delta : Mv \in \Delta^+, \langle Mv, v \rangle = 0\}.$$

Since $\text{Sol}(\text{AVI}(M, 0, \Delta))$ is a closed cone, Lemma 6.4 shows that (7.17) is equivalent to the following condition

$$q^T v > 0 \quad \forall v \in \text{Sol}(\text{AVI}(M, 0, \Delta)) \setminus \{0\}. \qquad (7.21)$$

CLAIM 1. *There exists $\varepsilon > 0$ such that if $(\widetilde{M}, \widetilde{q}) \in R^{n \times n} \times R^n$, $\widetilde{M}$ is copositive on Δ, and if (7.18) holds, then the set $\text{Sol}(\text{AVI}(\widetilde{M}, \widetilde{q}, \Delta))$ is nonempty.*

Suppose Claim 1 were false. Then we could find a sequence $\{(M^k, q^k)\}$ in $R^{n \times n} \times R^n$ such that M^k is copositive on Δ for every $k \in N$, $(M^k, q^k) \to (M, q)$ as $k \to \infty$, and $\text{Sol}(\text{AVI}(M^k, q^k, \Delta)) = \emptyset$ for every $k \in N$. According to Theorem 6.5, we must have

$$q^k \notin \text{int}\left([\text{Sol}(\text{AVI}(M^k, 0, \Delta))]^+ \right) \quad \forall k \in N.$$

Applying Lemma 6.4 we can assert that for each $k \in N$ there exists $v^k \in \text{Sol}(\text{AVI}(M^k, 0, \Delta)) \setminus \{0\}$ such that $(q^k)^T v^k \leq 0$. Then we have

$$v^k \in \Delta, \quad M^k v^k \in \Delta^+, \quad \langle M^k v^k, v^k \rangle = 0, \qquad (7.22)$$

for every $k \in N$. Without loss of generality we can assume that

$$\frac{v^k}{\|v^k\|} \to \bar{v} \in R^n, \quad \|\bar{v}\| = 1.$$

From (7.22) it follows that

$$\frac{v^k}{\|v^k\|} \in \Delta, \quad M^k \frac{v^k}{\|v^k\|} \in \Delta^+, \quad \left\langle M^k \frac{v^k}{\|v^k\|}, \frac{v^k}{\|v^k\|} \right\rangle = 0.$$

Taking limits as $k \to \infty$ we obtain

$$\bar{v} \in \Delta, \quad M\bar{v} \in \Delta^+, \quad \langle M\bar{v}, \bar{v} \rangle = 0.$$

This shows that $\bar{v} \in \text{Sol}(\text{AVI}(M, 0, \Delta))$. Since $(q^k)^T v^k \leq 0$, we see that $(q^k)^T \frac{v^k}{\|v^k\|} \leq 0$ for every $k \in N$. Letting $k \to \infty$ yields $q^T \bar{v} \leq 0$. Since $\bar{v} \in \text{Sol}(\text{AVI}(M, 0, \Delta)) \setminus \{0\}$, the last inequality contradicts (7.21). We have thus justified Claim 1.

CLAIM 2. *There exist $\varepsilon > 0$ and $\delta > 0$ such that if $(\widetilde{M}, \widetilde{q}) \in R^{n \times n} \times R^n$, $\widetilde{M}$ is copositive on Δ, and if (7.18) holds, then inclusion (7.19) is satisfied.*

To obtain a contradiction, suppose that there exist a sequence $\{(M^k, q^k)\}$ in $R^{n \times n} \times R^n$ and a sequence $\{x^k\}$ in R^n such that M^k is copositive on Δ for every $k \in N$, $x^k \in \text{Sol}(\text{AVI}(M^k, q^k, \Delta))$ for every $k \in N$, $(M^k, q^k) \to (M, q)$ as $k \to \infty$, and $\|x^k\| \to +\infty$ as $k \to \infty$. Since $x^k \in \text{Sol}(\text{AVI}(M^k, q^k, \Delta))$, we see that

$$x^k \in \Delta, \quad M^k x^k + q^k \in \Delta^+, \quad \langle M^k x^k + q^k, x^k \rangle = 0, \qquad (7.23)$$

for every $k \in N$. There is no loss of generality in assuming that

$$\frac{x^k}{\|x^k\|} \to \bar{v} \in R^n, \quad \|\bar{v}\| = 1.$$

From (7.23) it follows that

$$\bar{v} \in \Delta, \quad M\bar{v} \in \Delta^+, \quad \langle M\bar{v}, \bar{v} \rangle = 0.$$

From this we conclude that $\bar{v} \in \text{Sol}(\text{AVI}(M, 0, \Delta))$. Since

$$\langle M^k x^k + q^k, x^k \rangle = 0$$

and since $0^+\Delta = \Delta$ and M^k is copositive on Δ, we have

$$-(q^k)^T x^k = -\langle q^k, x^k \rangle = \langle M^k x^k, x^k \rangle \geq 0.$$

Then

$$q^T \bar{v} = \lim_{k \to \infty} \left((q^k)^T \frac{x^k}{\|x^k\|} \right) \leq 0.$$

This contradicts (7.21). Claim 2 has been proved.

Now we are in a position to show that there exist $\varepsilon > 0$, $\delta > 0$ and $\ell > 0$ such that if $(\widetilde{M}, \widetilde{q}) \in R^{n \times n} \times R^n$, $\widetilde{M}$ is copositive on Δ, and if (7.18) holds, then $\text{Sol}(\text{AVI}(\widetilde{M}, \widetilde{q}, \Delta)) \neq \emptyset$ and (7.19), (7.20) are satisfied.

Combining Claim 1 with Claim 2 we see that there exist $\varepsilon > 0$ and $\delta > 0$ such that if $(\widetilde{M}, \widetilde{q}) \in R^{n \times n} \times R^n$, $\widetilde{M}$ is copositive on Δ, and if (7.18) holds, then $\text{Sol}(\text{AVI}(\widetilde{M}, \widetilde{q}, \Delta)) \neq \emptyset$ and (7.19) is satisfied. According to Corollary 7.4, for the given matrix M and vector q, there exist a constant $\ell_M > 0$ and a neighborhood U_q of q such that

$$\text{Sol}(\text{AVI}(M, q', \Delta)) \subset \text{Sol}(\text{AVI}(M, q, \Delta)) + \ell_M \|q' - q\| \bar{B}_{R^n} \qquad (7.24)$$

for every $q' \in U_q$. Let $(\widetilde{M}, \widetilde{q}) \in R^{n \times n} \times R^n$ be such that $\widetilde{M}$ is copositive on Δ and (7.18) holds. Select any $\widetilde{x} \in \mathrm{Sol}(\mathrm{AVI}(\widetilde{M}, \widetilde{q}, \Delta))$. Setting

$$\bar{q} = \widetilde{q} + (\widetilde{M} - M)\widetilde{x} \tag{7.25}$$

we will show that

$$\widetilde{x} \in \mathrm{Sol}(\mathrm{AVI}(M, \bar{q}, \Delta)). \tag{7.26}$$

Since

$$\langle \widetilde{M}\widetilde{x} + \widetilde{q}, x - \widetilde{x} \rangle \geq 0 \quad \forall x \in \Delta,$$

using (7.25) we deduce that

$$\begin{aligned} 0 \leq \langle \widetilde{M}\widetilde{x} + \widetilde{q}, x - \widetilde{x} \rangle &= \langle \widetilde{M}\widetilde{x} + \bar{q} - \widetilde{M}\widetilde{x} + M\widetilde{x}, x - \widetilde{x} \rangle \\ &= \langle M\widetilde{x} + \bar{q}, x - \widetilde{x} \rangle \end{aligned}$$

for every $x \in \Delta$. This shows that (7.26) is valid. From (7.18), (7.19) and (7.25) it follows that

$$\|\bar{q} - q\| \leq \|\widetilde{q} - q\| + \|\widetilde{M} - M\|\|\widetilde{x}\| \leq \varepsilon(1 + \delta).$$

Consequently, choosing a smaller $\varepsilon > 0$ if necessary, we can assert that $\bar{q} \in U_q$ whenever $(\widetilde{M}, \widetilde{q}) \in R^{n \times n} \times R^n$, $\widetilde{M}$ is copositive on Δ, (7.18) holds. Hence from (7.24) and (7.26) we deduce that there exists $x \in \mathrm{Sol}(\mathrm{AVI}(M, q, \Delta))$ such that

$$\begin{aligned} \|\widetilde{x} - x\| &\leq \ell_M \|\bar{q} - q\| \\ &\leq \ell_M(\|\widetilde{q} - q\| + \|\widetilde{M} - M\|\|\widetilde{x}\|) \\ &\leq \ell_M(\|\widetilde{q} - q\| + \delta\|\widetilde{M} - M\|) \\ &\leq \ell(\|\widetilde{q} - q\| + \|\widetilde{M} - M\|), \end{aligned}$$

where $\ell = \max\{\ell_M, \delta\ell_M\}$. We have thus obtained (7.20). The proof is complete. $\square$

Our next goal is to establish the following interesting result on AVI problems with positive semidefinite matrices.

Theorem 7.5. (See Robinson (1979), Theorem 2) *Let $M \in R^{n \times n}$ be a positive semidefinite matrix, Δ a nonempty polyhedral convex set in R^n, and $q \in R^n$. Then the following two properties are equivalent:*

(i) *The solution set $\mathrm{Sol}(\mathrm{AVI}(M, q, \Delta))$ is nonempty and bounded;*

(ii) *There exists $\varepsilon > 0$ such that for each $\widetilde{M} \in R^{n \times n}$ and each $\widetilde{q} \in R^n$ with*

$$\max\{\|\widetilde{M} - M\|, \|\widetilde{q} - q\|\} < \varepsilon, \qquad (7.27)$$

the set $\mathrm{Sol}(\mathrm{AVI}(\widetilde{M}, \widetilde{q}, \Delta))$ is nonempty.

For proving the above theorem we shall need the following three auxiliary lemmas in which *it is assumed that $M \in R^{n \times n}$ is a positive semidefinite matrix, $\Delta \subset R^n$ is a nonempty polyhedral convex set, and $q \in R^n$.* We set $M\Delta = \{Mx : x \in \Delta\}$.

Lemma 7.1. (See, for instance, Best and Chakravarti (1992)) *For any $\bar{v} \in R^n$, if $\bar{v}^T M \bar{v} = 0$ then $(M + M^T)\bar{v} = 0$.*

Proof. Consider the unconstrained quadratic program

$$\min\left\{ f(x) := \frac{1}{2}x^T(M + M^T)x : x \in R^n \right\}.$$

From our assumptions it follows that

$$\begin{aligned}
\frac{1}{2}x^T(M + M^T)x = x^T M x &\geq 0 \\
&= \bar{v}^T M \bar{v} \\
&= \frac{1}{2}\bar{v}^T(M + M^T)\bar{v}
\end{aligned}$$

for every $x \in R^n$. Hence $\bar{v}$ is a global solution of the above problem. By Theorem 3.1 we have

$$0 = \nabla f(\bar{v}) = (M + M^T)\bar{v},$$

which completes the proof. $\square$

Lemma 7.2. *The inclusion*

$$q \in \mathrm{int}((0^+\Delta)^+ - M\Delta) \qquad (7.28)$$

holds if and only if

$$\forall v \in 0^+\Delta \setminus \{0\} \ \exists x \in \Delta \ \text{such that} \ \langle Mx + q, v \rangle > 0. \qquad (7.29)$$

Proof. *Necessity:* Suppose that (7.28) holds. Then there exists $\varepsilon > 0$ such that

$$B(q, \varepsilon) \subset (0^+\Delta)^+ - M\Delta. \qquad (7.30)$$

To obtain a contradiction, suppose that there exists $\bar{v} \in 0^+\Delta \setminus \{0\}$ such that

$$\langle Mx + q, \bar{v} \rangle \leq 0 \quad \forall x \in \Delta.$$

By (7.30), for every $q' \in B(q, \varepsilon)$ there exist $w \in (0^+\Delta)^+$ and $x \in \Delta$ such that $q' = w - Mx$. So we have

$$\langle q' - q, \bar{v} \rangle \geq \langle w, \bar{v} \rangle \geq 0 \quad \forall q' \in B(q, \varepsilon).$$

This clearly forces $\bar{v} = 0$, which is impossible.

Sufficiency: On the contrary, suppose that (7.29) is valid, but (7.28) is false. Then there exists a sequence $\{q^k\} \subset R^n$ such that $q^k \notin (0^+\Delta)^+ - M\Delta$ for all $k \in N$, and $q^k \to q$. From this we deduce that

$$(M\Delta + q^k) \cap (0^+\Delta)^+ = \emptyset \quad \forall k \in N.$$

Since $M\Delta + q^k$ and $(0^+\Delta)^+$ are two disjoint polyhedral convex sets, by Theorem 11.3 from Rockafellar (1970) there exists a hyperplane separating these sets properly. Since $(0^+\Delta)^+$ is a cone, by Theorem 11.7 from Rockafellar (1970) there exists a hyperplane which separates the above two sets properly and passes through the origin. So there exists $v^k \in R^n$ with $\|v^k\| = 1$ such that

$$\langle v^k, Mx + q^k \rangle \leq 0 \leq \langle v^k, w \rangle \quad \forall x \in \Delta, \ \forall w \in (0^+\Delta)^+. \tag{7.31}$$

(Actually, the above-mentioned hyperplane is defined by the formula $H = \{z \in R^n : \langle v^k, z \rangle = 0\}$). Without loss of generality we can assume that $v^k \to \bar{v} \in R^n$, $\|\bar{v}\| = 1$. From (7.31) it follows that

$$\langle \bar{v}, Mx + q \rangle \leq 0 \quad \forall x \in \Delta \tag{7.32}$$

and

$$\langle \bar{v}, w \rangle \geq 0 \quad \forall w \in (0^+\Delta)^+. \tag{7.33}$$

By Theorem 14.1 from Rockafellar (1970), from (7.33) it follows that $\bar{v} \in 0^+\Delta$. Combining this with (7.32) we see that (7.29) is false, which is impossible. □

Lemma 7.3. (See Gowda-Pang (1994a), Theorem 7) *The solution set* Sol(AVI(M, q, Δ)) *is nonempty and bounded if and only if* (7.28) *holds.*

Proof. *Necessity:* To obtain a contradiction, suppose that the set Sol(AVI(M, q, Δ)) is nonempty and bounded, but (7.28) does not hold. Then, by Lemma 7.2 there exists $\bar{v} \in 0^+\Delta \setminus \{0\}$ such that

(7.32) holds. Select a point $x^0 \in \text{Sol}(\text{AVI}(M, q, \Delta))$. For each $t > 0$, we set $x_t = x^0 + t\bar{v}$. Since $\bar{v} \in 0^+\Delta$, we have $x_t \in \Delta$ for every $t > 0$. Substituting x_t for x in (7.32) we get

$$\langle \bar{v}, Mx^0 + q \rangle + t\langle \bar{v}, M\bar{v} \rangle \leq 0 \quad \forall t > 0.$$

This implies that $\langle \bar{v}, M\bar{v} \rangle \leq 0$. Besides, since M is positive semidefinite, we have $\langle \bar{v}, M\bar{v} \rangle \geq 0$. So

$$\langle \bar{v}, M\bar{v} \rangle = 0. \tag{7.34}$$

By Lemma 7.1, from (7.34) we obtain

$$(M + M^T)\bar{v} = 0. \tag{7.35}$$

Fix any $x \in \Delta$. On account of (7.32), (7.34), (7.35) and the fact that $x^0 \in \text{Sol}(\text{AVI}(M, q, \Delta))$, we have

$$
\begin{aligned}
\langle Mx_t + q, x - x_t \rangle &= \langle Mx^0 + q + tM\bar{v}, x - x^0 - t\bar{v} \rangle \\
&= \langle Mx^0 + q, x - x^0 \rangle + t\langle M\bar{v}, x - x^0 \rangle \\
&\quad -t\langle Mx^0 + q, \bar{v} \rangle - t^2 \underbrace{\langle M\bar{v}, \bar{v} \rangle}_{=0} \\
&= \langle Mx^0 + q, x - x^0 \rangle - t\underbrace{\langle \bar{v}, Mx + q \rangle}_{\leq 0} \\
&\quad -t\underbrace{\langle (M + M^T)\bar{v}, x^0 \rangle}_{=0} \\
&\geq \langle Mx^0 + q, x - x^0 \rangle \\
&\geq 0.
\end{aligned}
$$

Since this holds for every $x \in \Delta$, $x_t \in \text{Sol}(\text{AVI}(M, q, \Delta))$. As the last inclusion is valid for each $t > 0$, we conclude that $\text{Sol}(\text{AVI}(M, q, \Delta))$ is unbounded, a contradiction.

Sufficiency: Suppose that (7.28) holds. We have to show that the set $\text{Sol}(\text{AVI}(M, q, \Delta))$ is nonempty and bounded. By (7.28),

$$q \in (0^+\Delta)^+ - M\Delta.$$

Hence there exist $w \in (0^+\Delta)^+$ and $\bar{x} \in \Delta$ such that $q = w - M\bar{x}$. Since $M\bar{x} + q = w \in (0^+\Delta)^+$, for every $v \in 0^+\Delta$ it holds

$$\langle M\bar{x} + q, v \rangle = \langle w, v \rangle \geq 0.$$

Since M is a positive semidefinite matrix, we see that both conditions (i) and (ii) in Theorem 6.1 are satisfied. Hence the set

Sol(AVI(M, q, Δ)) is nonempty. To show that Sol(AVI(M, q, Δ)) is bounded we suppose, contrary to our claim, that there exists a sequence $\{x^k\}$ in Sol(AVI(M, q, Δ)) such that $\|x^k\| \to +\infty$. There is no loss of generality in assuming that $x^k \neq 0$ for each $k \in N$, and

$$\frac{x^k}{\|x^k\|} \to \bar{v} \in R^n, \quad \|\bar{v}\| = 1.$$

Let $m \in N$, $A \in R^{m \times n}$ and $b \in R^m$ be such that $\Delta = \{x \in R^n : Ax \geq b\}$. Since $Ax^k \geq b$ for every $k \in N$, dividing the inequality by $\|x^k\|$ and letting $k \to \infty$ we obtain $A\bar{v} \geq 0$. This shows that $\bar{v} \in 0^+\Delta$. We have

$$\langle Mx^k + q, x - x^k \rangle \geq 0 \quad \forall x \in \Delta \; \forall k \in N.$$

Hence

$$\langle Mx^k + q, x \rangle \geq \langle Mx^k, x^k \rangle + \langle q, x^k \rangle \quad \forall x \in \Delta \; \forall k \in N. \qquad (7.36)$$

Dividing the last inequality by $\|x^k\|^2$ and letting $k \to \infty$ we get $0 \geq \langle M\bar{v}, \bar{v} \rangle$. Since M is positive semidefinite, from this we see that $\langle M\bar{v}, \bar{v} \rangle = 0$. Thus, by Lemma 7.1 we have

$$M\bar{v} = -M^T\bar{v}. \qquad (7.37)$$

Fix a point $x \in \Delta$. Since $\langle Mx^k, x^k \rangle \geq 0$ for every $k \in N$, (7.36) implies that

$$\langle Mx^k + q, x \rangle \geq \langle q, x^k \rangle \quad \forall k \in N.$$

Dividing the last inequality by $\|x^k\|$ and letting $k \to \infty$ we obtain

$$\langle M\bar{v}, x \rangle \geq \langle q, \bar{v} \rangle.$$

Combining this with (7.37) we can assert that

$$\langle Mx + q, \bar{v} \rangle \leq 0 \quad \forall x \in \Delta.$$

Since $\bar{v} \in (0^+\Delta) \setminus \{0\}$, from the last fact and Lemma 7.2 it follows that (7.28) does not hold. We have thus arrived at a contradiction. The proof is complete. □

Proof of Theorem 7.5.

We first prove the implication (i) $\Rightarrow$ (ii). To obtain a contradiction, suppose that Sol(AVI(M, q, Δ)) is nonempty and bounded, while there exists a sequence $(M^k, q^k) \in R^{n \times n} \times R^n$ such that $(M^k, q^k) \to (M, q)$ and

$$\text{Sol}(\text{AVI}(M^k, q^k, \Delta)) = \emptyset \quad \forall k \in N. \tag{7.38}$$

Since Δ is nonempty, for $j \in N$ large enough, the set

$$\Delta_j := \Delta \cap \{x \in R^n : \|x\| \leq j\}$$

is nonempty. Without restriction of generality we can assume that $\Delta_j \neq \emptyset$ for every $j \in N$. By the Hartman-Stampacchia Theorem (Theorem 5.1) we can find a point, denoted by $x^{k,j}$, in the solution set $\text{Sol}(\text{AVI}(M^k, q^k, \Delta_j))$. We have

$$\langle M^k x^{k,j} + q^k, x - x^{k,j} \rangle \geq 0 \quad \forall x \in \Delta_j. \tag{7.39}$$

Note that

$$\|x^{k,j}\| = j \quad \forall j \in N. \tag{7.40}$$

Indeed, if $\|x^{k,j}\| < j$ then there exists $\mu > 0$ such that $\bar{B}(x^{k,j}, \mu) \subset \bar{B}(0, j)$. Hence from (7.39) it follows that

$$\langle M^k x^{k,j} + q^k, x - x^{k,j} \rangle \geq 0 \quad \forall x \in \Delta \cap \bar{B}(x^{k,j}, \mu).$$

By Proposition 5.3, this implies that $x^{k,j} \in \text{Sol}(\text{AVI}(M^k, q^k, \Delta))$, which is impossible because (7.38) holds. Fixing an index $j \in N$ we consider the sequence $\{x^{k,j}\}_{k \in N}$. From (7.40) we deduce that this sequence has a convergent subsequence. There is no loss of generality in assuming that

$$\lim_{k \to \infty} x^{k,j} = x^j, \quad x^j \in R^n, \quad \|x^j\| = j. \tag{7.41}$$

Letting $k \to \infty$ we deduce from (7.39) that

$$\langle M x^j + q, x - x^j \rangle \geq 0 \quad \forall x \in \Delta_j. \tag{7.42}$$

On account of (7.41), without loss of generality we can assume that

$$\frac{x^j}{\|x^j\|} \to \bar{v} \in R^n, \quad \|\bar{v}\| = 1.$$

Let us fix a point $x \in \Delta$. It is clear that there exists an index $j_x \in N$ such that $x \in \Delta_j$ for every $j \geq j_x$. From (7.42) we deduce that

$$\langle M x^j + q, x - x^j \rangle \geq 0 \quad \forall j \geq j_x.$$

Hence

$$\langle Mx^j + q, x\rangle \geq \langle Mx^j, x^j\rangle + \langle q, x^j\rangle \quad \forall j \geq j_x. \qquad (7.43)$$

As in the last part of the proof of Lemma 7.3, we can show that $\bar{v} \in (0^+\Delta) \setminus \{0\}$ and deduce from (7.43) the following inequality

$$\langle Mx + q, \bar{v}\rangle \leq 0.$$

Since the latter holds for every $x \in \Delta$, applying Lemma 7.2 we see that the inclusion (7.28) cannot hold. According to Lemma 7.3, the last fact implies that the set $\mathrm{Sol}(\mathrm{AVI}(M, q, \Delta))$ cannot be nonempty and bounded. This contradicts our assumption.

We now prove the implication (ii) $\Rightarrow$ (i). Suppose that there exists $\varepsilon > 0$ such that if matrix $\widetilde{M} \in R^{n \times n}$ and vector $\widetilde{q} \in R^n$ satisfy condition (7.27) then the set $\mathrm{Sol}(\mathrm{AVI}(\widetilde{M}, \widetilde{q}, \Delta))$ is nonempty. Consequently, for any $\widetilde{q} \in R^n$ satisfying $\|\widetilde{q} - q\| < \varepsilon$, the set $\mathrm{Sol}(\mathrm{AVI}(M, \widetilde{q}, \Delta))$ is nonempty. Let $\widetilde{x} \in \mathrm{Sol}(\mathrm{AVI}(M, \widetilde{q}, \Delta))$. For any $v \in 0^+\Delta$ we have

$$(M\widetilde{x} + \widetilde{q})^T v = \langle M\widetilde{x} + \widetilde{q}, (\widetilde{x} + v) - \widetilde{x}\rangle \geq 0.$$

Hence $M\widetilde{x} + \widetilde{q} \in (0^+\Delta)^+$. So we have $\widetilde{q} \in (0^+\Delta)^+ - M\Delta$. Since this inclusion is valid for each $\widetilde{q}$ satisfying $\|\widetilde{q} - q\| < \varepsilon$, we conclude that

$$q \in \mathrm{int}((0^+\Delta)^+ - M\Delta).$$

By Lemma 7.3, the set $\mathrm{Sol}(\mathrm{AVI}(M, q, \Delta))$ is nonempty and bounded. The proof is complete. $\square$

Let us consider three illustrative examples.

Example 7.1. Setting $\Delta = [0, +\infty) \subset R^1$, $M = (-1)$, and $q = 0$, we have $\mathrm{Sol}(\mathrm{AVI}(M, q, \Delta)) = \{0\}$. Note that matrix M is not positive semidefinite. Taking $\widetilde{M} = M$ and $\widetilde{q} = -\theta$, where $\theta > 0$, we check at once that $\mathrm{Sol}(\mathrm{AVI}(\widetilde{M}, \widetilde{q}, \Delta)) = \emptyset$. So, for this AVI problem, property (i) in Theorem 7.5 holds, but property (ii) does not hold. This example shows that, in Theorem 7.5, one cannot omit the assumption that M is a positive semidefinite matrix.

Example 7.2. Setting $\Delta = (-\infty, +\infty) = R^1$, $M = (0)$, and $q = 0$, we have $\mathrm{Sol}(\mathrm{AVI}(M, q, \Delta)) = \Delta$. So property (i) in Theorem 7.5 does not hold for this example. Taking $\widetilde{M} = (0)$ and $\widetilde{q} = \theta$, where $\theta > 0$, we have $\mathrm{Sol}(\mathrm{AVI}(\widetilde{M}, \widetilde{q}, \Delta)) = \emptyset$. This shows that, for the

AVI problem under consideration, property (ii) in Theorem 7.5 fails to hold.

Example 7.3. Setting $\Delta = [1, +\infty) \subset R^1$, $M = (0)$, and $q = 0$, we have $\text{Sol}(\text{AVI}(M, q, \Delta)) = \Delta$. Taking $\widetilde{M} = M$ and $\widetilde{q} = \theta$, where $\theta > 0$, we see that $\text{Sol}(\text{AVI}(\widetilde{M}, \widetilde{q}, \Delta)) = \{1\}$. But taking $\widetilde{M} = (-\theta)$ and $\widetilde{q} = 0$, where $\theta > 0$, we see that $\text{Sol}(\text{AVI}(\widetilde{M}, \widetilde{q}, \Delta)) = \emptyset$. So, for this problem, both the properties (i) and (ii) in Theorem 7.5 do not hold.

In connection with Theorem 7.5, it is natural to raise the following open question.

QUESTION: Is it true that property (i) in Theorem 7.5 implies that there exists $\varepsilon > 0$ such that if matrix $\widetilde{M} \in R^{n \times n}$ and vector $\widetilde{q} \in R^n$ satisfy condition (7.27) then the set $\text{Sol}(\text{AVI}(\widetilde{M}, \widetilde{q}, \Delta))$ is bounded (may be empty)?

The next example shows that property (i) in Theorem 7.5 does not imply that the solution sets $\text{Sol}(\text{AVI}(\widetilde{M}, \widetilde{q}, \Delta))$, where $(\widetilde{M}, \widetilde{q})$ is taken from a neighborhood of (M, q), are uniformly bounded.

Example 7.4. (See Robinson (1979), pp. 139–140) Let $\Delta = [0, +\infty) \subset R^1$, $M = (0)$, and $q = 1$. It is clear that

$$\text{Sol}(\text{AVI}(M, q, \Delta)) = \{0\}.$$

Taking $\widetilde{M} = (-\mu)$ and $\widetilde{q} = 1$, where $\mu > 0$, we have

$$\text{Sol}(\text{AVI}(\widetilde{M}, \widetilde{q}, \Delta)) = \left\{ 0, \frac{1}{\mu} \right\}.$$

From this we conclude that there exist no $\varepsilon > 0$ and $\delta > 0$ such that if matrix $\widetilde{M} \in R^{1 \times 1}$ and vector $\widetilde{q} \in R^1$ satisfy condition (7.27) then $\text{Sol}(\text{AVI}(\widetilde{M}, \widetilde{q}, \Delta)) \subset \delta \bar{B}_{R^1}$.

The following theorem is one of the main results on solution stability of AVI problems. One can observe that this theorem and Theorem 7.4 are independent results.

Theorem 7.6. (See Robinson (1979), Theorem 2) *Suppose that $\Delta \subset R^n$ is a nonempty polyhedral convex set. Suppose that $M \in R^{n \times n}$ is a given matrix and $q \in R^n$ is a given vector. If M is a positive semidefinite matrix and if the solution set $\text{Sol}(\text{AVI}(M, q, \Delta))$ is nonempty and bounded, then there exist constants $\varepsilon > 0$, $\delta > 0$ and $\ell > 0$ such that if $(\widetilde{M}, \widetilde{q}) \in R^{n \times n} \times R^n$, $\widetilde{M}$ is positive semidefinite, and if*

$$\max\{\|\widetilde{M} - M\|, \|\widetilde{q} - q\|\} < \varepsilon, \tag{7.44}$$

then the set $\mathrm{Sol}(\mathrm{AVI}(\widetilde{M}, \widetilde{q}, \Delta))$ *is nonempty,*

$$\mathrm{Sol}(\mathrm{AVI}(\widetilde{M}, \widetilde{q}, \Delta)) \subset \delta \bar{B}_{R^n}, \tag{7.45}$$

and

$$\mathrm{Sol}(\mathrm{AVI}(\widetilde{M}, \widetilde{q}, \Delta)) \subset \mathrm{Sol}(\mathrm{AVI}(M, q, \Delta)) + \ell(\|\widetilde{M} - M\| + \|\widetilde{q} - q\|)\bar{B}_{R^n}. \tag{7.46}$$

Proof. Since M is positive semidefinite and $\mathrm{Sol}(\mathrm{AVI}(M, q, \Delta))$ is nonempty and bounded, by Lemmas 7.2 and 7.3 we have

$$\forall v \in 0^+\Delta \setminus \{0\} \ \exists x \in \Delta \ \text{ such that } \ \langle Mx + q, v \rangle > 0. \tag{7.47}$$

Moreover, according to Theorem 7.5, there exists $\varepsilon_0 > 0$ such that for each matrix $\widetilde{M} \in R^{n \times n}$ and each $\widetilde{q} \in R^n$ satisfying

$$\max\{\|\widetilde{M} - M\|, \ \|\widetilde{q} - q\|\} < \varepsilon_0,$$

the set $\mathrm{Sol}(\mathrm{AVI}(\widetilde{M}, \widetilde{q}, \Delta))$ is nonempty. We claim that there exist constants $\varepsilon > 0$ and $\delta > 0$ such that (7.45) holds for every $(\widetilde{M}, \widetilde{q}) \in R^{n \times n} \times R^n$ satisfying condition (7.44) and the requirement that $\widetilde{M}$ is a positive semidefinite matrix. Indeed, if the claim were false we would find a sequence $\{(M^k, q^k)\}$ in $R^{n \times n} \times R^n$ and a sequence $\{x^k\}$ in R^n such that M^k is positive semidefinite for every $k \in$, $(M^k, q^k) \to (M, q)$, $x^k \in \mathrm{Sol}(\mathrm{AVI}(M^k, q^k, \Delta))$ for every $k \in N$, and $\|x^k\| \to +\infty$ as $k \to \infty$. For each $x \in \Delta$, we have

$$\langle M^k x^k + q^k, x - x^k \rangle \geq 0 \quad \forall k \in N. \tag{7.48}$$

Without loss of generality we can assume that $x^k \neq 0$ for every $k \in N$, and

$$\frac{x^k}{\|x^k\|} \to \bar{v} \in R^n, \quad \|\bar{v}\| = 1.$$

It is easily seen that $\bar{v} \in (0^+\Delta)^+$. From (7.48) it follows that

$$\langle M^k x^k + q^k, x \rangle \geq \langle M^k x^k, x^k \rangle + \langle q^k, x^k \rangle \quad \forall k \in N. \tag{7.49}$$

Dividing the last inequality by $\|x^k\|^2$ and letting $k \to \infty$ we get $0 \geq \langle M\bar{v}, \bar{v} \rangle$. Since M is positive semidefinite, from this we see that $\langle M\bar{v}, \bar{v} \rangle = 0$. By Lemma 7.1 we have

$$M\bar{v} = -M^T \bar{v}. \tag{7.50}$$

Fix a point $x \in \Delta$. Since M^k is positive semidefinite, we have $\langle M^k x^k, x^k \rangle \geq 0$ for every $k \in N$. Hence (7.49) implies that

$$\langle M^k x^k + q^k, x \rangle \geq \langle q^k, x^k \rangle \quad \forall k \in N.$$

Dividing the last inequality by $\|x^k\|$ and letting $k \to \infty$ we obtain

$$\langle M\bar{v}, x \rangle \geq \langle q, \bar{v} \rangle.$$

Combining this with (7.50) we get

$$\langle Mx + q, \bar{v} \rangle \leq 0 \quad \forall x \in \Delta.$$

Since $\bar{v} \in (0^+\Delta)^+ \setminus \{0\}$, the last fact contradicts (7.47). Our claim has been proved. We can now proceed analogously to the proof of Claim 3 in the proof of Theorem 7.4 to find the required constants $\varepsilon > 0$, $\delta > 0$ and $\ell > 0$. $\square$

7.4   Commentaries

As it has been noted in Robinson (1981), p. 206, the class of polyhedral multifunctions is closed under finite addition, scalar multiplication, and finite composition. This means that if $\Phi : R^n \to 2^{R^m}$, $\Psi : R^m \to 2^{R^s}$, $\Phi_j : R^n \to 2^{R^m}$ $(j = 1, \ldots, m)$ are some given polyhedral multifuntions and $\lambda \in R$ is a given scalar, then the formulae

$$(\lambda\Phi)(x) = \lambda\Phi(x) \quad (\forall x \in R^n),$$

$$(\Phi_1 + \ldots + \Phi_k)(x) = \Phi_1(x) + \ldots + \Phi_k(x) \quad (\forall x \in R^n),$$

and

$$(\Psi \circ \Phi)(x) = \Psi(\Phi(x)) \quad (\forall x \in R^n),$$

create new polyhedral multifunctions which are denoted by $\lambda\Phi$, $\Phi_1 + \ldots + \Phi_k$ and $\Psi \circ \Phi$, respectively.

The proof of Theorem 7.4 is similar in spirit to the proof of Theorem 7.5.1 in Cottle et al. (1992).

The 'elementary' proof of the results of Robinson (see Theorems 7.5 and 7.6) on the solution stability of AVI problems with positive semidefinite matrices given in this chapter is new. We hope that it can expose furthermore the beauty of these results. The original proof of Robinson is based on a general solution stability theorem for variational inequalities in Banach spaces (see Robinson (1979), Theorem 1).

Results presented in this chapter deal only with upper-Lipschitz continuity properties of the solution map of parametric AVI problems. For multifunctions, the lower semicontinuity, the upper semicontinuity, the openness, the Aubin property, the metric regularity, and the single-valuedness are other interesting properties which have many applications (see Aubin and Frankowska (1990), Mordukhovich (1993), Rockafellar and Wets (1998), and references therein). It is of interest to characterize these properties of the solution map in parametric AVI problems (in particular, of the solution map in parametric LCP problems). Some results in this direction have been obtained (see, for instance, Jansen and Tijs (1987), Gowda (1992), Donchev and Rockafellar (1996), Oettli and Yen (1995), Gowda and Sznajder (1996)). We will study the lower semicontinuity and the upper semicontinuity the solution map of parametric AVI problems in Chapter 18.

Chapter 8

Linear Fractional Vector Optimization Problems

Linear Fractional Vector Optimization (LFVO) is an interesting area in the wider theory of vector optimization (see, for example, Choo and Atkins (1982, 1983), Malivert (1995), Malivert and Popovici (2000), and Steuer (1986)). LFVO problems have applications in finance and production management (see Steuer (1986)). In a LFVO problem, any point satisfying the first-oder necessary optimality condition is a solution. Therefore, solving a LFVO problem is to solve a monotone affine Vector Variational Inequality (VVI). The original concept of VVI was proposed by Giannessi (1980). In this chapter we will apply the results of the preceding two chapters to establish some facts about connectedness and stability of the solution sets in LFVO problems. In particular, we will prove that the efficient solution set of a LFVO problem with a bounded constraint set is connected.

8.1 LFVO Problems

Let $f_i : R^n \to R$ $(i = 1, 2, \cdots, m)$ be m *linear fractional functions*, that is

$$f_i(x) = \frac{a_i^T x + \alpha_i}{b_i^T x + \beta_i}$$

for some $a_i \in R^n, b_i \in R^n, \alpha_i \in R$, and $\beta_i \in R$. Let $\Delta = \{x \in R^n : Cx \leq d\}$, where $C \in R^{r \times n}$ and $d \in R^r$, be a nonempty polyhedral convex set. We assume that $b_i^T x + \beta_i > 0$ for all $i \in \{1, \cdots, m\}$ and for all $x \in \Delta$. Define

$$f(x) = (f_1(x), \ldots, f_m(x)), \quad A = (a_1, \ldots, a_m), \quad B = (b_1, \ldots, b_m),$$

$$\alpha = (\alpha_1, \ldots, \alpha_m), \quad \beta = (\beta_1, \ldots, \beta_m), \quad \omega = (A, B, \alpha, \beta).$$

Thus A and B are $n \times m$-matrices, α and β are vectors of R^m, and ω is a parameter containing all the data related to the vector function f.

Consider the following vector optimization problem

(VP)   Minimize $f(x)$ subject to $x \in \Delta$.

Definition 8.1. One says that $x \in \Delta$ is an *efficient solution* of (VP) if there exists no $y \in \Delta$ such that $f(y) \leq f(x)$ and $f(y) \neq f(x)$. If there exists no $y \in \Delta$ such that $f(y) < f(x)$, then one says that $x \in \Delta$ is a *weakly efficient solution* of (VP).

Let us denote the *efficient solution set* and the *weakly efficient solution set*) of (VP) by Sol(VP) and Sol^w(VP), respectively.

The following lemma will be useful for obtaining necessary and sufficient optimality conditions for (VP).

Lemma 8.1. (See Malivert (1995)) *Let* $\varphi(x) = (a^T x + \alpha)/(b^T x + \beta)$ *be a linear fractional function. Suppose that* $b^T x + \beta \neq 0$ *for every* $x \in \Delta$. *Then for any* $x, y \in \Delta$, *it holds*

$$\varphi(y) - \varphi(x) = \frac{b^T x + \beta}{b^T y + \beta} \langle \nabla\varphi(x), y - x \rangle, \tag{8.1}$$

where $\nabla\varphi(x)$ *denotes the gradient of* φ *at* x.

Proof. By the definition of gradient,

$$\langle \nabla\varphi(x), y - x \rangle$$
$$= \lim_{t \downarrow 0} \frac{1}{t} [\varphi(x + t(y - x)) - \varphi(x)]$$
$$= \lim_{t \downarrow 0} \frac{1}{t} \left[\frac{a^T(x + t(y - x)) + \alpha}{b^T(x + t(y - x)) + \beta} - \frac{a^T x + \alpha}{b^T x + \beta} \right]$$
$$= \frac{a^T(y - x)(b^T x + \beta) - b^T(y - x)(a^T x + \alpha)}{(b^T x + \beta)^2}.$$

Hence we obtain

$$\frac{b^T x + \beta}{b^T y + \beta} \langle \nabla\varphi(x), y - x \rangle$$
$$= \frac{a^T(y - x)(b^T x + \beta) - b^T(y - x)(a^T x + \alpha)}{(b^T x + \alpha)(b^T x + \beta)}$$
$$= \varphi(y) - \varphi(x),$$

which completes the proof.   □

Given any $x, y \in \Delta$, $x \neq y$, we consider two points belonging to the line segment $[x, y]$:

$$z_t = x + t(y - x), \quad z_{t'} = x + t'(y - x) \quad (t \in [0, 1], \ t' \in [0, t)).$$

By (8.1),

$$
\begin{aligned}
\varphi(z_t) &- \varphi(z_{t'}) \\
&= (\varphi(z_t) - \varphi(x)) - (\varphi(z_{t'} - \varphi(x)) \\
&= \frac{b^T + \beta}{b^T z_t + \beta} \langle \nabla \varphi(x), z_t - x \rangle - \frac{b^T x + \beta}{b^T z_{t'} + \beta} \langle \nabla \varphi(x), z_{t'} - x \rangle \\
&= \langle \nabla \varphi(x), y - x \rangle \frac{b^T x + \beta)^2}{(b^T z_t + \beta)(b^T z_{t'} + \beta)} (t - t').
\end{aligned}
$$

From this we can conclude that: (i) If $\langle \nabla \varphi(x), y - x \rangle > 0$ then $\varphi(z_{t'}) < \varphi(z_t)$ for every $t' \in [0, t)$; (ii) If $\langle \nabla \varphi(x), y - x \rangle < 0$ then $\varphi(z_{t'}) > \varphi(z_t)$ for every $t' \in [0, t)$; (iii) If $\langle \nabla \varphi(x), y - x \rangle = 0$ then $\varphi(z_{t'}) = \varphi(z_t)$ for every $t' \in [0, t)$. Hence the function φ is monotonic on every line segment or ray contained in Δ.

By definition, a function $\varphi : \Delta \to R$ is *quasiconcave* on Δ if

$$\varphi((1 - t)x + ty) \geq \min\{\varphi(x), \varphi(y)\} \quad \forall x, y \in \Delta, \forall t \in (0, 1). \quad (8.2)$$

We say that φ is *semistrictly quasiconcave* on Δ if φ is quasiconcave on Δ and the inequality in (8.2) is strict whenever $\varphi(x) \neq \varphi(y)$. If φ is quasiconcave on Δ and the inequality in (8.2) is strict whenever $x \neq y$, then we say that φ is *strictly quasiconcave* on Δ. Function φ is said to be *quasiconvex* (resp., *semistrictly quasiconvex, strictly quasiconcave*)on Δ if the function $-\varphi$ is quasiconcave (resp., semistrictly quasiconcave, strictly quasiconcave) on Δ.

From the above-mentioned monotonicity of linear fractional functions it follows that if $\varphi : \Delta \to R$ is a linear fractional function then it is, at the same time, semistrictly quasiconcave and semistrictly quasiconvex on Δ. In general, linear fractional functions are not strictly quasiconcave on their effective domain. Indeed, for any $\mu \in R$, $t \in (0, 1)$, and x^1, x^2 from the set

$$\{x \in \Delta : \varphi(x) = \mu\} = \{x \in \Delta : a^T x + \alpha = b^T x + \beta\},$$

we have $\varphi((1 - t)x^1 + tx^2) = \min\{\varphi(x^1), \varphi(x^2)\}$.

Let

$$\Lambda = \{\lambda \in R_+^m : \sum_{i=1}^m \lambda_i = 1\}.$$

Then

$$\text{ri}\Lambda = \{\lambda \in R_+^m : \sum_{i=1}^m \lambda_i = 1, \; \lambda_i > 0 \text{ for all } i\}.$$

Theorem 8.1. (See Malivert (1995)) *Let* $x \in \Delta$. *The following assertions hold:*

(i) $x \in \text{Sol}(\text{VP})$ *if and only if there exists* $\lambda = (\lambda_1, \dots, \lambda_m) \in \text{ri}\Lambda$ *such that*

$$\left\langle \sum_{i=1}^m \lambda_i \left[(b_i^T x + \beta_i)a_i - (a_i^T x + \alpha_i)b_i \right], y - x \right\rangle \geq 0, \quad \forall y \in \Delta. \tag{8.3}$$

(ii) $x \in \text{Sol}^w(\text{VP})$ *if and only if there exists* $\lambda = (\lambda_1, \dots, \lambda_m) \in \Lambda$ *such that* (8.3) *holds.*

(iii) *Condition* (8.3) *is satisfied if and only if there exists* $\mu = (\mu_1, \dots, \mu_r)$, $\mu_j \geq 0$ *for all* $j = 1, \dots, r$, *such that*

$$\sum_{i=1}^m \lambda_i \left[(b_i^T x + \beta_i)a_i - (a_i^T x + \alpha_i)b_i \right] + \sum_{j \in I(x)} \mu_j C_j^T = 0, \tag{8.4}$$

where C_j *denotes the* j-*th row of the matrix* C *and*

$$I(x) = \{j : C_j x = d_j\}.$$

Proof. (i) We claim that $x \in\in \text{Sol}(\text{VP})$ if and only if

$$Q_x(\Delta - x) \cap (-R_+^m) = \{0\}, \tag{8.5}$$

where

$$Q_x = \begin{pmatrix} (b_1^T x + \beta_1)a_1^T - (a_1^T x + \alpha_1)b_1^T \\ \vdots \\ (b_m^T x + \beta_m)a_m^T - (a_m^T x + \alpha_m)b_m^T \end{pmatrix}$$

is an $(m \times n)$-matrix and $Q_x(\Delta - x) = \{Q_x(y - x) : y \in \Delta\}$. Indeed, $x \notin \text{Sol}(\text{VP})$ if and only if there exist $y \in \Delta$ and i_0 such that

$$f_i(y) \leq f_i(x) \; \forall i \in \{1, \dots, m\}, \quad f_{i_0}(y) < f_{i_0}(x).$$

By Lemma 8.1, the last system of inequalities is equivalent to the following one:

$$\langle \nabla f_i(x), y - x \rangle \leq 0 \ \ \forall i \in \{1, \ldots, m\}, \quad \langle \nabla f_{i_0}(x), y - x \rangle < 0.$$
$$(8.6)$$

Since

$$\langle \nabla f_i(x), y - x \rangle = \frac{a_i^T(y - x)(b_i^T x + \beta_i) - b_i^T(y - x)(a_i^T x + \alpha_i)}{(b_i^T x + \beta_i)^2},$$

we can rewrite (8.6) as follows

$$\begin{cases} \left[(b_i^T x + \beta_i)a_i^T - (a_i^T x + \alpha_i)b_i^T \right](y - x) \leq 0 \ \ \forall i \in \{1, \ldots, m\}, \\ \left[(b_{i_0}^T x + \beta_{i_0})a_{i_0}^T - (a_{i_0}^T x + \alpha_{i_0})b_{i_0}^T \right](y - x) < 0. \end{cases}$$

Therefore, $x \notin \text{Sol(VP)}$ if and only if there exists $y \in \Delta$ such that

$$Q_x(y - x) \in -R_+^m \quad \text{and} \quad Q_x(y - x) \neq 0.$$

Our claim has been proved.

It is clear that $D := Q_x(\Delta - x)$ is a polyhedral convex set. Hence, by Corollary 19.7.1 from Rockafellar (1970), $K := \text{cone} D$ is a polyhedral convex cone. In particular, K is a closed convex cone. It is easily seen that (8.5) is equivalent to the property $K \cap (-R_+^m) = \{0\}$. Setting

$$K^+ = \{z \in R^m : \langle z, v \rangle \geq 0 \ \ \forall v \in K\},$$

we have $K^+ \cap \text{int} R_+^k \neq \emptyset$. Indeed, on the contrary, suppose that $K^+ \cap \text{int} R_+^k = \emptyset$. Then, by the separation theorem, there exists $\xi \in R^m \setminus \{0\}$ such that

$$\langle \xi, u \rangle \leq 0 \leq \langle \xi, z \rangle \ \ \forall u \in \text{int} R_+^m, \forall z \in K^+.$$

This implies that $\xi \in -R_+^m$ and $\xi \in (K^+)^+ = K$. So we get $\xi \in K \cap (-R_+^m) = \{0\}$, a contradiction.

Fix any $\tilde{\lambda} \in K^+ \cap \text{int} R_+^k$. For $\lambda := \tilde{\lambda}/(\tilde{\lambda}_1 + \ldots + \tilde{\lambda}_m)$, we have $\lambda \in K^+ \cap \text{ri} \Delta$. Since $\langle \lambda, v \rangle \geq 0$ for every $v \in K$, we deduce that

$$\langle Q_x^T \lambda, (y - x) \rangle = \langle \lambda, Q_x(y - x) \rangle \geq 0$$

for every $x \in \Delta$. Hence (8.3) is valid.

(ii) It is easily seen that $x \in \text{Sol}^w(\text{VP})$ if and only if

$$Q_x(\Delta - x) \cap (-\text{int} R_+^m) = \emptyset.$$

Using the separation theorem we find a multiplier $\lambda = (\lambda_1, \ldots, \lambda_m) \in \Lambda$ satisfying (8.3).

(iii) It suffices to apply the Farkas Lemma. $\square$

Condition (8.3) can be rewritten in the form

(VI)$_\lambda$ $\qquad\qquad \langle M(\lambda)y + q(\lambda), y - x \rangle \geq 0 \quad \forall y \in \Delta,$

where

$$M(\lambda) = (M_{kj}(\lambda)),$$

$$M_{kj}(\lambda) = \sum_{i=1}^{m} \lambda_i \left(b_{i,j} a_{i,k} - a_{i,j} b_{i,k} \right), \quad 1 \leq k \leq n, \ 1 \leq j \leq n,$$

$$q(\lambda) = (q_k(\lambda)), \quad q_k(\lambda) = \sum_{i=1}^{m} \lambda_i \left(\beta_i a_{i,k} - \alpha_i b_{i,k} \right), \quad 1 \leq k \leq n,$$

$a_{i,k}$ and $b_{i,k}$ are the k–th components of a_i and b_i, respectively.

It is clear that $(M(\lambda))^T = -M(\lambda)$. Therefore, $\langle M(\lambda)v, v \rangle = 0$ for every $v \in R^n$. In particular, $M(\lambda)$ is a positive semidefinite matrix.

8.2   Connectedness of the Solution Sets

Let X, Y be two topological spaces and $G : X \to 2^Y$ be a multifunction.

Definition 8.2. One says that G is *upper semicontinuous* (usc) at $a \in X$ if f for every open set $V \subset Y$ satisfying $G(a) \subset V$ there exists a neighborhood U of a, such that $G(a') \subset V$ for all $a' \in U$. One says that G is *lower semicontinuous* (lsc) at $a \in X$ if $G(a) \neq \emptyset$ and for every open set $V \subset Y$ satisfying $G(a) \cap V \neq \emptyset$ there exists a neighborhood U of a such that $G(a') \cap V \neq \emptyset$ for all $a' \in U$.

Multifunction G is said to be *continuous* at $a \in X$ if it is simultaneously upper semicontinuous and lower semicontinuous at that point.

Definition 8.3. A topological space Z is said to be *connected* if there exists no pair (Z_1, Z_2) of disjoint nonempty open subsets Z_1, Z_2 of Z, such that $Z = Z_1 \cup Z_2$. The space Z is *arcwise connected* if for any $a, b \in Z$ there exits a continuous mapping $\gamma : [0, 1] \to Z$ such that $\gamma(0) = a, \gamma(1) = b$. One says that Z is *contractible* if there exist a point $z^0 \in Z$ and a continuous function $H : Z \times [0, 1] \to Z$ such that $H(z, 0) = z$ and $H(z, 1) = z^0$ for every $z \in Z$.

It is clear that if Z is contractible then it is arcwise connected. It is also clear that if Z is arcwise connected then it is connected. The reverse implications are not true in general.

The simple proof of the following result is left after the reader.

Theorem 8.2. (See Warburton (1983), and Hirriart-Urruty (1985)) *Assume that X is connected. If*

(i) *for every $x \in X$ the set $G(x)$ is nonempty and connected, and*

(ii) *G is upper semicontinuous at every $a \in X$ (or G is lower semicontinuous at every $a \in X$),*

then the image set

$$G(X) := \bigcup_{x \in X} G(x),$$

which is equipped with the induced topology, is connected.

Remark 8.1. Let X and Y be two normed spaces, $G : X \to 2^Y$ a multifunction. If G is upper-Lipschitz at $a \in X$ and $G(a)$ is a compact set, then G is usc at a. So, according to Theorem 7.6, if M is a positive semidefinite matrix and if the solution set $\text{Sol}(\text{AVI}(M, q, \Delta))$ of (6.1) is nonempty and bounded, then the solution map $\text{Sol}(\text{AVI}(\cdot, \cdot, \Delta) : \mathcal{P}_n \times R^n \to 2^{R^n}$ is usc at (M, q). Here the symbol $\mathcal{P}_n$ stands for the set of all positive semidefinite $n \times n$-matrices.

We now turn our attention back to problem (VP). Denote by $F(\lambda)$ the solution set of the problem $(\text{VI})_\lambda$ described in Section 8.1. By Proposition 5.4, $F(\lambda)$ is closed and convex. If Δ is compact then, by Theorem 5.1, $F(\lambda)$ is nonempty and bounded. Consider the set-valued map $F : \Lambda \to 2^{R^n}$, $\lambda \mapsto F(\lambda)$. According to Theorems 8.1 and 8.2,

$$\text{Sol}(\text{VP}) = \bigcup_{\lambda \in \text{ri}\Lambda} F(\lambda) = F(\text{ri}\Lambda), \tag{8.7}$$

$$\text{Sol}^w(\text{VP}) = \bigcup_{\lambda \in \Lambda} F(\lambda) = F(\Lambda). \tag{8.8}$$

Remark 8.2. Using the results and the terminology in Lee et al. (1998) and Lee and Yen (2001) we can say that solving problem (VP) is equivalent to solve the monotone affine VVI defined by Δ and the affine functions

$$g_i(x) = (b_i^T x + \beta_i) a_i - (a_i^T x + \alpha_i) b_i \quad (i = 1, 2, \dots, m).$$

Thus the first-order optimality condition of a LFVO problem can be treated as a special vector variational inequality.

Theorem 8.3. (Benoist (1998), Yen and Phuong (2000)) *If Δ is compact, then* Sol(VP) *is a connected set.*

Proof. Since riΛ is a convex set (so it is connected), $F(\lambda)$ is nonempty and connected for every $\lambda \in$ riΛ, and the map $F(\cdot)$ is upper semicontinuous at every $\bar{\lambda} \in$ riΛ, Theorem 8.3 can be applied to the set-valued map $F :$ ri$\Lambda \to 2^{R^n}$. As a consequence, $F($ri$\Lambda)$ is connected. Hence, by (8.7), Sol(VP) is connected. $\square$

Theorem 8.4. (Choo and Atkins (1983)) *If Δ is compact then* Sol^w(VP) *is a connected set.*

Proof. Apply Theorem 8.3 and formula (8.8) to the set-valued map $F : \Lambda \to 2^{R^n}$. $\square$

In fact, Choo and Atkins (1983) established the following stronger result: *If Δ is compact then* Sol^w(VP) *is connected by line segments.* The latter means that for any points $x, \bar{y} \in$ Sol^w(VP) there exists a finite sequence of points $x_0 = x, x_1, \ldots, x_k = \bar{y}$ such that each line segment $[x_j, x_{j+1}]$ $(j = 0, 1, \ldots, k - 1)$ is a subset of Sol^w(VP).

If Δ is unbounded, then Sol(VP) and Sol$(VP)^w$ may be disconnected.

Example 8.1. (Choo and Atkins (1983)) Consider problem (VP) with

$$\Delta = \{x = (x_1, x_2) \in R^2 : x_1 \geq 2, \ 0 \leq x_2 \leq 4\},$$
$$f_1(x) = \frac{-x_1}{x_1 + x_2 - 1}, \quad f_2(x) = \frac{-x_1}{x_1 - x_2 + 3}.$$

Using Theorem 8.1, one can verify that

$$\text{Sol(VP)} = \text{Sol}^w(\text{VP}) = \{(x_1, 0) : x_1 \geq 2\} \cup \{(x_1, 4) : x_1 \geq 2\}.$$

Recall that a *component* of a topological space is a maximal connected subset (that is, a connected subset which is properly contained in no other connected subset).

Example 8.2. (Hoa et al. (2004)) Consider problem (VP) where $n = m$, $m \geq 2$,

$$\Delta = \{x \in R^m : x_1 \geq 0, x_2 \geq 0, \ldots, x_m \geq 0, \sum_{k=1}^{m} x_k \geq 1\},$$

and

$$f_i(x) = \frac{-x_i + \dfrac{1}{2}}{\sum_{k=1}^{m} x_k - \dfrac{3}{4}} \qquad (i = 1, \ldots, m).$$

Using Theorem 8.1 one can show that

$$\text{Sol}(VP) = \text{Sol}^w(VP) = \{(x_1, 0, \ldots, 0)^T : x_1 \geq 1\}$$
$$\cup \{(0, x_2, \ldots, 0)^T : x_2 \geq 1\}$$
$$\cdots \cdots \cdots$$
$$\cup \{(0, \ldots, 0, x_m)^T : x_m \geq 1\}.$$

Hence each of the sets $\text{Sol}(VP)$ and $\text{Sol}^w(VP)$ has exactly m components.

The following question (see Yen and Phuong (2000)) remains open: *Is it true that every component of Sol(VP) is connected by line segments?*

The following possible estimates have been mentioned in Hoa et al. (2004):

$$\chi(\text{Sol}(VP)) \leq \min\{m, n\}, \quad \chi(\text{Sol}^w(VP)) \leq \min\{m, n\}.$$

Here $\chi(M)$ denotes the number of components of a subset $M \subset R^n$ in its induced topology. So far, the above estimates have not been proved even for bicriteria LFVO problems.

It is worthy to observe that the sets $\text{Sol}(VP)$ and $\text{Sol}^w(VP)$ may be not contractible, even in the case they are arcwise connected.

Example 8.3. (Huy and Yen (2004b)) Consider problem (VP), where

$$\Delta = \{x = (x_1, x_2, x_3) : \quad x_1 + x_2 + x_3 \geq 2, \ x_1 + x_2 - 2x_3 \leq 2,$$
$$x_1 - 2x_2 + x_3 \leq 2, \ -2x_1 + x_2 + x_3 \leq 2\},$$

$$f_i(x) = \frac{-x_i}{x_1 + x_2 + x_3 - 1} \qquad (i = 1, 2, 3).$$

Using Theorem 8.1 one can prove that the sets $\text{Sol}(VP)$ and $\text{Sol}^w(VP)$ coincide with the surrounding surface of the parallelepiped Δ; that is

$$\text{Sol}^w(VP) = \text{Sol}(VP) = F_1 \cup F_2 \cup F_3,$$

where

$$F_1 = \{x : x_1 = 2 - x_2 + 2x_3, \ x_3 \geq 0, \ x_3 \leq x_2 \leq 2 + x_3\}$$
$$F_2 = \{x : x_2 = 2 - x_3 + 2x_1, \ x_1 \geq 0, \ x_1 \leq x_3 \leq 2 + x_1\}$$
$$F_3 = \{x : x_3 = 2 - x_1 + 2x_2, \ x_2 \geq 0, \ x_2 \leq x_1 \leq 2 + x_2\}.$$

8.3   Stability of the Solution Sets

Since problem (VP) depends on the parameter $\omega = (A, B, \alpha, \beta)$, in this section it is convenient for us to rename the sets Sol(VP) and Sol^w(VP) into $E(\omega)$ and $E^w(\omega)$, respectively. The solution set of the problem $(VI)_\lambda$ is now denoted by $F(\omega, \lambda)$.

Theorem 8.5. *If Δ is compact, then the multifunction $\omega' \mapsto E^w(\omega')$ is upper semicontinuous at ω.*

Proof. Since Δ is compact and nonempty, $E^w(\omega)$ is a nonempty compact subset of Δ. Suppose that $\Omega \subset R^n$ is an open set containing $E^w(\omega)$. Choose $\delta > 0$ so that

$$E^w(\omega) + \delta \bar{B}_{R^n} \subset \Omega. \tag{8.9}$$

For each $y \in E^w(\omega)$, by Theorem 8.2 there exists $\lambda \in \Delta$ such that $y \in F(\omega, \lambda)$. By Theorem 7.6, there exist constants $\ell(\lambda) > 0$ and $\epsilon_1(\lambda) > 0$, such that $\ell(\lambda)\epsilon_1(\lambda) < 2^{-1/2}\delta$ and

$$F(\omega', \lambda') \subset F(\omega, \lambda) + \ell(\lambda)\|(\omega', \lambda') - (\omega, \lambda)\|\bar{B}_{R^n} \tag{8.10}$$

for all $\omega' = (A', B', \alpha', \beta')$ and $\lambda' \in \Lambda$ satisfying

$$\|\omega' - \omega\| < \epsilon_1(\lambda), \quad \|\lambda' - \lambda\| < \epsilon_1(\lambda).$$

By definition,

$$\|A\| = \max\{\|Av\| \, : \, v \in \bar{B}_{R^m}\}$$

and

$$\begin{aligned}
\|(\omega', \lambda') - (\omega, \lambda)\| &= (\|\omega' - \omega\|^2 + \|\lambda' - \lambda\|^2)^{1/2} \\
&= \Big(\|A' - A\|^2 + \|B' - B\|^2 \\
&\quad + \|\alpha' - \alpha\|^2 + \|\beta' - \beta\|^2 + \|\lambda' - \lambda\|^2\Big)^{1/2}.
\end{aligned}$$

Since Λ is compact and the family $\{U(\lambda)\}_{\lambda \in \Lambda}$, where

$$U(\lambda) := \{\lambda' \in \Lambda \, : \, \|\lambda' - \lambda\| < \epsilon_1(\lambda)\},$$

is an open covering of Λ, there exists a finite sequence $\lambda^{(1)}, \ldots, \lambda^{(k)} \in \Lambda$ such that

$$\Lambda \subset U(\lambda^{(1)}) \cup \ldots \cup U(\lambda^{(k)}). \tag{8.11}$$

Let

$$\varepsilon = \min\{\varepsilon_1(\lambda^{(i)}) \, : \, i = 1, \ldots, k\}.$$

Fix any $\omega' = (A', B', \alpha', \beta')$ satisfying $\|\omega' - \omega\| < \epsilon$. For every $y' \in E^w(\omega')$ there exists $\lambda' \in \Lambda$ such that $y' \in F(\omega', \lambda')$. By (8.11), there exists $i_0 \in \{1, \ldots, k\}$ such that $\lambda' \in U(\lambda^{(i_0)})$. By (8.10),

$$
\begin{aligned}
F(\omega', \lambda') \; &\subset F(\omega, \lambda^{(i_0)}) + \ell(\lambda^{(i_0)})\|(\omega', \lambda') - (\omega, \lambda^{(i_0)})\|\bar{B}_{R^n} \\
&\subset F(\omega, \lambda^{(i_0)}) + 2^{1/2}\ell(\lambda^{(i_0)})\epsilon_1(\lambda^{(i_0)})\bar{B}_{R^n} \\
&\subset F(\omega, \lambda^{(i_0)}) + \delta\bar{B}_{R^n}.
\end{aligned}
$$

Combining this with (8.9) we can deduce that $E(\omega') \subset \Omega$ for every ω' satisfying $\|\omega' - \omega\| < \delta$. The proof is complete. $\square$

If one can prove that there exists a finite upper bound for the family $\{\ell(\lambda)\}_{\lambda \in \Lambda}$ (see the preceding proof), then the multifunction $\omega' \mapsto E^w(\omega')$ is upper-Lipschitz at ω.

By giving a counterexample, Kum, Lee and Yen (2004) have shown that, even in the case where Δ is a compact polyhedral convex set, the multifunction $\omega' \mapsto E(\omega')$ may be not upper semicontinuous at ω. Nevertheless, from Theorem 8.5 it is easy to derive the following sufficient condition for the usc property of this multifunction.

Theorem 8.6. *If Δ is compact and if $E(\omega) = E^w(\omega)$, then the multifunction $\omega' \mapsto E(\omega')$ is upper semicontinuous at ω.*

8.4 Commentaries

The material presented in this chapter is adapted from Yen and Phuong (2000). We have seen that the results on the solution existence and the stability of monotone AVI problems in Chapters 6 and 7 are very useful for studying LFVO problems on compact constraint sets.

Theorem 8.3 solves a question discussed in the final part of Choo and Atkins (1983). This result is a special corollary of the theorem of Benoist (1998) which asserts that the efficient solution set of the problem of maximizing a vector function with strictly quasiconcave components over a convex compact set is connected. The proof given here is due to Yen and Phuong (2000). Note that Warburton (1983) has extended the result in Theorem 8.4 by proving that, in a vector maximization problem with continuous quasiconvex functions and a compact convex constraint set, the weakly efficient solution set is connected. Connectedness and contractibility of the efficient sets of quasiconcave vector maximization problems have been discussed intensively in the literature (see Warburton (1983), Schaible (1983), Choo et al. (1985), Luc (1987), Hu and Sun (1993), Wantao and Kunping (1993), Benoist (2001), Huy and Yen (2004a, 2004b), and the references therein).

The reader is referred to Bednardczuck (1995) and Penot and Sterna-Karwat (1986) for some general results on stability of vector optimization problems. Stability of the solution maps of LFVO problems on noncompact sets deserves further investigations. It seems to us that some ideas and techniques related to asymptotic cones and asymptotic functions from Auslender and Teboulle (2003) can be useful for this purpose.

Chapter 9

The Traffic Equilibrium Problem

A traffic network can be modelled in a form of a variational inequality. Solutions of the variational inequality correspond to the equilibrium flows on the traffic network. Variational inequality can be also a suitable model for studying other kinds of economic equilibria. The aim of this chapter is to discuss the variational inequality model of the traffic equilibrium problem. Later on, in Chapter 17, by using some results on solution sensitivity of convex QP problems we will establish a fact about the Lipschitz continuity of the equilibrium flow in a traffic network where the travel costs and the demands are subject to change.

9.1 Traffic Networks Equilibria

Consider a traffic system with several cities and many roads connecting them. Suppose that the technical conditions (capacity and quality of roads, etc.) are established. Assume that we know the demands for transportation of some kind of materials or goods between each pair of two cities. The system is well functioning if all these demands are satisfied. The aim of the owner of the network is to keep the system well functioning. The users (drivers, passengers, etc.) do not behave blindly. To go from A to B they will choose one of the roads leading them from A to B with the minimum cost. This natural law is known as the user-optimizing principle or the Wardrop principle. The traffic flow satisfying demands and this law is said to be an *equilibrium flow* of the network. By using this principle, in most of the cases, the owner can compute or estimate the traffic flow on every road. The owner can affect on the network, for example, by requiring high fees from the users of the good roads to force them to use also some roads of lower quality. In this way, a new equilibrium flow, which is more suitable in the opinion of the owner, can be reached.

Traffic network is an example of networks acting in accordance with the Wardrop equilibrium principle. Other examples can be telephone networks or computer networks.

As it was proved by Smith (1979) and Dafermos (1980), a traffic network can be modelled in a variational inequality.

Consider a graph $\mathcal{G}$ consisting of a set $\mathcal{N}$ of nodes and a set $\mathcal{A}$ of arcs. Every *arc* is a pair of two nodes. The inclusion $a \in \mathcal{A}$ means that a is an arc. A *path* is an ordered family of arcs $a_1, \ldots, a_m$, where the second node of a_s coincides with the first node of a_{s+1} for $s = 1, \ldots, m - 1$. We say that the path $\{a_1, \ldots, a_m\}$ connects the first node of a_1 with the second node of a_m.

Let I be a given set of the *origin-destination pairs* (OD−pairs, for brevity). Each OD−pair consists of two nodes: the origin (the first node of the pair) and the destination (the second node of the pair). Denote by P_i the family of all paths connecting the origin with the destination of an OD−pair $i \in I$. Let $P = \bigcup_{i \in I} P_i$ and let $|P|$ denote the number of elements of P.

A vector $v = (v_a : a \in \mathcal{A})$, where $v_a \geq 0$ for all $a \in \mathcal{A}$, is said to be a *flow* (or *flow on arcs*) on the graph. Each v_a indicates the amount of material flow on arc a.

Let there be given a vector function

$$c(v) = (c_a(v) : a \in \mathcal{A}),$$

where $c_a(v) \geq 0$ for all $a \in \mathcal{A}$. This function $c(\cdot)$, which maps $R^{|\mathcal{A}|}$ to $R^{|\mathcal{A}|}$, is called the *travel cost function*. Each number $c_a(v)$ is interpreted as the travel cost for *one unit* of material flow to go through an arc a provided that the flow v exists on the network. There are many examples explaining why the travel cost on one arc should depend on the flows on other arcs.

The travel cost on a path $p \in P_i$ ($i \in I$) is given by the formula

$$C_p(v) = \sum_{a \in p} c_a(v)$$

Let $C(v) = (C_p(v) : p \in P)$. For each $i \in I$, define the *minimum travel cost* $u_i(v)$ for the $OD-$pair i by setting

$$u_i(v) = \min\{C_p(v) : p \in P_i\}.$$

Obviously, $C_p(v) - u_i(v) \geq 0$ for each $i \in I$ and for each $p \in P_i$. Let $D = (\delta_{ap})$ be *the incidence matrix* of the relations "arcs-paths"; that is

$$\delta_{ap} = \begin{cases} 0 & \text{if } p \notin P \\ 1 & \text{if } p \in P \end{cases}$$

for all $a \in \mathcal{A}$ and $p \in P$.

It is natural to assume the fulfilment of the following *flow-invariant law:*

$$v_a = \sum_{p \in P} \delta_{ap} v_p. \tag{9.1}$$

Let there be given also a *vector of demands* $g = (g_i : i \in I)$. Every component g_i indicates the *demand* for an $OD-$pair i, that is the amount of the material flow going from the origin to the destination of the pair i. We say that a flow v on the network satisfies demands if

$$\sum_{p \in P_i} v_p = g_i \quad \forall i \in I. \tag{9.2}$$

Note that

$$\Delta := \left\{ v = (v_p : p \in P) \in R_+^{|P|} : \sum_{p \in P_i} v_p = g_i \quad \forall i \in I \right\} \tag{9.3}$$

(the set of flows satisfying demands) is a polyhedral convex set.

If there are given upper bounds $(\gamma_p : p \in P)$, $\gamma_p > 0$ for all $p \in P$, for the capacities of the arcs, then the set of flows satisfying demands is given by the formula

$$\Delta = \left\{ v \in R_+^{|P|} : \sum_{p \in P_i} v_p = g_i \ \forall i \in I, \ 0 \leq v_p \leq \gamma_p \ \forall p \in P \right\}. \tag{9.4}$$

In this case, Δ is a compact polyhedral convex set.

Definition 9.1. *A traffic network* $\{\mathcal{G}, I, c(v), g\}$ *consists of a graph* $\mathcal{G} = (\mathcal{N}, \mathcal{A})$, *a set* I *of* $OD-$*pairs, a travel cost function* $c(v) = (c_a(v) : a \in \mathcal{A})$, *and a vector of demands* $g = (g_i : i \in I)$.

The following *user-optimizing principle* was introduced by Wardrop (1952). This equilibrium condition explains how the equilibrium flow must depend on the travel cost function.

Definition 9.2. (The Wardrop principle) A flow $\bar{v}$ on the network $\{\mathcal{G}, I, c(v), g\}$ is said to be *an equilibrium flow* if it satisfies demands and, for each $i \in I$ and for each $p \in P_i$, it holds

$$C_p(\bar{v}) - u_i(\bar{v}) = 0 \quad \text{if} \quad \bar{v}_p > 0.$$

The above principle can be stated equivalently as follows: If $C_p(\bar{v})$ (the travel cost on path $p \in P_i$) is greater than $u_i(\bar{v})$ (the minimum travel cost for the $OD-$pair i) then $\bar{v}_p = 0$ (the flow on p is zero). It is important to stress that *the fact that the flow on p is zero does not imply that the flows on all the arcs of p are zeros!*

The problem of finding an equilibrium flow $\bar{v}$ on the given network $\{\mathcal{G}, I, c(v), g\}$ is called *the network equilibrium problem.*

9.2 Reduction of the Network Equilibrium Problem to a Complementarity Problem

Let

$$S(v) = \{C_p(v) - u_i(v) : p \in P_i, \ i \in I\}.$$

We see that the number of components of vector $S(v)$ is equal to $|P|$. Since $C_p(v) - u_i(v) \geq 0$ for all $p \in P_i$ and $i \in I$, we have $S(v) \geq 0$. Note that $v = (v_p : p \in P)$ is also a nonnegative vector.

Proposition 9.1. *A flow $\bar{v} \in \Delta$ on a network $\{\mathcal{G}, I, c(v), g\}$ is an equilibrium flow if and only if*

$$\begin{cases} \bar{v} \geq 0, \quad S(\bar{v}) \geq 0, \\ \sum_{p \in P_i} \bar{v}_p = g_i \quad \forall i \in I, \\ \langle S(\bar{v}), \bar{v} \rangle = 0. \end{cases} \tag{9.5}$$

Proof. Suppose that $\bar{v} = (\bar{v}_p : p \in P)$ is a flow satisfying (9.5). Since $\bar{v} \geq 0$ and $S(\bar{v}) \geq 0$, the equality $\langle S(\bar{v}), \bar{v} \rangle = 0$ is equivalent to

$$\bar{v}_p(C_p(\bar{v}) - u_i(\bar{v})) = 0 \quad \forall p \in P_i, \quad \forall i \in I.$$

Therefore, for each $i \in I$ and for each $p \in P_i$, if $\bar{v}_p > 0$ then $C_p(\bar{v}) - u_i(\bar{v}) = 0$. This means that the Wardrop principle is satisfied. Conversely, if $\bar{v} = (\bar{v}_p : p \in P)$ is an equilibrium flow then it is easy to show that (9.5) holds. $\square$

Note that (9.5) is a (generalized) nonlinear complementarity problem under a polyhedral convex set constraint of the form

$$v \in \Delta, \quad f(v) \geq 0, \quad v^T f(v) = 0,$$

where $f(v) := S(v)$ and $\Delta \subset R_+^{|P|}$.

9.3 Reduction of the Network Equilibrium Problem to a Variational Inequality

Consider the incidence matrix $B = (\beta_{ip})$ of the relations "path-OD-pair", where

$$\beta_{ip} = \begin{cases} 1 & \text{if} \quad p \in P_i \\ 0 & \text{if} \quad p \notin P_i. \end{cases}$$

Note that a flow v satisfies demands if and only if

$$Bv = g. \tag{9.6}$$

Indeed, (9.6) means that $(Bv)_i = g_i$ for all $i \in I$. The latter is equivalent to (9.2). Let Δ be defined by (9.3) or (9.4).

The next proposition, which is due to Smith (1979) and Dafermos (1980), reduces the network equilibrium problem to a variational inequality. The proof of below is taken from De Luca and Maugeri (1989).

Proposition 9.2. *A flow $\bar{v} \in \Delta$ is an equilibrium flow of the network $\{\mathcal{G}, I, c(v), g\}$ if and only if*

$$\langle C(\bar{v}), v - \bar{v} \rangle \geq 0 \quad \forall v \in \Delta. \tag{9.7}$$

Proof. *Necessity:* Let $\bar{v} \in \Delta$ be an equilibrium flow. Let $v \in \Delta$. Define

$$P_i^{(1)} = \{p \in P_i : C_p(\bar{v}) = u_i(\bar{v})\},$$

$$P_i^{(2)} = \{p \in P_i \; : \; C_p(\bar{v}) > u_i(\bar{v})\}.$$

According to the Wardrop principle, we have

$$
\begin{aligned}
&\langle C(v), v - \bar{v} \rangle \\
&= \sum_{p \in P} C_p(\bar{v})(v_p - \bar{v}_p) \\
&= \sum_{i \in I} \left(\sum_{p \in P_i} C_p(\bar{v})(v_p - \bar{v}_p) \right) \\
&= \sum_{i \in I} \left(\sum_{p \in P_i^{(1)}} C_p(\bar{v})(v_p - \bar{v}_p) + \sum_{p \in P_i^{(2)}} C_p(\bar{v})(v_p - \bar{v}_p) \right) \\
&\geq \sum_{i \in I} \left(\sum_{p \in P_i^{(1)}} u_i(\bar{v})(v_p - \bar{v}_p) + \sum_{p \in P_i^{(2)}} u_i(\bar{v})v_p \right) \\
&= \sum_{i \in I} u_i(\bar{v}) \left(\sum_{p \in P_i^{(1)}} (v_p - \bar{v}_p) + \sum_{p \in P_i^{(2)}} v_p \right) \\
&= \sum_{i \in I} u_i(v)(g_i - g_i) \\
&= 0.
\end{aligned}
$$

So, (9.7) is valid.

Sufficiency: Let $\bar{v} \in \Delta$ be such that (9.7) holds. It suffices to show that if for an $OD-$pair $i_0 \in I$ there exist two paths $\hat{p} \in P_{i_0}, \widetilde{p} \in P_{i_0}$ such that

$$C_{\hat{p}}(\bar{v}) > C_{\widetilde{p}}(\bar{v}),$$

then $\bar{v}_{\widetilde{p}} = 0$. Consider vector $v = (v_p \; : \; p \in P)$ defined by

$$
v_p = \begin{cases}
\bar{v}_p & \text{if} \quad p \notin \{\hat{p}, \widetilde{p}\}, \\
\bar{v}_{\hat{p}} + \bar{v}_{\widetilde{p}} & \text{if} \quad p = \widetilde{p}, \\
0 & \text{if} \quad p = \hat{p}.
\end{cases}
$$

We have $v \geq 0$ and

$$\sum_{i \in P_i} v_p = g_i \quad \forall i \in I.$$

Indeed, if $i \neq i_0$ then the equality is obvious. If $i = i_0$ then we have

$$
\begin{aligned}
\sum_{p \in P_{i_0}} v_p &= \sum_{p \in P_{i_0} \setminus \{\hat{p}, \tilde{p}\}} + (\bar{v}_{\hat{p}} + \bar{v}_{\tilde{p}}) \\
&= \sum_{p \in P_{i_0}} \bar{v}_p \\
&= g_{i_0}.
\end{aligned}
$$

Hence $v \in \Delta$. From (9.7) it follows that

$$
\begin{aligned}
0 \leq \langle C(\bar{v}), v - \bar{v} \rangle &= \sum_{p \in P} C_p(\bar{v})(v_p - \bar{v}_p) \\
&= C_{\hat{p}}(\bar{v})(v_{\hat{p}} - \bar{v}_{\hat{p}}) + C_{\tilde{p}}(\bar{v})(v_{\tilde{p}} - \bar{v}_{\tilde{p}}) \\
&= -C_{\hat{p}}(\bar{v})\bar{v}_{\hat{p}} + C_{\tilde{p}}(\bar{v})\bar{v}_{\hat{p}} \\
&= -\bar{v}_{\hat{p}}(C_{\hat{p}}(\bar{v}) - C_{\tilde{p}}(\bar{v})).
\end{aligned}
$$

Since $C_{\hat{p}}(\bar{v}) - C_{\tilde{p}}(\bar{v}) > 0$, we have $\bar{v}_{\hat{p}} = 0$. $\quad\square$

We proceed to show that (9.7) can be expressed as a variational inequality on the set of flows on arcs.

According to (9.1), $v_A = Dv$ for every $v \in \Delta$. Therefore, the set Z of flows on arcs can be defined as follows

$$
\begin{aligned}
Z &= \{z : z = Dv, \ v \in \Delta\} \\
&= \{z : z = Dv, \ Bv = g, \ v \geq 0\}.
\end{aligned} \tag{9.8}
$$

Proposition 9.3. *A flow* $\bar{v}_A = (\bar{v}_a : a \in A) \in Z$ *is corresponding to an equilibrium flow of the network* $\{\mathcal{G}, I, c(v), g\}$ *if and only if*

$$
\langle c(\bar{v}), v_A - \bar{v}_A \rangle \geq 0, \tag{9.9}
$$

for all $v_A = (v_a : a \in A) \in Z$.

Proof. Recall that $c(v)$ is the vector of travel costs on arcs. By definition of the matrix $D = (\delta_{ap})$, we have

$$
\begin{aligned}
\langle c(\bar{v}), v_{\mathcal{A}} - \bar{v}_{\mathcal{A}} \rangle &= \sum_{a \in \mathcal{A}} c_a(v)(v_a - \bar{v}_a) \\
&= \sum_{a \in \mathcal{A}} c_a(\bar{v}) \left(\sum_{p \in P} \delta_{ap} v_p - \sum_{p \in P} \delta_{ap} \bar{v}_p \right) \\
&= \sum_{a \in \mathcal{A}} \left(\sum_{p \in P} (c_a(\bar{v}) \delta_{ap})(v_p - \bar{v}_p) \right) \\
&= \sum_{p \in P} \left(\sum_{a \in \mathcal{A}} (c_a(\bar{v}) \delta_{ap})(v_p - \bar{v}_p) \right) \\
&= \sum_{p \in P} C_p(\bar{v})(v_p - \bar{v}_p) \\
&= \langle C(\bar{v}), v - \bar{v} \rangle.
\end{aligned}
$$

The proof is completed by using (9.8), Proposition 9.2, and the equality

$$
\langle c(\bar{v}), v_{\mathcal{A}} - \bar{v}_{\mathcal{A}} \rangle = \langle C(\bar{v}), v - \bar{v} \rangle
$$

which holds for every $v \in \Delta$. □

The variational inequality in (9.9), in some sense, is simpler than that one in (9.7). Both of them are variational inequalities on polyhedral convex sets, but the constraint set of (9.9) usually has a smaller dimension. Besides, in most of the cases we can assume that $c(v)$ is a *locally strongly monotone function* (see Chapter 17), while we cannot do so for $C(v)$.

9.4   Commentaries

The material of this chapter is adapted from the reports of Yen and Zullo (1992), Chen and Yen (1993).

Numerical methods for solving the traffic equilibrium problem can be seen in Patriksson (1999). Various network equilibrium problems leading to finite-dimensional variational inequalities are discussed in Nagurney (1993).

Chapter 10

Upper Semicontinuity of the KKT Point Set Mapping

We have studied QP problems in Chapters 1–4. Studying various stability aspects of QP programs is an interesting topic. Although the general stability theory in nonlinear mathematical programming is applicable to convex and nonconvex QP problems, the specific structure of the latter allows one to have more complete results.

In this chapter we obtain some conditions which ensure that a small perturbation in the data of a quadratic programming problem can yield only a small change in its Karush-Kuhn-Tucker point set. Convexity of the objective function and boundedness of the constraint set are not assumed. Obtaining *necessary* conditions for the upper semicontinuity of the KKT point set mapping will be our focus point. Sufficient conditions for the upper semicontinuity of the mapping will be developed on the framework of the obtained necessary conditions.

10.1 KKT Point Set of the Canonical QP Problems

Here we study QP problems of the canonical form:

$$\begin{cases} \text{Minimize} \quad f(x) := \dfrac{1}{2}x^T D x + c^T x \\ \text{subject to} \quad x \in \Delta(A; b) := \{x \in R^n \ : \ Ax \geq b, \ x \geq 0\}, \end{cases} \tag{10.1}$$

where $D \in R_S^{n \times n}$, $A \in R^{m \times n}$, $c \in R^n$ and $b \in R^m$ are given data. In the sequel, sometime problem (10.1) will be referred to as $QP(D, A, c, b)$.

Recall that $\bar{x} \in R^n$ is a *Karush-Kuhn-Tucker point* of (10.1) if there exists a vector $\bar{\lambda} \in R^m$ such that

$$\begin{cases} D\bar{x} - A^T\bar{\lambda} + c \geq 0, & A\bar{x} - b \geq 0, \\ \bar{x} \geq 0, & \bar{\lambda} \geq 0, \\ \bar{x}^T(D\bar{x} - A^T\bar{\lambda} + c) + \bar{\lambda}^T(A\bar{x} - b) = 0. \end{cases} \tag{10.2}$$

The set of all the Karush-Kuhn-Tucker points of (10.1) is denoted by $S(D, A, c, b)$. In Chapter 3, we have seen that if $\bar{x}$ is a local solution of (10.1) then $\bar{x} \in S(D, A, c, b)$. This fact leads to the following standard way to solve (10.1): *Find first the set $S(D, A, c, b)$ then compare the values $f(x)$ among the points $x \in S(D, A, c, b)$.* Hence, one may wish to have some criteria for the (semi)continuity of the following multifunction

$$(D, A, c, b) \mapsto S(D, A, c, b). \tag{10.3}$$

In Section 10.2 we will obtain a necessary condition for the *upper semicontinuity* of the multifunction $S(\cdot, \cdot, c, b)$ at a given point $(D, A) \in R_S^{n \times n} \times R^{m \times n}$. In Section 10.3 we study a special class of QP problems for which the necessary condition obtained in this section is also a sufficient condition for the usc property of the multifunction in (10.3). This class contains some nonconvex QP problems. Sections 10.4 and 10.5 are devoted to sufficient conditions for the usc property of the multifunction in (10.3). In Section 10.5 we will investigate some questions concerning the usc property of the KKT point set mapping in a general QP problem.

Note that the upper Lipschitz property of the multifunction $S(D, A, \cdot, \cdot)$ with respect to the parameters (c, b) is a direct consequence of Theorem 7.3 in Chapter 7.

Since (10.2) can be rewritten as a linear complementarity problem, the study of continuity of the multifunction (10.3) is closely related to the study of continuity and stability of the solution map in linear complementarity theory (see Jansen and Tijs (1987), Cottle et al. (1992), Gowda (1992), Gowda and Pang (1992, 1994a)). However, when the data of (10.1) are perturbed, only some components of the matrix $M = M(D, A)$ (see formula (10.18) below) are perturbed. So, necessary conditions for (semi)continuity and stability of the Karush-Kuhn-Tucker point set cannot be derived from the corresponding results in linear complementarity theory (see, for example, Gowda and Pang (1992)) where all the components of M are perturbed.

10.2 A Necessary Condition for the usc Property of $S(\cdot)$

We now obtain a necessary condition for $S(\cdot, \cdot, c, b)$ to be upper semicontinuous at a given pair $(D, A) \in R_S^{n \times n} \times R^{m \times n}$.

Theorem 10.1. *Assume that the set $S(D, A, c, b)$ is bounded. If the multifunction $S(\cdot, \cdot, c, b)$ is upper semicontinuous at (D, A), then*

$$S(D, A, 0, 0) = \{0\}. \tag{10.4}$$

Proof. Arguing by contradiction, we assume that $S(D, A, c, b)$ is bounded, the multifunction $S(\cdot, \cdot, c, b)$ is usc at (D, A), but (10.4) is violated. The latter means that there is a nonzero vector $\hat{x} \in S(D, A, 0, 0)$. Hence there exists $\hat{\lambda} \in R^m$ such that

$$D\hat{x} - A^T \hat{\lambda} \geq 0, \quad A\hat{x} \geq 0, \tag{10.5}$$

$$\hat{x} \geq 0, \quad \hat{\lambda} \geq 0, \tag{10.6}$$

$$\hat{x}^T D \hat{x} = 0. \tag{10.7}$$

Setting

$$x_t = \frac{1}{t}\hat{x}, \quad \lambda_t = \frac{1}{t}\hat{\lambda}, \quad \text{for every } t \in (0, 1), \tag{10.8}$$

we claim that there exist matrices $D_t \in R_S^{n \times n}$ and $A_t \in R^{m \times n}$ such that $D_t \to D$, $A_t \to A$ as $t \to 0$, and

$$D_t x_t - A_t^T \lambda_t + c \geq 0, \quad A_t x_t - b \geq 0, \tag{10.9}$$

$$x_t > 0, \quad \lambda_t \geq 0, \tag{10.10}$$

$$x_t^T (D_t x_t - A_t^T \lambda_t + c) + \lambda_t^T (A_t x_t - b) = 0. \tag{10.11}$$

Matrices D_t and A_t will be of the form

$$D_t = D + tD_0, \quad A_t = A + tA_0, \tag{10.12}$$

where matrices D_0 and A_0 are to be constructed. Since

$$
\begin{aligned}
D_t x_t - A_t^T \lambda_t + c &= \frac{1}{t}(D + tD_0)\hat{x} - \frac{1}{t}(A^T + tA_0^T)\hat{\lambda} + c \\
&= \frac{1}{t}(D\hat{x} - A^T\hat{\lambda}) + D_0\hat{x} - A_0^T\hat{\lambda} + c,
\end{aligned}
$$

and

$$
\begin{aligned}
A_t x_t - b &= \frac{1}{t}(A + tA_0)\hat{x} - b \\
&= \frac{1}{t}A\hat{x} + A_0\hat{x} - b,
\end{aligned}
$$

the following conditions, due to (10.5), imply (10.9):

$$
D_0\hat{x} - A_0^T\hat{\lambda} + c \geq 0, \quad A_0\hat{x} - b \geq 0. \tag{10.13}
$$

As $x_t = \frac{1}{t}\hat{x}$ and $\lambda_t = \frac{1}{t}\hat{\lambda}$, (10.6) implies (10.10). Taking account of (10.7), we have

$$
\begin{aligned}
& x_t^T(D_t x_t - A_t^T \lambda_t + c) + \lambda_t^T(A_t x_t - b) \\
&= \frac{1}{t}\hat{x}^T\left[\frac{1}{t}(D\hat{x} - A^T\hat{\lambda}) + D_0\hat{x} - A_0^T\hat{\lambda} + c\right] \\
&\quad + \frac{1}{t}\hat{\lambda}^T\left(\frac{1}{t}A\hat{x} + A_0\hat{x} - b\right) \\
&= \frac{1}{t^2}\left(\hat{x}^T D\hat{x} - \hat{x}^T A^T\hat{\lambda} + \hat{\lambda}^T A\hat{x}\right) \\
&\quad + \frac{1}{t}\hat{x}^T\left(D_0\hat{x} - A_0^T\hat{\lambda} + c\right) + \frac{1}{t}\hat{\lambda}^T(A_0\hat{x} - b) \\
&= \frac{1}{t}\left[\hat{x}^T(D_0\hat{x} - A_0^T\hat{\lambda} + c) + \hat{\lambda}^T(A_0\hat{x} - b)\right].
\end{aligned}
$$

So the following equality implies (10.11):

$$
\hat{x}^T(D_0\hat{x} - A_0^T\hat{\lambda} + c) + \hat{\lambda}^T(A_0\hat{x} - b) = 0. \tag{10.14}
$$

Let $\hat{x} = (\hat{x}_1, \ldots, \hat{x}_n)$, where $\hat{x}_i > 0$ for $i \in I \subset \{1, \ldots, n\}$, and $\hat{x}_i = 0$ for $i \notin I$. Since $\hat{x} \neq 0$, I must be nonempty. Fixing an $i_0 \in I$, we define A_0 as the $m \times n$-matrix whose i_0-th column is $\hat{x}_0^{-1}b$, and whose other columns consist solely of zeros. For this A_0 we have $A_0\hat{x} - b = 0$, hence the second inequality in (10.13) is satisfied, and condition (10.14) becomes the following one:

$$
\hat{x}^T(D_0\hat{x} - A_0^T\hat{\lambda} + c) = 0.
$$

We have to find a matrix $D_0 \in R_S^{n \times n}$ such that this condition and the first inequality in (10.13) are valid. For this purpose it is enough to find a symmetric matrix D_0 such that

$$D_0 \hat{x} - w = 0, \tag{10.15}$$

where $w := A_0^T \hat{\lambda} - c \in R^n$.

If $w = (w_1, \ldots, w_n)$, then we put $D_0 = (d_{ij})$, $1 \leq i, j \leq n$, where

$$\begin{aligned} d_{ii} &:= \hat{x}_i^{-1} w_i & \text{for all } i \in I, \\ d_{i_0 j} = d_{j i_0} &:= \hat{x}_{i_0}^{-1} w_j & \text{for all } j \in \{1, 2, \ldots, n\} \setminus I, \end{aligned}$$

and

$$d_{ij} := 0 \quad \text{for other pairs } (i, j), \ 1 \leq i, j \leq n.$$

A simple direct computation shows that this symmetric matrix D_0 satisfies (10.15).

We have thus constructed matrices A_0 and D_0 such that for x_t, λ_t, D_t and A_t defined by (10.8) and (10.12), the conditions (10.9)–(10.11) are satisfied. As a consequence, $x_t \in S(D_t, A_t, c, b)$. Since $S(D, A, c, b)$ is a bounded set, there exists a bounded open set Ω such that $S(D, A, c, b) \subset \Omega$. Since $D_t \to D$ and $A_t \to A$ as $t \to 0$, and the multifunction $S(\cdot, \cdot, c, b)$ is usc at (D, A), we have $x_t \in \Omega$ for all t sufficiently small. This is a contradiction, because $\|x_t\| = \dfrac{1}{t} \|\hat{x}\| \to \infty$ as $t \to 0$. The proof is complete. $\square$

Observe also that, in general, (10.4) is not a sufficient condition for the upper semicontinuity of $S(\cdot)$ at (D, A, c, b).

Example 10.1. Consider the problem $QP(D, A, c, b)$ where

$$D = \begin{bmatrix} -1 & 0 \\ 0 & -1 \end{bmatrix}, \quad A = [0, \ -1], \quad b = (-1), \quad c = (0, 0).$$

For each $t \in (0, 1)$, let $A_t = [-t, \ -1]$. By direct computation using (10.2) we obtain

$$S(D, A, 0, 0) = \{0\}, \quad S(D, A, c, b) = \{(0, 0), \ (0, 1)\},$$
$$S(D, A_t, c, b) = \left\{ (0, 0), \ (0, 1), \ \left(\frac{1}{t}, 0 \right), \ \left(\frac{t}{t^2 + 1}, \frac{1}{t^2 + 1} \right) \right\}.$$

Thus, for any bounded open set $\Omega \subset R^2$ containing $S(D, A, c, b)$, the inclusion

$$S(D, A_t, c, b) \subset \Omega$$

fails to hold for $t > 0$ small enough. Since $A_t \to A$ as $t \to 0$, $S(\cdot)$ cannot be usc at (D, A, c, b).

In the next section we will study a special class of quadratic programs for which (10.4) is not only a necessary but also a sufficient condition for the upper semicontinuity of $S(\cdot)$ at a given point (D, A, c, b).

10.3   A Special Case

We now study those canonical QP problems for which the following condition (H) holds:

(H) *There exists $\bar{x} \in R^n$ such that $A\bar{x} > 0$, $\bar{x} \geq 0$.*

Denote by $\mathcal{H}$ the set of all the matrices $A \in R^{m \times n}$ satisfying (H).

The next statement can be proved easily by applying Lemma 3 from Robinson (1977) and the Farkas Lemma (Theorem 3.2).

Lemma 10.1. *Each one of the following two conditions is equivalent to* (H):

 (i) *There exists $\delta > 0$ such that, for every matrix A' satisfying $\|A' - A\| < \delta$ and for every $b \in R^n$, the system $A'x \geq b$, $x \geq 0$ is solvable.*

 (ii) *For any $\lambda \in R^n$, if*

$$-A^T\lambda \geq 0, \quad \lambda \geq 0, \tag{10.16}$$

  then $\lambda = 0$.

Obviously, (H) implies the existence of an $\hat{x} \in R^n$ satisfying $A\hat{x} > 0$, $\hat{x} > 0$. Thus $\Delta(A, 0)$ has nonempty interior. Now suppose that (H) is fulfilled and $b \in R^n$ is an arbitrarily chosen vector. Since $\Delta(A, b) + \Delta(A, 0) \subset \Delta(A, b)$ and, by Lemma 10.1, $\Delta(A, b)$ is nonempty, we conclude that $\Delta(A, b)$ is an unbounded set with nonempty interior. Besides, it is clear that there exists $\tilde{x} \in R^n$ satisfying

$$A\tilde{x} > b, \quad \tilde{x} > 0.$$

The latter property is a specialization of the Slater constraint qualification (Mangasarian (1969), p. 78), and the Mangasarian-Fromovitz constraint qualification (called by Mangasarian the modified Arrow-Hurwicz-Uzawa constraint qualification) (Mangasarian

(1969), pp. 172-173). These well-known constraint qualifications play an important role in the stability analysis of nonlinear optimization problems.

In the sequence, the inequality system $Ax \geq b$, where $A \in R^{m \times n}$ and $b \in R^m$, is said to be *regular* if there exists $x^0 \in R^n$ such that $Ax^0 > b$.

As it has been noted in Section 5.4, a pair $(\bar{x}, \bar{\lambda}) \in R^n \times R^m$ satisfies (10.2) if and only if $\bar{z} := \begin{pmatrix} \bar{x} \\ \bar{\lambda} \end{pmatrix}$ is a solution to the following linear complementarity problem

$$Mz + q \geq 0, \quad z \geq 0, \quad z^T(Mz + q) = 0, \tag{10.17}$$

where

$$M = M(D, A) := \begin{pmatrix} D & -A^T \\ A & 0 \end{pmatrix}, \quad q := \begin{pmatrix} c \\ -b \end{pmatrix}, \quad z = \begin{pmatrix} x \\ \lambda \end{pmatrix} \in R^{n+m}. \tag{10.18}$$

Denoting by $\mathrm{Sol}(M, q)$ the solution set of (10.17), we have

$$S(D, A, c, b) = \pi_1\left(\mathrm{Sol}(M, q)\right), \tag{10.19}$$

where $\pi_1 : R^{n+m} \to R^n$ is the linear operator defined by setting $\pi_1 \begin{pmatrix} x \\ \lambda \end{pmatrix} := x$ for every $\begin{pmatrix} x \\ \lambda \end{pmatrix} \in R^{m+n}$.

The notion of R_0-matrix, which is originated to Garcia (1973), has proved to be useful in characterizing the upper semicontinuity property of the solution set of linear complementarity problems (see Jansen and Tijs (1987), Cottle et al. (1992), Gowda (1992), Gowda and Pang (1992), Oettli and Yen (1995, 1996a, 1996b)), and in studying other questions concerning these problems (see Cottle et al. (1992)). R_0-matrices are called also pseudo-regular matrices (Gowda and Pang (1992), p. 78).

Definition 10.1. (See Cottle et al. (1992), Definition 3.8.7) A matrix $M \in R^{p \times p}$ is called an R_0-matrix if the linear complementarity problem

$$Mz \geq 0, \quad z \geq 0, \quad z^T Mz = 0, \quad (z \in R^p),$$

has the unique solution $z = 0$.

Theorem 10.2. *Assume that $A \in \mathcal{H}$ and that $S(D, A, c, b)$ is bounded. If the multifunction $S(\cdot, \cdot, c, b)$ is upper semicontinuous at (D, A), then $M(D, A)$ is an R_0-matrix.*

Proof. Since $S(D, A, c, b)$ is bounded and $S(\cdot, \cdot, c, b)$ is usc at (D, A), by Theorem 10.1, (10.4) holds. Let $\hat{z} = \begin{pmatrix} \hat{x} \\ \hat{\lambda} \end{pmatrix}$ be such that

$$M\hat{z} \geq 0, \quad \hat{z} \geq 0, \quad \hat{z}^T M \hat{z} = 0, \tag{10.20}$$

where $M = M(D, A)$. This means that the system (10.5)–(10.7) is satisfied. Hence, $\hat{x} \in S(D, A, 0, 0)$. Then $\hat{x} = 0$ by (10.4), and the system (10.5)–(10.7) implies

$$-A^T \hat{\lambda} \geq 0, \quad \hat{\lambda} \geq 0.$$

Since $A \in \mathcal{H}$, $\hat{\lambda} = 0$. Thus any $\hat{z}$ satisfying (10.20) must be zero. So $M(D, A)$ is an $\mathbf{R}_0$-matrix. □

Corollary 10.1. *Let $A \in \mathcal{H}$. If for every $(c, b) \in R^n \times R^m$ the multifunction $S(\cdot, \cdot, c, b)$ is upper semicontinuous at (D, A), then $M(D, A)$ is an $\mathbf{R}_0$-matrix.*

Proof. Consider problem (10.17), where $M = M(D, A)$ and q are defined via D, A, c, b by (10.18). Lemma 1 from Oettli and Yen (1995) shows that there exists $\bar{q} \in R^{n+m}$ such that $\text{Sol}(M, \bar{q})$ is bounded. If $(\bar{c}, \bar{b}) \in R^n \times R^m$ is the pair satisfying $\bar{q} = \begin{pmatrix} \bar{c} \\ -\bar{b} \end{pmatrix}$, then it follows from (10.19) that $S(D, A, \bar{c}, \bar{b})$ is bounded. Since $S(\cdot, \cdot, \bar{c}, \bar{b})$ is usc at (D, A), $M(D, A)$ is an $\mathbf{R}_0$-matrix by Theorem 10.2. □

The following statement gives a sufficient condition for the usc property of the multifunction $S(\cdot)$.

Theorem 10.3. *If $M(D, A)$ is an $\mathbf{R}_0$-matrix, then for any $(c, b) \in R^n \times R^m$ the set $S(D, A, c, b)$ is bounded, and the multifunction $S(\cdot)$ is upper semicontinuous at (D, A, c, b). If, in addition, $S(D, A, c, b)$ is nonempty, then there exist constants $\gamma > 0$ and $\delta > 0$ such that*

$$\begin{aligned} S(D', A', c', b') &\subset S(D, A, c, b) \\ &+\gamma(\|D' - D\| + \|A' - A\| + \|c' - c\| + \|b' - b\|)B_{R^n}, \end{aligned} \tag{10.21}$$

for all $(c', b') \in R^n \times R^m$, $D' \in R^{n \times n}$ and $A' \in R^{m \times n}$ satisfying $\|D' - D\| < \delta$, $\|A' - A\| < \delta$.

Proof. Since $M(D, A)$ is an $\mathbf{R}_0$-matrix, by Proposition 5.1 and Theorem 5.6 in Jansen and Tijs (1987) and the remarks before Theorem 2 of in Gowda (1992), $\text{Sol}(M, q)$ is a bounded set, and the solution map $\text{Sol}(\cdot)$ is usc at (M, q). It follows from (10.19) that

$S(D, A, c, b)$ is bounded. Let $\Omega \subset R^n$ be an arbitrary open set containing $S(D, A, b, c)$. By the upper semicontinuity of $\text{Sol}(\cdot)$ at (M, q), we have

$$\text{Sol}(M', q') \subset \Omega \times R^m, \qquad (10.22)$$

for all (M', q') in a neighborhood of (M, q). Using (10.19) and (10.22) we get $S(D', A', c', b') \subset \Omega$, for all (D', A', c', b') in a neighborhood of (D, A, c, b).

The upper Lipschitz property described in (10.21) follows from a result of Gowda (1992). Indeed, since $S(D, A, c, b)$ is nonempty, $\text{Sol}(M, q)$ is nonempty. Since M is an $\mathbf{R}_0$-matrix, by Theorem 9 of Gowda (1992) there exist γ_0 and δ_0 such that

$$\text{Sol}(M', q') \subset \text{Sol}(M, q) + \gamma_0(\|M' - M\| + \|q' - q\|) B_{R^{n+m}} \quad (10.23)$$

for all $q' \in R^{n+m}$ and for all $M' \in R^{(n+m) \times (n+m)}$ satisfying $\|M' - M\| < \delta_0$. The inclusion (10.21) follows easily from (10.23) and (10.19). $\quad \square$

Combining Theorem 10.3 with Corollary 10.1 we get the following result.

Corollary 10.2. *If $A \in \mathcal{H}$, then for every $(c, b) \in R^n \times R^m$ the multifunction $S(\cdot, \cdot, c, b)$ is upper semicontinuous at (D, A) if and only if $M(D, A)$ is an $\mathbf{R}_0$-matrix.*

We now find necessary and sufficient conditions for $M(D, A)$ to be an $\mathbf{R}_0$-matrix. By definition, $M = M(D, A)$ is an $\mathbf{R}_0$-matrix if and only if the system

$$D\hat{x} - A^T\hat{\lambda} \geq 0, \quad A\hat{x} \geq 0, \qquad (10.24)$$

$$\hat{x} \geq 0, \quad \hat{\lambda} \geq 0, \qquad (10.25)$$

$$\hat{x}^T D\hat{x} = 0 \qquad (10.26)$$

has the unique solution $(\hat{x}, \hat{\lambda}) = (0, 0)$.

Proposition 10.1. *If $M = M(D, A)$ is an $\mathbf{R}_0$-matrix then $A \in \mathcal{H}$ and the following condition holds:*

$$\left[D\hat{x} \geq 0, \ A\hat{x} \geq 0, \ \hat{x} \geq 0, \ \hat{x}^T D\hat{x} = 0 \right] \implies \hat{x} = 0. \qquad (10.27)$$

Proof. If $\hat{\lambda} \in R^m$ is such that $-A^T\hat{\lambda} \geq 0$, $\hat{\lambda} \geq 0$, then $(0, \hat{\lambda})$ is a solution of the system (10.24)–(10.26). If M is an $\mathbf{R}_0$-matrix then we must have $\hat{\lambda} = 0$. By Lemma 10.1, $A \in \mathcal{H}$. Furthermore, for any

$\hat{x} \in R^n$ satisfying $D\hat{x} \geq 0$, $A\hat{x} \geq 0$, $\hat{x} \geq 0$ and $\hat{x}^T D\hat{x} = 0$, it is clear that $(\hat{x}, 0)$ is a solution of (10.24)–(10.26). If M is an $\mathbf{R}_0$-matrix then $(\hat{x}, 0) = (0, 0)$. We have thus proved (10.27). □

The above proposition shows that the inclusion $A \in \mathcal{H}$ and the property (10.27) are necessary conditions for $M = M(D, A)$ to be an $\mathbf{R}_0$-matrix. Sufficient conditions for $M = M(D, A)$ to be an $\mathbf{R}_0$-matrix are given in the following proposition. Recall that a matrix is said to be *nonnegative* if each of its elements is a nonnegative real number.

Proposition 10.2. *Assume that $A \in \mathcal{H}$. The following properties hold:*

(i) *If A is a nonnegative matrix and D is an $\mathbf{R}_0$-matrix then $M(D, A)$ is an $\mathbf{R}_0$-matrix.*

(ii) *If D a positive definite or a negative definite matrix, then $M(D, A)$ is an $\mathbf{R}_0$-matrix.*

Proof. For proving (i), let D be an $\mathbf{R}_0$-matrix and let $(\hat{x}, \hat{\lambda})$ be a pair satisfying (10.24)–(10.26). Since A is a nonnegative matrix, the inequalities $D\hat{x} - A^T\hat{\lambda} \geq 0$ and $\hat{\lambda} \geq 0$ imply $D\hat{x} \geq A^T\hat{\lambda} \geq 0$. Hence (10.24)–(10.26) yield $D\hat{x} \geq 0$, $\hat{x} \geq 0$, $\hat{x}^T D\hat{x} = 0$. Since D is an $\mathbf{R}_0$-matrix, $\hat{x} = 0$. This fact and (10.24)–(10.26) imply $-A^T\hat{\lambda} \geq 0$, $\hat{\lambda} \geq 0$. Since $A \in \mathcal{H}$, $\hat{\lambda} = 0$ by Lemma 10.1. Thus $(\hat{x}, \hat{\lambda}) = (0, 0)$ is the unique solution of (10.24)–(10.26). Hence M is an $\mathbf{R}_0$-matrix. We omit the easy proof of (ii). □

Observe that in Proposition 10.2(i) the condition that A is a nonnegative matrix cannot be dropped.

Example 10.2. Let $n = 2$, $m = 1$, $D = \mathrm{diag}(1, -1)$, $A = (1, -1)$. It is clear that D is an $\mathbf{R}_0$-matrix and the condition $A \in \mathcal{H}$ is satisfied with $\bar{x} = \begin{pmatrix} 1 \\ 1/2 \end{pmatrix}$. Meanwhile, M is not an $\mathbf{R}_0$-matrix. Indeed, one can verify that the pair $(\hat{x}, \hat{\lambda})$, where $\hat{x} = (1, 1)$ and $\hat{\lambda} = 1$, is a solution of the system (10.24)–(10.26).

Definition 10.2 (Murty (1972), p. 67). We say that $D = (d_{ij}) \in R^{n \times n}$ is a *nondegenerate matrix* if, for any nonempty subset $\alpha \subset \{1, \dots, n\}$, the determinant of the principal submatrix $D_{\alpha\alpha}$ consisting of the elements d_{ij} $(i \in \alpha, j \in \alpha)$ of D is nonzero.

Every nondegenerate matrix is an $\mathbf{R}_0$-matrix (see Cottle et al. (1992), p. 180). It can be proved that the set of nondegenerate

$n \times n$-matrices is open and dense in $R^{n \times n}$. From the following simple observation it follows that symmetric nondegenerate $\mathbf{R}_0$-matrices form a dense subset in the set of all symmetric matrices.

Proposition 10.3. *For any matrix $D \in R^{n \times n}$ and for any $\epsilon > 0$ there exists a nonnegative diagonal matrix U^ϵ such that $D + U^\epsilon$ is a nondegenerate matrix, and $\|U^\epsilon\| \leq \epsilon$.*

Proof. The proposition is proved by induction on n. For $n = 1$, if $D = [d]$, $d \neq 0$, then we set $U^\epsilon = [0]$. If $D = [0]$ then we set $U^\epsilon = [\epsilon]$. Assume that the conclusion of the proposition is true for all indexes $n \leq k - 1$. Let $D = (d_{ij})$ be a $k \times k$-matrix which is *not* nondegenerate. Denote by D_{k-1} the left-top submatrix of the order $(k - 1) \times (k - 1)$ of D. By induction, there is a diagonal matrix $U^\epsilon_{k-1} = \text{diag}(\alpha_1, \ldots, \alpha_{k-1})$ such that every principal minor of the matrix $D_{k-1} + U^\epsilon_{k-1}$ is nonzero, and $\|U^\epsilon_{k-1}\| \leq \epsilon$. The required matrix U^ϵ is sought in the form

$$U^\epsilon = \text{diag}(\alpha_1, \ldots, \alpha_{k-1}, y),$$

where $y \in R$ is a parameter.

From the construction of U^ϵ it follows that all the determinants of the principal submatrices of $D + U^\epsilon$ which do not contain the element $d_{kk} + y$, are nonzero. Obviously, there are 2^{k-1} principal submatrices of $D + U^\epsilon$ containing the element $d_{kk} + y$. The determinant of each one of these submatrices has the form $\alpha_i y + \beta_i$, $1 \leq i \leq 2^{k-1}$, where α_i and β_i are certain real numbers. Moreover, α_i equals 1 or equals one of the principal minors of $D_{k-1} + U^\epsilon_{k-1}$. So $\alpha_i \neq 0$ for all i. Since the numbers $-\dfrac{\beta_i}{\alpha_i}$, $1 \leq i \leq 2^{k-1}$, cannot cover the segment $[0, \epsilon]$, there exists $\bar{y} \in [0, \epsilon]$ such that $\bar{y} \neq -\dfrac{\beta_i}{\alpha_i}$ for all i. From what has already been said, we conclude that for $U^\epsilon := \text{diag}(\alpha_1, \ldots, \alpha_{k-1}, \bar{y})$ the matrix $D + U^\epsilon$ is nondegenerate. In addition, it is clear that $\|U^\epsilon\| \leq \epsilon$. The proof is complete. $\square$

Remark 10.1. The property of being a nondegenerate matrix is not invariant under the operation of matrix conjugation. This means that even if D is nondegenerate and P is nonsingular, the matrix $P^{-1}DP$ still may have zero principal minors. Examples can be found even in $R^{2 \times 2}$. Consequently, a linear operator with a nondegenerate matrix in one basis may have a degenerate matrix in another basis.

It follows from Theorem 10.3 and Proposition 10.2 that the multifunction $S(\cdot)$ is usc at (D, A, c, b) if $A \in \mathcal{H}$, A is a nonnegative matrix and D is an $\mathbf{R}_0$-matrix. There are many nonconvex QP problems fulfilling these conditions. For example, in the quadratic programs whose objective functions are given by the formula

$$f(x) = c^T x + \sum_{i=1}^{s} \mu_i x_i^2 - \sum_{i=s+1}^{n} \mu_i x_i^2,$$

where $c \in R^n$, $1 \le s < n$ and $\mu_i > 0$ for all i, D is an $\mathbf{R}_0$-matrix. Proposition 10.3 shows that the set of symmetric $\mathbf{R}_0$-matrices is dense in $R_S^{n \times n}$.

10.4 Sufficient Conditions for the usc Property of $S(\cdot)$

Consider problem (10.1) whose Karush-Kuhn-Tucker point set is denoted by $S(D, A, c, b)$. A necessary condition for the usc property of $S(\cdot)$ was obtained in Section 10.2. Sufficient conditions for having that property were given in Section 10.3 only for a special class of QP problems. Our aim in this section is to find sufficient conditions for the usc property of the multifunction $S(\cdot)$ which are applicable for larger classes of QP problems.

For a matrix $A \in R^{m \times n}$, the dual of the cone

$$\Lambda[A] := \{\lambda \in R^m : -A^T \lambda \ge 0, \ \lambda \ge 0\}$$

is denoted by $(\Lambda[A])^*$. By definition, $(\Lambda[A])^* = \{\xi \in R^m : \lambda^T \xi \le 0 \ \forall \lambda \in \Lambda[A]\}$. The interior of $(\Lambda[A])^*$ is denoted by int $(\Lambda[A])^*$. By Lemma 6.4,

$$\text{int}\,(\Lambda[A])^* = \{\xi \in R^m : \lambda^T \xi < 0 \quad \forall \lambda \in \Lambda[A] \setminus \{0\}\}.$$

The proofs of Theorems 10.4–10.6 below are based on some observations concerning the structure of the Karush-Kuhn-Tucker system (10.2). It turns out that the desired stability property of the set $S(D, A, c, b)$ depends greatly on the behavior of the quadratic form $x^T D x$ on the recession cone of $\Delta(A, b)$ and also on the position of b with respect to the set int $(\Lambda[A])^*$.

One can note that in Example 10.1 the solution set Sol $(D, A, 0, 0)$ is empty. In the following theorem, such "abnormal" situation will be excluded.

Theorem 10.4. *If* $\mathrm{Sol}\,(D, A, 0, 0) = \{0\}$ *and if* $b \in \mathrm{int}\,(\Lambda[A])^*$ *then, for any* $c \in R^n$, *the multifunction* $S(\cdot)$ *is upper semicontinuous at* (D, A, c, b).

Proof. Suppose the theorem were false. Then we could find an open set Ω containing $S(D, A, c, b)$, a sequence $\{(D^k, A^k, c^k, b^k)\}$ converging to (D, A, c, b) in $R_S^{n \times n} \times R^{m \times n} \times R^n \times R^m$, a sequence $\{x^k\}$ with the property that $x^k \in S(D^k, A^k, c^k, b^k)$ and $x^k \notin \Omega$ for every k. By the definition of KKT point, there exists a sequence $\{\lambda^k\} \subset R^m$ such that

$$D^k x^k - (A^k)^T \lambda^k + c^k \geq 0, \quad A^k x^k - b^k \geq 0, \qquad (10.28)$$

$$x^k \geq 0, \quad \lambda^k \geq 0, \qquad (10.29)$$

$$(x^k)^T (D^k x^k - (A^k)^T \lambda^k + c^k) + (\lambda^k)^T (A^k x^k - b^k) = 0. \qquad (10.30)$$

We first consider the case where the sequence of norms $\{\|(x^k, \lambda^k)\|\}$ is bounded. As the sequences $\{\|x^k\|\}$ and $\{\|\lambda^k\|\}$ are also bounded, from $\{x^k\}$ and $\{\lambda^k\}$, respectively, one can extract converging subsequences $\{x^{k_i}\}$ and $\{\lambda^{k_i}\}$. Assume that $x^{k_i} \to x^0 \in R^n$ and $\lambda^{k_i} \to \lambda^0 \in R^m$ as $i \to \infty$. From (10.28)–(10.30) it follows that

$$\begin{aligned} &Dx^0 - A^T \lambda^0 + c \geq 0, \quad Ax^0 - b \geq 0, \\ &x^0 \geq 0, \quad \lambda^0 \geq 0, \\ &(x^0)^T (Dx^0 - A^T \lambda^0 + c) + (\lambda^0)^T (Ax^0 - b) = 0. \end{aligned}$$

Hence $x^0 \in S(D, A, c, b) \subset \Omega$. On the other hand, since $x^{k_i} \notin \Omega$ for all i and Ω is open, we have $x^0 \notin \Omega$, a contradiction.

We now turn to the case where the sequence $\{\|(x^k, \lambda^k)\|\}$ is unbounded. In this case, there exists a subsequence, which is denoted again by $\{\|(x^k, \lambda^k)\|\}$, such that $\|(x^k, \lambda^k)\| \to \infty$ and $\|(x^k, \lambda^k)\| \neq 0$ for every k. Let

$$z^k := \frac{(x^k, \lambda^k)}{\|(x^k, \lambda^k)\|} = \left(\frac{x^k}{\|(x^k, \lambda^k)\|}, \frac{\lambda^k}{\|(x^k \lambda^k)\|} \right). \qquad (10.31)$$

Since $\|z^k\| = 1$, there is a subsequence of $\{z^k\}$, which is denoted again by $\{z^k\}$, such that $z^k \to \bar{z} \in R^n \times R^m$, $\|\bar{z}\| = 1$. Let $\bar{z} = (\bar{x}, \bar{\lambda})$. By (10.31),

$$\frac{x^k}{\|(x^k, \lambda^k)\|} \to \bar{x}, \quad \frac{\lambda^k}{\|(x^k, \lambda^k)\|} \to \bar{\lambda}.$$

Dividing both sides of (10.28) and (10.29) by $\|(x^k, \lambda^k)\|$, both sides of (10.30) by $\|(x^k, \lambda^k)\|^2$, and taking limits as $k \to \infty$, we obtain

$$D\bar{x} - A^T \bar{\lambda} \geq 0, \quad A\bar{x} \geq 0, \qquad (10.32)$$

$$\bar{x} \geq 0, \quad \bar{\lambda} \geq 0, \tag{10.33}$$

$$\bar{x}^T (D\bar{x} - A^T\bar{\lambda}) + \bar{\lambda}^T A\bar{x} = 0. \tag{10.34}$$

By (10.32) and (10.33), $\bar{x} \in \Delta(A, 0) = \{x \in R^n : Ax \geq 0, x \geq 0\}$. Let us suppose for the moment that $\bar{x} \neq 0$. It is obvious that (10.34) can be rewritten as $\bar{x}^T D\bar{x} = 0$. If $x^T Dx \geq 0$ for all $x \in \Delta(A, 0)$ then $\bar{x} \in \mathrm{Sol}\,(D, A, 0, 0)$, contrary to the assumption $\mathrm{Sol}\,(D, A, 0, 0) = \{0\}$. If there exists $\hat{x} \in \Delta(A, 0)$ such that $\hat{x}^T D\hat{x} < 0$ then

$$\inf\{x^T Dx : x \in \Delta(A, 0)\} = -\infty,$$

because $\Delta(A, 0)$ is a cone. Thus $\mathrm{Sol}\,(D, A, 0, 0) = \emptyset$, contrary to the condition $\mathrm{Sol}\,(D, A, 0, 0) = \{0\}$. Therefore $\bar{x} = 0$.

As $\|(\bar{x}, \bar{\lambda})\| = 1$, from (10.32) and (10.33) it follows that $\bar{\lambda} \in \Lambda[A] \setminus \{0\}$. The assumption $b \in \mathrm{int}\,(\Lambda[A])^*$ implies

$$\bar{\lambda}^T b < 0. \tag{10.35}$$

Since $\|(x^k, \lambda^k)\| \to \infty$, $\dfrac{\lambda^k}{\|(x^k, \lambda^k)\|} \to \bar{\lambda}$ and $\|\bar{\lambda}\| = \|(\bar{x}, \bar{\lambda})\| = 1$, $\|\lambda^k\| \to \infty$. Using the obvious identity $(x^k)^T (A^k)^T \lambda^k = (\lambda^k)^T A^k x^k$ we can rewrite (10.30) as the following

$$(x^k)^T D^k x^k + (x^k)^T c^k = (\lambda^k)^T b^k. \tag{10.36}$$

If the sequence $\{x^k\}$ is bounded, then dividing both sides of (10.36) by $\|(x^k, \lambda^k)\|$ and letting $k \to \infty$ we obtain $\bar{\lambda}^T b = 0$, contrary to (10.35). So the sequence $\{x^k\}$ must be unbounded, and it has a subsequence, denoted again by $\{x^k\}$, such that $\|x^k\| \to \infty$, $\|x^k\| \neq 0$ for all k, and $\dfrac{x^k}{\|x^k\|} \to \hat{x}$ with $\|\hat{x}\| = 1$. For the sequence $\{(\lambda^k)^T b^k\}$ there are only two possibilities:

(α) There exists an integer i_0 such that

$$(\lambda^k)^T b^k \leq 0 \tag{10.37}$$

for all $k \geq i_0$, and

(β) For each i there exists an integer $k_i > i$ such that

$$(\lambda^{k_i})^T b^{k_i} > 0. \tag{10.38}$$

If case (α) arises, then (10.36) implies

$$(x^k)^T D^k x^k + (x^k)^T c^k \leq 0 \tag{10.39}$$

for all $k \geq i_0$. Dividing both sides of (10.39) by $\|x^k\|^2$ and letting $k \to \infty$ we get

$$\hat{x}^T D \hat{x} \leq 0. \tag{10.40}$$

By (10.28) and (10.29),

$$A^k x^k \geq b^k, \quad x^k \geq 0.$$

Dividing both sides of each of the last two inequalities by $\|x^k\|$ and letting $k \to \infty$ we obtain

$$A\hat{x} \geq 0, \quad \hat{x} \geq 0. \tag{10.41}$$

Since $0 \in \mathrm{Sol}\,(D, A, 0, 0)$, by (10.40) and (10.41) we have $\hat{x} \in \mathrm{Sol}\,(D, A, 0, 0)$, contrary to the condition $\mathrm{Sol}\,(D, A, 0, 0) = \{0\}$. Thus case (α) is impossible. If case (β) happens, then by dividing both sides of (10.38) by $\|(x^{k_i}, \lambda^{k_i})\|$ and letting $i \to \infty$ we obtain $\bar{\lambda}^T b \geq 0$, contrary to (10.35). The proof is complete, because neither (α) nor (β) can occur. $\square$

Theorem 10.5. *If* $\mathrm{Sol}\,(-D, A, 0, 0) = \{0\}$ *and* $b \in -\mathrm{int}\,(\Lambda[A])^*$ *then, for any* $c \in R^n$, *the multifunction* $S(\cdot)$ *is upper semicontinuous at* (D, A, c, b).

Proof. Except for several small changes, this proof is very similar to the proof of Theorem 10.4. Suppose, contrary to our claim, that there is an open set $\Omega \subset R^n$ containing $S(D, A, c, b)$, a sequence $\{(D^k, A^k, c^k, b^k)\}$ converging to (D, A, c, b) in $R_S^{n \times n} \times R^{m \times n} \times R^n \times R^m$, a sequence $\{x^k\}$ with $x^k \in S(D^k, A^k, c^k, b^k)$ and $x^k \notin \Omega$ for every k. By the definition of KKT point, there is a sequence $\{\lambda^k\}$ satisfying (10.28)–(10.30). If the sequence of $\{\|(x^k, \lambda^k)\|\}$ is bounded then, arguing similarly as in the preceding proof, we will arrive at a contradiction. If the sequence $\{\|(x^k, \lambda^k)\|\}$ is unbounded then, without any loss of generality, we can assume that the sequence $\left\{ \dfrac{(x^k, \lambda^k)}{\|(x^k, \lambda^k)\|} \right\}$ converges to a certain pair $(\bar{x}, \bar{\lambda})$ with $\|(\bar{x}, \bar{\lambda})\| = 1$. Dividing both sides of (10.28) and of (10.29) by $\|(x^k, \lambda^k)\|$, both sides of (10.30) by $\|(x^k, \lambda^k)\|^2$ and letting $k \to \infty$ we obtain (10.32)–(10.34). From (10.34) we have $\bar{x}^T(-D)\bar{x} = 0$. The assumption $\mathrm{Sol}\,(-D, A, 0, 0) = \{0\}$ forces $\bar{x} = 0$. Thus $\bar{\lambda} \in \Lambda[A] \setminus \{0\}$. Since $b \in -\mathrm{int}\,(\Lambda[A])^*$, we have

$$\bar{\lambda}^T b > 0. \tag{10.42}$$

Since $\|(x^k, \lambda^k)\| \to \infty$, $\dfrac{\lambda^k}{\|(x^k, \lambda^k)\|} \to \bar{\lambda}$, and $\|\bar{\lambda}\| = 1$, we must have $\|\lambda^k\| \to \infty$. Again, rewrite (10.30) in the form (10.36). If the sequence $\{x^k\}$ is bounded, we can divide both sides of (10.36) by $\|(x^k, \lambda^k)\|$ and let $k \to \infty$ to obtain $\bar{\lambda}^T b = 0$, which contradicts (10.42). Thus the sequence $\{x^k\}$ must be bounded, and it has a subsequence, denoted again by $\{x^k\}$, such that $\|x^k\| \to \infty$, $\|x^k\| \neq 0$ for all k, and $\dfrac{x^k}{\|x^k\|} \to \hat{x}$ with $\|\hat{x}\| = 1$.

If there exists an index i_0 such that (10.37) holds, then dividing both sides of (10.36) by $\|(x^k, \lambda^k)\|$ and taking limit as $k \to \infty$ we have $\bar{\lambda}^T b = 0$, contrary to (10.42).

Assume that for each i, there exists an integer $k_i > i$ such that (10.38) holds. From (10.36) and (10.38) it follows that

$$(x^{k_i})^T D_{k_i} x^{k_i} + (x^{k_i})^T c^{k_i} \geq 0 \tag{10.43}$$

for all i. Dividing both sides of (10.43) by $\|x^{k_i}\|^2$ and taking limit as $i \to \infty$ we get $\hat{x}^T D \hat{x} \geq 0$ or, equivalently,

$$\hat{x}^T(-D)\hat{x} \leq 0. \tag{10.44}$$

By (10.28) and (10.29), $A_{k_i} x^{k_i} \geq b^{k_i}$, $x^{k_i} \geq 0$. Dividing both sides of each of the last two inequalities by $\|x^{k_i}\|$ and taking limits we obtain (10.41). Properties (10.41), (10.44), and the inclusion $0 \in \text{Sol}(-D, A, 0, 0)$ yield $\hat{x} \in \text{Sol}(-D, A, 0, 0)$, contrary to the condition $\text{Sol}(-D, A, 0, 0) = \{0\}$. Thus, in all possible cases we have arrived at a contradiction. The proof is complete. $\square$

Our third sufficient condition for the stability of the Karush-Kuhn-Tucker point set can be formulated as follows.

Theorem 10.6. *If $S(D, A, 0, 0) = \{0\}$ and $\Lambda[A] = \{0\}$ then, for any $(c, b) \in R^n \times R^m$, the multifunction $S(\cdot)$ is upper semicontinuous at (D, A, c, b).*

Proof. Repeat the arguments in the proof of Theorem 10.4 until reaching the system (10.32)–(10.34). Since $S(D, A, 0, 0) = \{0\}$, we have $\bar{x} = 0$, hence (10.32)–(10.34) imply $-A^T \bar{\lambda} \geq 0$, $\bar{\lambda} \geq 0$. By $\|\bar{\lambda}\| = \|(\bar{x}, \bar{\lambda})\| = 1$, one has $\bar{\lambda} \in \Lambda[A] \setminus \{0\}$, contrary to the assumption that $\Lambda[A] = \{0\}$. $\square$

10.5 Corollaries and Examples

We now consider some corollaries of the results established in the preceding section and give several illustrative examples.

Corollary 10.3. *If $\Lambda[A] = \{0\}$ and if the matrix D is a positive definite (or negative definite) then, for any pair $(c, b) \in R^n \times R^m$, the multifunction $S(\cdot)$ is upper semicontinuous at (D, A, c, b).*

Proof. If D is positive definite, then $S(D, A, 0, 0) = \mathrm{Sol}(D, A, 0, 0) = \{0\}$. So our assertion follows from Theorem 10.6.

If D is negative definite, then $S(D, A, 0, 0) = \mathrm{Sol}(-D, A, 0, 0) = \{0\}$, and again the assertion follows from Theorem 10.6. $\square$

We proceed to show that the condition $b \in \mathrm{int}\,(\Lambda[A])^*$ in Theorem 10.4 is equivalent to the regularity of the following system of linear inequalities

$$Ax \geq b, \quad x \geq 0. \tag{10.45}$$

Lemma 10.2. *System (10.45) is regular if and only if $b \in \mathrm{int}\,(\Lambda[A])^*$.*

Proof. Assume (10.45) is regular, i.e. there exists x^0 such that $Ax^0 > b$, $x^0 > 0$. Let $q := Ax^0 - b > 0$ and let $\bar{\lambda}$ be any vector from $\Lambda[A] \setminus \{0\}$, that is $A^T \bar{\lambda} \leq 0, \bar{\lambda} \geq 0$, and $\bar{\lambda} \neq 0$. Then

$$\bar{\lambda}^T b = \bar{\lambda}^T (Ax^0 - q) = (x^0)^T A^T \bar{\lambda} - \bar{\lambda}^T q \leq -\lambda^T q < 0.$$

Hence $b \in \mathrm{int}\,(\Lambda[A])^*$.

Conversely, assume that $b \in \mathrm{int}\,(\Lambda[A])^*$. Suppose for a moment that (10.45) is irregular. Since the system $Ax > b$, $x \geq 0$ has no solutions, for any sequence $b^k \to b$ with $b^k > b$ for all k, the systems

$$Ax \geq b^k, \quad x \geq 0$$

have no solutions. By Theorem 2.7.9 from Cottle et al. (1992), which is a corollary of the Farkas Lemma, there exists $\lambda^k \in R^m$ such that

$$-A^T \lambda^k \geq 0, \quad \lambda^k \geq 0, \quad (\lambda^k)^T b^k > 0. \tag{10.46}$$

Since $\lambda^k \neq 0$, without loss of generality, we can assume that $\|\lambda^k\| = 1$ for every k, and $\lambda^k \to \bar{\lambda}$ with $\|\bar{\lambda}\| = 1$. Taking limits in (10.46) as $k \to \infty$ we get

$$-A^T \bar{\lambda} \geq 0, \quad \bar{\lambda} \geq 0, \quad \bar{\lambda}^T b \geq 0.$$

Hence $\bar{\lambda} \in \Lambda[A] \setminus \{0\}$, and the inequality $\bar{\lambda}^T b \geq 0$ contradicts the assumption $b \in \mathrm{int}\,(\Lambda[A])^*$. We have thus proved that (10.45) is regular. $\square$

Corollary 10.4. *If (10.45) is regular and if $\Delta(A, b)$ is bounded, then the multifunction $S(\cdot)$ is upper semicontinuous at (D, A, c, b).*

Proof. Since $\Delta(A, b)$ is nonempty, bounded, and $\Delta(A, b) + \Delta(A, 0) \subset \Delta(A, b)$, we have $\Delta(A, 0) = \{0\}$. Since (10.45) is regular, by Lemma 10.2 we have $b \in \text{int}\,(\Lambda[A])^*$. Applying Theorem 10.4 we get the desired result. $\square$

We have the following sufficient condition for stability of the KKT point set in QP problems with bounded constraint sets.

Corollary 10.5. *If $\Delta(A, 0) = \{0\}$ and $\lambda^T b \neq 0$ for all $\lambda \in \Lambda[A] \setminus \{0\}$ then, for any $c \in R^n$, the multifunction $S(\cdot)$ is upper semicontinuous at (D, A, c, b).*

Proof. Obviously, the condition $\Delta(A, 0) = \{0\}$ implies

$$S(D, A, 0, 0) = \text{Sol}\,(D, A, 0, 0) = \text{Sol}\,(-D, A, 0, 0) = \Delta(A, 0) = \{0\}. \tag{10.47}$$

Since $\Lambda[A]$ is a convex cone, the assumption $\lambda^T b \neq 0$ for all $\lambda \in \Lambda[A] \setminus \{0\}$ implies that one of the following two cases must occur:

(i) $\lambda^T b < 0$ for all $\lambda \in \Lambda[A] \setminus \{0\}$,
(ii) $\lambda^T b > 0$ for all $\lambda \in \Lambda[A] \setminus \{0\}$.

In the first case, the desired conclusion follows from (10.47) and Theorem 10.4. In the second case, the conclusion follows from (10.47) and Theorem 10.5. $\square$

The following two examples show that the obtained sufficient conditions for stability can be applied to nonconvex QP problems.

Example 10.3. Consider problem (10.1) where $n = 2$, $m = 1$,

$$D = \begin{bmatrix} -1 & 0 \\ 0 & -1 \end{bmatrix} \text{ or } D = \begin{bmatrix} 1 & 0 \\ 0 & -1 \end{bmatrix},$$

$$A = \begin{bmatrix} -\frac{1}{2}, -1 \end{bmatrix}, \ b = (-1), \ c = \begin{pmatrix} 0 \\ 0 \end{pmatrix}.$$

We have $\Delta(A, 0) = \{0\}$, $\text{Sol}\,(D, A, 0, 0) = \{0\}$ and $b \in \text{int}\,(\Lambda[A])^*$. By Theorem 10.4, $S(\cdot)$ is usc at (D, A, c, b).

Example 10.4. Consider problem (10.1) where $n = 2$, $m = 1$,

$$D = \begin{bmatrix} 1 & 0 \\ 0 & -1 \end{bmatrix}, \ A = [-1, 0], \ b = (-1), \ c = \begin{pmatrix} 0 \\ 0 \end{pmatrix}.$$

An easy computation shows that

$$S(D, A, 0, 0) = \{0\}, \ \text{Sol}\,(-D, A, 0, 0) = \{0\}, \ \text{and } b \in -\text{int}\,(\Lambda[A])^*.$$

The multifunction $S(\cdot)$ is usc at (D, A, c, b) by Theorem 10.5.

The next two examples show that when the condition $b \in \operatorname{int}(\Lambda[A])^*$ is violated the conclusion of Theorem 10.4 may hold or may not hold, as well.

Example 10.5. Let $D = [1]$, $A = [0]$, $b = (1)$, $c = (0)$, $A_t = [-t]$, where $t \in (0, 1)$. It is easily seen that

$$S(D, A, 0, 0) = \{0\}, \quad \operatorname{Sol}(D, A, 0, 0) = \{0\}, \quad S(D, A, c, b) = \emptyset,$$
$$S(D, A_t, c, b) = \left\{\frac{1}{t}\right\}, \quad \Lambda[A] = R_+, \quad b \notin \operatorname{int}(\Lambda[A])^*.$$

We have $S(D, A, c, b) \subset \Omega$, where $\Omega = \emptyset$. Since $A_t \to A$ and the inclusion $S(D, A_t, c, b) \subset \Omega$ cannot hold for sufficiently small $t > 0$, $S(\cdot)$ cannot be usc at (D, A, c, b).

Example 10.6. Let $D = [-1]$, $A = [-1]$, $b = (1)$, $c = (0)$. It is easy to verify that

$$S(D, A, 0, 0) = \{0\}, \quad \operatorname{Sol}(D, A, 0, 0) = \{0\}, \quad S(D, A, c, b) = \emptyset,$$
$$\Lambda[A] = R_+, \quad b \notin \operatorname{int}(\Lambda[A])^*.$$

The map $S(\cdot)$ is usc at (D, A, c, b). Indeed, since $S(-D, A, 0, 0) = \{0\}$ and $b \in -\operatorname{int}(\Lambda[A])^*$, Theorem 10.5 can be applied.

The following two examples show that if $b \notin -\operatorname{int}(\Lambda[A])^*$ then the conclusion of Theorem 10.5 may hold or may not hold, as well.

Example 10.7. Let D, A, c, b be defined as in Example 10.5. In this case we have

$$S(D, A, 0, 0) = \{0\}, \quad \operatorname{Sol}(-D, A, 0, 0) = \{0\},$$
$$\Lambda[A] = R_+, \quad b \notin -\operatorname{int}(\Lambda[A])^*.$$

As it has been shown in Example 10.5, the map $S(\cdot)$ is not usc at (D, A, c, b).

Example 10.8. Let $D = [1]$, $A = [-1]$, $b = (-1)$, $c = (0)$. It is a simple matter to verify that

$$S(D, A, 0, 0) = \{0\}, \quad \operatorname{Sol}(-D, A, 0, 0) = \{0\},$$
$$\Lambda[A] = R_+, \quad b \notin -\operatorname{int}(\Lambda[A])^*.$$

The fact that $S(\cdot)$ is usc at (D, A, c, b) follows from Theorem 10.4, because $\operatorname{Sol}(D, A, 0, 0) = \{0\}$ and $b \in \operatorname{int}(\Lambda[A])^*$.

10.6   USC Property of $S(\cdot)$: The General Case

In this section we obtain necessary and sufficient conditions for the stability of the Karush-Kuhn-Tucker point set in a general QP problem.

Given matrices $A \in R^{m \times n}$, $F \in R^{s \times n}$, $D \in R_S^{n \times n}$, and vectors $c \in R^n$, $b \in R^m$, $d \in R^s$, we consider the following general indefinite QP problem $QP(D, A, c, b, F, d)$:

$$\begin{cases} \text{Minimize } f(x) := \frac{1}{2}x^T D x + c^T x \\ \text{subject to } x \in R^n, \ Ax \geq b, \ Fx \geq d \end{cases} \qquad (10.48)$$

In what follows, *the pair (F, d) is not subject to change.* So the set $\Delta(F, d) := \{x \in R^n : Fx \geq d\}$ is fixed. Define $\Delta(A, b) = \{x \in R^n : Ax \geq b\}$ and recall (see Definition 3.1 and Corollary 3.2) that $\bar{x} \in \Delta(A, b) \cap \Delta(F, d)$ is said to be a Karush-Kuhn-Tucker point of $QP(D, A, c, b, F, d)$ if there exists a pair $(\bar{u}, \bar{v}) \in R^m \times R^s$ such that

$$\begin{aligned} & D\bar{x} - A^T\bar{u} - F^T\bar{v} + c = 0, \\ & A\bar{x} \geq b, \quad \bar{u} \geq 0, \\ & F\bar{x} \geq d, \quad \bar{v} \geq 0, \\ & \bar{u}^T(A\bar{x} - b) + \bar{v}^T(F\bar{x} - d) = 0. \end{aligned}$$

The KKT point set and the solution set of (10.48) are denoted, respectively, by $S(D, A, c, b, F, d)$ and $\text{Sol}(D, A, c, b, F, d)$.

If $s = n$, $d = 0$, and F is the unit matrix in $R^{n \times n}$, then problem (10.48) has the following canonical form (10.1). In agreement with the notation of the preceding sections, we write $S(D, A, c, b)$ instead of $S(D, A, c, b, F, d)$, and $\text{Sol}(D, A, c, b)$ instead of $\text{Sol}(D, A, c, b, F, d)$ if (10.48) has the canonical form. The upper semicontinuity of the multifunction

$$p' \mapsto S(p'), \ p' = (D', A', c', b') \in R_S^{n \times n} \times R^{m \times n} \times R^n \times R^m, \ (10.49)$$

has been studied in Sections 10.3–10.5. This property can be interpreted as the *stability* of the KKT point set $S(D, A, c, b)$ with respect to the change in the problem parameters. In this section we are interested in finding out how the results proved in Sections 10.3–10.5 can be extended to the case of problem (10.48). We will obtain some necessary and sufficient conditions for the upper semi-continuity of the multifunction

$$p' \mapsto S(p', F, d), \ p' = (D', A', c', b') \in R_S^{n \times n} \times R^{m \times n} \times R^n \times R^m.$$
$$(10.50)$$

As for the canonical problem, the obtained results can be interpreted as the necessary and sufficient *conditions for the stability* of the Karush-Kuhn-Tucker point set $S(D, A, c, b, F, d)$ with respect to the change in the problem parameters.

Our proofs are based on several observations concerning the system of equalities and inequalities defining the KKT point set. It is worthy to stress that the proofs in the preceding sections cannot be applied to the case of problem (10.48). This is because, unlike the case of the canonical problem (10.1), $\Delta(F, d)$ may fail to be a cone with nonempty interior and the vertex 0. So we have to use some new arguments. Fortunately, the proof schemes in the preceding sections will be useful also for the case of problem (10.48).

Theorem 10.7 below deals with the case where $\Delta(F, d)$ is a polyhedral cone with a vertex x^0, where $x^0 \in R^n$ is an arbitrarily given vector. Theorem 10.8 works for the case where $\Delta(F, d)$ is an arbitrary polyhedral set, but the conclusion is weaker than that of Theorem 10.7.

For any $M \in R^{r \times n}$ and $q \in R^r$, the set $\{x \in R^n : Mx \geq q\}$ is denoted by $\Delta(M, q)$. For $F \in R^{s \times n}$ and $A \in R^{m \times n}$, we abbreviate the set

$$\{(u, v) \in R^m \times R^s : A^T u + F^T v = 0, \ u \geq 0, \ v \geq 0\}$$

to $\Lambda[A, F]$. Note that

$$\mathrm{int}(\Lambda[A, F])^*$$
$$= \{(\xi, \eta) : \xi^T u + \eta^T v < 0 \ \forall (u, v) \in \Lambda[A, F] \setminus \{(0, 0)\}\}.$$

The next two remarks clarify some points in the assumption and conclusion of Theorem 10.7 below.

Remark 10.2. If there is a point $x^0 \in R^n$ such that $F(x^0) = d$ then $\Delta(F, d) = x^0 + \Delta(F, 0)$. Conversely, for any $x^0 \in R^n$ and any polyhedral cone K, there exists a positive integer s and a matrix $F \in R^{s \times n}$ such that $x^0 + K = \Delta(F, d)$, where $d := F(x^0)$.

Remark 10.3. If $\Delta(F, d)$ and $\Delta(A, b)$ are nonempty, then $\Delta(F, 0)$ and $\Delta(A, 0)$, respectively, are the recession cones of $\Delta(F, d)$ and

$\Delta(A, b)$. By definition, $S(D, A, 0, 0, F, 0)$ is the Karush-Kuhn-Tucker point set of the following QP problem

     Minimize $x^T D x$ subject to $x \in R^n$, $Ax \geq 0$, $Fx \geq 0$,

whose constraint set is the intersection $\Delta(A, 0) \cap \Delta(F, 0)$.

Theorem 10.7. *Assume that the set* $S(p, F, d)$, *where* $p = (D, A, c, b)$, *is bounded and there exists* $x^0 \in R^n$ *such that* $F(x^0) = d$. *If the multifunction (10.50) is upper semicontinuous at* p *then*

$$S(D, A, 0, 0, F, 0) = \{0\}. \tag{10.51}$$

Proof. Suppose, contrary to our claim, that there is a nonzero vector $\bar{x} \in S(D, A, 0, 0, F, 0)$. By definition, there exists a pair $(\bar{u}, \bar{v}) \in R^m \times R^s$ such that

$$D\bar{x} - A^T \bar{u} - F^T \bar{v} = 0, \tag{10.52}$$

$$A\bar{x} \geq 0, \quad \bar{u} \geq 0, \tag{10.53}$$

$$F\bar{x} \geq 0, \quad \bar{v} \geq 0, \tag{10.54}$$

$$\bar{u}^T A\bar{x} + \bar{v}^T F\bar{x} = 0. \tag{10.55}$$

For every $t \in (0, 1)$, we set

$$x_t = x^0 + \frac{1}{t}\bar{x}, \quad u_t = \frac{1}{t}\bar{u}, \quad v_t = \frac{1}{t}\bar{v}, \tag{10.56}$$

where x^0 is given by our assumptions. We claim that there exist matrices $D_t \in R_S^{n \times n}$, $A_t \in R^{m \times n}$ and vectors $c_t \in R^n$, $b_t \in R^m$ such that

$$\max\{\|D_t - D\|, \|A_t - A\|, \|c_t - c\|, \|b_t - b\|\} \to 0 \quad \text{as } t \to 0,$$

and

$$D_t x_t - A_t^T u_t - F^T v_t + c_t = 0, \tag{10.57}$$

$$A_t x_t \geq b_t, \quad u_t \geq 0, \tag{10.58}$$

$$F x_t \geq d, \quad v_t \geq 0, \tag{10.59}$$

$$u_t^T (A_t x_t - b_t) + v_t^T (F x_t - d) = 0. \tag{10.60}$$

The matrices D_t, A_t and the vectors c_t, b_t will have the following representations

$$D_t = D + t D_0, \quad A_t = A + t A_0 \tag{10.61}$$

$$c_t = c + tc_0, \quad b_t = b + tb_0, \tag{10.62}$$

where the matrices D_0, A_0 and the vectors c_0, b_0 are to be constructed. First we observe that, due to (10.54) and (10.56), (10.59) holds automatically. Clearly,

$$
\begin{aligned}
A_t x_t - b_t &= (A + tA_0)\left(x^0 + \frac{\bar{x}}{t}\right) - (b + tb_0) \\
&= t(A_0 x^0 - b_0) + \frac{1}{t} A\bar{x} + A_0\bar{x} + Ax^0 - b
\end{aligned}
$$

and

$$
\begin{aligned}
& u_t^T(A_t x_t - b_t) + v_t^T(Fx_t - d) \\
&= \frac{\bar{u}^T}{t}\left[t(A_0 x^0 - b_0) + \frac{1}{t}A\bar{x} + A_0\bar{x} + Ax^0 - b\right] \\
&\quad + \frac{\bar{v}^T}{t}\left[F\left(x^0 + \frac{\bar{x}}{t}\right) - d\right] \\
&= \bar{u}^T(A_0 x^0 - b_0) + \frac{1}{t^2}(\bar{u}^T A\bar{x} + \bar{v}^T F\bar{x}) + \frac{\bar{u}^T}{t}(A_0\bar{x} + Ax^0 - b).
\end{aligned}
$$

Therefore, by (10.53) and (10.55), if we have

$$A_0\bar{x} + Ax^0 - b = 0 \tag{10.63}$$

and

$$A_0 x^0 - b_0 = 0, \tag{10.64}$$

then (10.58) and (10.60) will be fulfilled. By (10.52),

$$
\begin{aligned}
D_t x_t & - A_t^T u_t - F^T v_t + c_t \\
&= (D + tD_0)\left(x^0 + \frac{\bar{x}}{t}\right) - (A + tA_0)^T\frac{\bar{u}}{t} - F^T\frac{\bar{v}}{t} + c + tc_0 \\
&= \frac{1}{t}(D\bar{x} - A^T\bar{u} - F^T\bar{v}) + t(D_0 x^0 + c_0) + Dx^0 \\
&\quad + D_0\bar{x} - A_0^T\bar{u} + c, \\
&= t(D_0 x^0 + c_0) + Dx^0 + D_0\bar{x} - A_0^T\bar{u} + c.
\end{aligned}
$$

Hence, if we have

$$Dx^0 + D_0\bar{x} - A_0^T\bar{u} + c = 0 \tag{10.65}$$

and

$$D_0 x^0 + c_0 = 0, \tag{10.66}$$

then (10.57) will be fulfilled.

Let $\bar{x} = (\bar{x}_1, \ldots, \bar{x}_n)$, where $\bar{x}^i \neq 0$ for $i \in I$ and $\bar{x}_i = 0$ for $i \notin I$, $I \subset \{1, \ldots, n\}$. Since $\bar{x} \neq 0$, I is nonempty. Fixing an index

$i_0 \in I$, we define A_0 as the $m \times n$-matrix in which the i_0-th column is $\bar{x}_{i_0}^{-1}(b - Ax^0)$, and the other columns consist solely of zeros. Let $b_0 = A_0 x^0$. One can verify immediately that (10.63) and (10.64) are satisfied; hence conditions (10.58) and (10.60) are fulfilled. From what has been said it follows that our claim will be proved if we can construct a matrix $D_0 \in R_S^{n \times n}$ and a vector c_0 satisfying (10.65) and (10.66). Let $D_0 = (d_{ij})$, where d_{ij} $(1 \le i, j \le n)$ are defined by the following formulae:

$$d_{ii} = \bar{x}_i^{-1}\left(A_0^T \bar{u} - Dx^0 - c\right)_i \quad \forall i \in I,$$
$$d_{i_0 j} = d_{j i_0} = \bar{x}_{i_0}^{-1}\left(A_0^T \bar{u} - Dx^0 - c\right)_j \quad \forall j \in \{1, \ldots, n\} \setminus I,$$

and $d_{ij} = 0$ for other pairs (i, j), $1 \le i, j \le n$. Here $(A_0^T \bar{u} - Dx^0 - c)_k$ denotes the k-th component of the vector $A_0^T \bar{u} - Dx^0 - c$. Since D_0 is a symmetric matrix, $D_0 \in R_S^{n \times n}$. If we define $c_0 = -D_0 x^0$ then (10.66) is satisfied. A direct computation shows that (10.65) is also satisfied.

We have thus constructed matrices D_0, A_0 and vectors c_0, b_0 such that for $x_t, u_t, v_t, D_t, A_t, c_t, b_t$ defined by (10.56), (10.61) and (10.62), conditions (10.57)–(10.60) are satisfied. Consequently, $x_t \in S(D_t, A_t, c_t, b_t, F, d)$. Since $S(p, F, d)$ is bounded, there is a bounded open set $\Omega \subset R^n$ such that $S(p, F, d) \subset \Omega$. Since

$$\max\{\|D_t - D\|, \|A_t - A\|, \|c_t - c\|, \|b_t - b\|\} \to 0$$

as $t \to 0$ and the multifunction $p' \mapsto S(p', F, d)$ is usc at $p = (D, A, c, b)$, $x_t \in \Omega$ for all sufficiently small t. This is impossible, because $\|x_t\| = \|x^0 + \dfrac{\bar{x}}{t}\| \to \infty$ as $t \to 0$. The proof is complete. $\square$

Remark 10.4. If $d = 0$ then $\Delta(F, d)$ is a cone with the vertex 0. In order to verify the assumptions of Theorem 10.7, one can choose $x^0 = 0$. In particular, this is the case of the canonical problem (10.1). Applying Theorem 10.7 we obtain the following necessary condition for the upper semicontinuity of the multifunction (10.49): *If $S(p)$, $p = (D, A, c, b)$ is bounded and if the multifunction $p' \mapsto S(p')$, $p' = (D', A', c', b')$, is usc at p, then $S(D, A, 0, 0) = \{0\}$.* Thus Theorem 10.8 above extends Theorem 10.1 to the case where $\Delta(F, d)$ can be any polyhedral convex cone in R^n, merely the standard cone R_+^n.

In the sequel, $S(D, A)$ denotes the set of all $x \in R^n$ such that there exists $u = u(x) \in R^m$ satisfying the following system:

$$Dx - A^T u = 0, \quad Ax \ge 0, \quad u \ge 0, \quad u^T A x = 0.$$

Remark 10.5. From the definition it follows that $S(D, A) = S(D, A, 0, 0, F, 0)$, where $s = n$ and $F = 0 \in R^{n \times n}$.

Theorem 10.8. *Assume that $\Delta(F, d)$ is nonempty and $S(p, F, d)$, where $p = (D, A, c, b)$, is bounded. If the multifunction (10.50) is upper semicontinuous at p then*

$$S(D, A) \cap \Delta(F, 0) = \{0\}. \tag{10.67}$$

Remark 10.6. Observe that (10.51) implies (10.67). Indeed, suppose that (10.51) holds. The fact that $0 \in S(D, A) \cap \Delta(F, 0)$ is obvious. So, if (10.67) does not hold then there exists $\bar{x} \in S(D, A) \cap \Delta(F, 0)$, $\bar{x} \neq 0$. Taking $\bar{u} = u(\bar{x})$, $\bar{v} = 0 \in R^s$, we see at once that the system (10.52)–(10.55) is satisfied. This means that $\bar{x} \in S(D, A, 0, 0, F, 0) \setminus \{0\}$, contrary to (10.51). Note that, in general, (10.67) does not imply (10.51).

Remark 10.7. If there exists x^0 such that $Fx^0 = d$ then, of course, $x^0 \in \Delta(F, d) = \{x \in R^n : Fx \geq d\}$. In particular, $\Delta(F, d) \neq \emptyset$. Thus Theorem 10.8 can be applied to a larger class of problems than Theorem 10.7. However, Remark 10.6 shows that the conclusion of Theorem 10.8 is weaker than that of Theorem 10.7. One question still unanswered is whether the assumptions of Theorem 10.8 always imply (10.51).

Proof of Theorem 10.8.

Assume that $\Delta(F, d)$ is nonempty, $S(D, A, c, b, F, d)$ is bounded and the multifunction $S(\cdot, F, d)$ is usc at p but (10.67) is violated. Then, there is a nonzero vector $\bar{x} \in S(D, A) \cap \Delta(F, 0)$. Hence there exists $\bar{u} \in R^m$ such that

$$D\bar{x} - A^T\bar{u} = 0, \tag{10.68}$$

$$A\bar{x} \geq 0, \quad \bar{u} \geq 0, \tag{10.69}$$

$$\bar{u}^T A\bar{x} = 0, \tag{10.70}$$

$$F\bar{x} \geq 0. \tag{10.71}$$

Let x^0 be an arbitrary point of $\Delta(F, d)$. Setting

$$x_t = x^0 + \frac{1}{t}\bar{x}, \quad u_t = \frac{1}{t}\bar{u}$$

for every $t \in (0, 1)$, we claim that there exist matrices

$$D_t \in R_S^{n \times n}, \quad A_t \in R^{m \times n}$$

and vectors $c_t \in R^n$, $b_t \in R^m$ such that

$$\max\{\|D_t - D\|, \|A_t - A\|, \|c_t - c\|, \|b_t - b\|\} \to 0$$

as $t \to 0$, and

$$D_t x_t - A_t^T u_t - F^T 0 + c_t = 0,$$
$$A_t x_t \geq b_t, \quad u_t \geq 0,$$
$$F x_t \geq d,$$
$$u_t^T (A_t x_t - b_t) + 0^T (F x_t - d) = 0.$$

The matrices D_t, A_t and vectors c_t, b_t are defined by (10.61) and (10.62), where D_0, A_0, c_0, b_0 are constructed as in the proof of Theorem 10.7. Using (10.68)–(10.71) and arguing similarly as in the preceding proof we shall arrive at a contradiction. $\square$

The following theorem gives three sufficient conditions for the upper semicontinuity of the multifunction (10.50). These conditions express some requirements on the behavior of the quadratic form $x^T D x$ on the cone $\Delta(A, 0) \cap \Delta(F, 0)$ and the position of the vector (b, d) relatively to the set $\text{int}(\Lambda[A, F])^*$.

Theorem 10.9. *Suppose that one of the following three pairs of conditions*

$$\text{Sol}(D, A, 0, 0, F, 0) = \{0\}, \quad (b, d) \in \text{int}\,(\Lambda[A, F])^*, \qquad (10.72)$$

$$\text{Sol}(-D, A, 0, 0, F, 0) = \{0\}, \quad (b, d) \in -\text{int}\,(\Lambda[A, F])^*, \qquad (10.73)$$

and

$$S(D, A, 0, 0, F, 0) = \{0\}, \quad \text{int}\,(\Lambda[A, F])^* = R^m \times R^s, \qquad (10.74)$$

is satisfied. Then, for any $c \in R^n$ (and also for any $b \in R^m$ if (10.74) takes place), the multifunction $p' \mapsto S(p', F, d)$, where $p' = (D', A', c', b')$, is upper semicontinuous at $p = (D, A, c, b)$.

Proof. On the contrary, suppose that one of the three pairs of conditions (10.72)–(10.74) is satisfied but, for some $c \in R^n$ (and also for some $b \in R^m$ if (10.74) takes place), the multifunction $p' \mapsto S(p', F, d)$ is not usc at $p = (D, A, c, b)$. Then there exist an open subset $\Omega \subset R^n$ containing $S(p, F, d)$, a sequence $p^k = (D^k, A^k, c^k, b^k)$ converging to p in $R_S^{n \times n} \times R^{m \times n} \times R^n \times R^m$, and a sequence $\{x^k\}$ such that, for each k, $x^k \in S(p^k, F, d)$ and $x^k \notin \Omega$. By the definition of KKT point, for each k there exists a pair $(u^k, v^k) \in R^m \times R^s$ such that

$$D^k x^k - (A^k)^T u^k - F^T v^k + c^k = 0, \qquad (10.75)$$

$$A^k x^k \geq b^k, \quad u^k \geq 0, \tag{10.76}$$

$$F x^k \geq d, \quad v^k \geq 0, \tag{10.77}$$

$$(u^k)^T (A^k x^k - b^k) + (v^k)^T (F x^k - d) = 0. \tag{10.78}$$

If the sequence $\{(x^k, u^k, v^k)\}$ is bounded, then the sequences $\{x^k\}$, $\{u^k\}$ and $\{v^k\}$ are also bounded. Therefore, without loss of generality, we can assume that the sequences $\{x^k\}$, $\{u^k\}$ and $\{v^k\}$ converge, respectively, to some points $x^0 \in R^n$, $u^0 \in R^m$ and $v^0 \in R^s$, as $k \to \infty$. Letting $k \to \infty$, from (10.75)–(10.78) we get

$$
\begin{aligned}
&D x^0 - A^T u - F^T v + c = 0, \\
&A x^0 \geq b, \quad u^0 \geq 0, \\
&F x^0 \geq d, \quad v^0 \geq 0, \\
&(u^0)^T (A x^0 - b) + (v^0)^T (F x^0 - d) = 0.
\end{aligned}
$$

Hence $x^0 \in S(p, F, d) \subset \Omega$. On the other hand, since $x^k \notin \Omega$ for each k, we must have $x^0 \notin \Omega$, a contradiction. We have thus shown that the sequence $\{(x^k, u^k, v^k)\}$ must be unbounded. By considering a subsequence, if necessary, we can assume that $\|(x^k, u^k, v^k)\| \to \infty$ and, in addition, $\|(x^k, u^k, v^k)\| \neq 0$ for all k. Since the sequence of vectors

$$
\frac{(x^k, u^k, v^k)}{\|(x^k, u^k, v^k)\|} = \left(\frac{x^k}{\|(x^k, u^k, v^k)\|}, \frac{u^k}{\|(x^k, u^k, v^k)\|}, \frac{v^k}{\|(x^k, u^k, v^k)\|} \right)
$$

is bounded, it has a convergent subsequence. Without loss of generality, we can assume that

$$
\frac{(x^k, u^k, v^k)}{\|(x^k, u^k, v^k)\|} \to (\bar{x}, \bar{u}, \bar{v}) \in R^n \times R^m \times R^s, \quad \|(\bar{x}, \bar{u}, \bar{v})\| = 1. \tag{10.79}
$$

Dividing both sides of (10.75), (10.76) and (10.77) by $\|(x^k, u^k, v^k)\|$, both sides of (10.78) by $\|(x^k, u^k, v^k)\|^2$, and letting $k \to \infty$, by (10.79) we obtain

$$D \bar{x} - A^T \bar{u} - F^T \bar{v} = 0, \tag{10.80}$$

$$A \bar{x} \geq 0, \ \bar{u} \geq 0, \tag{10.81}$$

$$F \bar{x} \geq 0, \ \bar{v} \geq 0, \tag{10.82}$$

$$\bar{u}^T A \bar{x} + \bar{v}^T F \bar{x} = 0. \tag{10.83}$$

We first consider the case where (10.72) is fulfilled. It is evident that (10.80)–(10.83) imply

$$\bar{x}^T D \bar{x} = 0, \quad A \bar{x} \geq 0, \quad F \bar{x} \geq 0. \tag{10.84}$$

If $\bar{x} \neq 0$ then, taking account of the fact that the constraint set $\Delta(A, 0) \cap \Delta(F, 0)$ of $QP(D, A, 0, 0, F, 0)$ is a cone, one can deduce from (10.84) that either $\mathrm{Sol}(D, A, 0, 0, F, 0) = \emptyset$ or

$$\bar{x} \in \mathrm{Sol}(D, A, 0, 0, F, 0).$$

This contradicts the first condition in (10.72). Thus $\bar{x} = 0$. Then it follows from (10.80)–(10.83) that $(\bar{u}, \bar{v}) \in \Lambda[A, F] \setminus \{(0, 0)\}$. Since $(b, d) \in \mathrm{int}\,(\Lambda[A, F])^*$ by (10.72),

$$\bar{u}^T b + \bar{v}^T d < 0. \tag{10.85}$$

Consider the sequence $\{(u^k)^T b^k + (v^k)^T d\}$. By (10.75) and (10.78),

$$(x^k)^T D^k x^k + (c^k)^T x^k = (u^k)^T b^k + (v^k)^T d. \tag{10.86}$$

If for each positive integer i there exists an integer k_i such that $k_i > i$ and

$$(u^{k_i})^T b^{k_i} + (v^{k_i})^T d > 0 \tag{10.87}$$

then, dividing both sides of (10.87) by $\|(x^{k_i}, u^{k_i}, v^{k_i})\|$ and letting $i \to \infty$, we have

$$\bar{u}^T b + \bar{v}^T d \geq 0,$$

contrary to (10.85). Consequently, there must exist a positive integer i_0 such that

$$(u^k)^T b^k + (v^k)^T d \leq 0 \quad \text{for every } k \geq i_0. \tag{10.88}$$

If the sequence $\{x^k\}$ is bounded then, dividing both sides of (10.86) by $\|(x^k, u^k, v^k)\|$ and letting $k \to \infty$, we get $\bar{u}^T b + \bar{v}^T d = 0$, contrary to (10.85). Thus $\{x^k\}$ is unbounded. We can assume that $\|x^k\| \to \infty$ and $\|x^k\| \neq 0$ for each k. Then $\left\{\dfrac{x^k}{\{x^k\}}\right\}$ is bounded. We can assume that

$$\frac{x^k}{\|x^k\|} \to \hat{x} \quad \text{with } \|\hat{x}\| = 1.$$

Combining (10.86) with (10.88) gives

$$(x^T)^k D^k x^k + (c^k)^T x^k \leq 0 \quad \text{for every } k \geq i_0. \tag{10.89}$$

Dividing both sides of (10.89) by $\|x^k\|^2$ and letting $k \to \infty$, we obtain

$$\hat{x}^T D \hat{x} \leq 0. \tag{10.90}$$

By (10.76) and (10.77),

$$A^k x^k \geq b^k, \quad F x^k \geq d.$$

Dividing both sides of each of the last inequalities by $\|x^k\|$ and letting $k \to \infty$, we have

$$A\hat{x} \geq 0, \quad F\hat{x} \geq 0. \tag{10.91}$$

Combining (10.90) with (10.91), we can assert that

$$\text{Sol}(D, A, 0, 0, F, 0) \neq \{0\},$$

contrary to the first condition in (10.72). Thus we have proved the theorem for the case where (10.72) is fulfilled.

Now we turn to the case where condition (10.73) is fulfilled. We deduce (10.84) from (10.80)–(10.83). If $\bar{x} \neq 0$ then from (10.84) we get $\text{Sol}(-D, A, 0, 0, F, 0) \neq \{0\}$, which contradicts the first condition in (10.73). Thus $\bar{x} = 0$. From (10.80)–(10.83) it follows that $(\bar{u}, \bar{v}) \in \Lambda[A, F] \setminus \{(0, 0)\}$. By the second condition in (10.73),

$$\bar{u}^T b + \bar{v}^T d > 0. \tag{10.92}$$

Consider the sequence $\{(u^k)^T b^k + (v^k)^T d\}$. We have (10.86). If there exists a positive integer i_0 such that (10.88) is valid then, dividing both sides of (10.88) by $\|(x^k, u^k, v^k)\|$ and letting $k \to \infty$, we obtain $\bar{u}^T b + \bar{v}^T d \leq 0$, contrary to (10.92). Therefore, for each positive integer i, one can find an integer $k_i > i$ such that (10.87) holds. If the sequence $\{x^k\}$ is bounded then, dividing both sides of (10.86) by $\|(x^k, u^k, v^k)\|$ and letting $k \to \infty$, we have $\bar{u}^T b + \bar{v}^T d = 0$, contrary to (10.92). Thus the sequence $\{x^k\}$ is unbounded. We can assume that $\|x^k\| \to \infty$ and $\|x^k\| \neq 0$ for all k. Since the sequence $\left\{\dfrac{x^k}{\|x^k\|}\right\}$ is well defined and bounded, without loss of generality, we can assume that

$$\frac{x^k}{\|x^k\|} \to \hat{x} \quad \text{with } \|\hat{x}\| = 1.$$

Combining (10.86) with (10.87) gives

$$(x^{k_i})^T D^{k_i} x^{k_i} + (c^{k_i})^T x^{k_i} > 0 \quad \text{for all } i. \tag{10.93}$$

Dividing both sides of (10.93) by $\|x^{k_i}\|^2$ and letting $i \to \infty$, we obtain $\hat{x}^T D \hat{x} \geq 0$ or, equivalently,

$$\hat{x}^T (-D) \hat{x} \leq 0. \tag{10.94}$$

By (10.76) and (10.77),

$$A^{k_i} x^{k_i} \geq b^{k_i}, \quad F x^{k_i} \geq d. \tag{10.95}$$

Dividing both sides of each of the inequalities in (10.95) by $\|x^{k_i}\|$ and letting $i \to \infty$, we have

$$A\hat{x} \geq 0, \quad F\hat{x} \geq 0. \tag{10.96}$$

Combining (10.94) with (10.96) yields $\mathrm{Sol}(-D, A, 0, 0, F, 0) \neq \{0\}$, contrary to the first condition in (10.73). This proves the theorem in the case where (10.73) is fulfilled.

Now let us consider the last case where (10.74) is assumed. From (10.80)–(10.83) we have $\bar{x} \in S(D, A, 0, 0, F, 0)$. By the first condition in (10.73), $\bar{x} = 0$. Then it follows from (10.80)–(10.83) that

$$A^T \bar{u} + F^T \bar{v} = 0, \quad \bar{u} \geq 0, \quad \bar{v} \geq 0, \quad \|(0, \bar{u}, \bar{v})\| = 1.$$

Therefore, $(\bar{u}, \bar{v}) \in \Lambda[A, F] \setminus \{(0, 0)\}$. Since $\bar{u}^T \bar{u} + \bar{v}^T \bar{v} > 0$, then $(\bar{u}, \bar{v}) \notin \mathrm{int}(\Lambda[A, F])^*$. This contradicts the second condition in (10.74).

We have thus proved that if one of the pairs of conditions (10.72)–(10.74) is fulfilled, then the conclusion of the theorem must hold true. □

We now proceed to show how the sufficient conditions (10.72) and (10.73) look like in the case of the canonical problem (10.1). As in Section 10.4, for any $A \in R^{n \times n}$, $\Lambda[A] = \{\lambda \in R^m : -A^T \lambda \geq 0, \ \lambda \geq 0\}$. We have

$$\mathrm{int}(\Lambda[A])^* = \{\xi \in R^m : \lambda^T \xi < 0 \ \ \forall \lambda \in \Lambda[A] \setminus \{0\}\}.$$

Lemma 10.3. *Suppose that, in problem (10.48), $s = n$, $d = 0$, and F is the unit matrix in $R^{n \times n}$. Then the following statements hold:*
   (a_1) If $b \in \mathrm{int}\,(\Lambda[A])^*$ then $(b, 0) \in \mathrm{int}\,(\Lambda[A, F])^*$;
   (a_2) If $\mathrm{Sol}\,(D, A, 0, 0) = \{0\}$ then $\mathrm{Sol}\,(D, A, 0, 0, F, 0) = \{0\}$;
   (a_3) If $b \in -\mathrm{int}\,(\Lambda[A])^*$ then $(b, 0) \in -\mathrm{int}\,(\Lambda[A, F])^*$;
   (a_4) If $\mathrm{Sol}\,(-D, A, 0, 0) = \{0\}$ then $\mathrm{Sol}\,(-D, A, 0, 0, F, 0) = \{0\}$.
Proof. If $b \in \mathrm{int}(\Lambda[A])^*$ then

$$\lambda^T b < 0 \quad \text{for all } \lambda \in \Lambda[A] \setminus \{0\}. \tag{10.97}$$

For any $(u, v) \in \Lambda[A, F] \setminus \{0\}$ we have

$$A^T u + F^T v = 0, \quad u \geq 0, \quad v \geq 0.$$

This yields
$$-A^T u = v \geq 0, \quad u \geq 0, \quad u \neq 0,$$
hence $u \in \Lambda[A] \setminus \{0\}$. By (10.97), $b^T u + 0^T v = b^T u = u^T b < 0$. This shows that $(b, 0) \in \text{int}(\Lambda[A, F])^*$. Statement (a_1) has been proved. It is clear that (a_3) follows from (a_1).

For proving (a_2) and (a_4) it suffices to note that, under our assumptions,

$$\text{Sol}(D, A, 0, 0) = \text{Sol}(D, A, 0, 0, F, 0)$$

and

$$\text{Sol}(-D, A, 0, 0) = \text{Sol}(-D, A, 0, 0, F, 0).$$

□

We check at once that Theorems 10.4 and 10.5 follow from Theorem 10.10 and Lemma 10.3.

10.7 Commentaries

The material of this chapter is taken from Tam and Yen (1999, 2000), Tam (2001a).

Several authors have made efforts in studying stability properties of the QP problems. Daniel (1973) established some basic facts about the solution stability of a QP problem whose objective function is a positive definite quadratic form. Guddat (1976) studied continuity properties of the solution set of a convex QP problem. Robinson (1979) obtained a fundamental result (see Theorem 7.6 in Chapter 7) on the stable behavior of the solution set of a monotone affine generalized equation (an affine variational inequality in the terminology of Gowda and Pang (1994), which yields a fact on the Lipschitz continuity of the solution set of a convex QP problem. Best and Chakravarti (1990) obtained some results on the continuity and differentiability of the optimal value function in a perturbed convex QP problem. By using the linear complementarity theory, Cottle, Pang and Stone (1992), studied in detail the stability of convex QP problems. Best and Ding (1995) proved a result on the continuity of the optimal value function in a convex QP problem. Auslender and Coutat (1996) established some results on stability and differential stability of generalized linear-quadratic programs, which include convex QP problems as a special case. Several attempts have been made to study the stability of nonconvex

QP problems (see, for instance, Klatte (1985), Tam (1999), Tam (2001a, 2001b, 2002)).

The proof of Theorem 10.1 is based on a construction developed by Oettli and Yen (1995, 1996a) for linear complementarity problems, homogeneous equilibrium problems, and quasi-complementarity problems.

Chapter 11

Lower Semicontinuity of the KKT Point Set Mapping

Our aim in this chapter is to characterize the lower semicontinuity of the Karush-Kuhn-Tucker point set mapping in quadratic programming. Necessary and sufficient conditions for the lsc property of the KKT point set mapping in canonical QP problems are obtained in Section 11.1. The lsc property of the KKT point set mapping in standard QP problems under linear perturbations is studied in Section 11.2.

11.1 The Case of Canonical QP Problems

Consider the canonical QP problem of the form (10.1). The following statement gives a necessary condition for the lower semicontinuity of the multifunction (10.3).

Theorem 11.1. *Let $D \in R_S^{n \times n}$ and $A \in R^{m \times n}$ be given. If the multifunction $S(D, A, \cdot, \cdot)$ is lower semicontinuous at $(c, b) \in R^n \times R^m$, then the set $S(D, A, c, b)$ is finite.*

Proof. Setting

$$M = \begin{pmatrix} D & -A^T \\ A & 0 \end{pmatrix}, \quad \bar{q} = \begin{pmatrix} c \\ -b \end{pmatrix},$$

and $s = n + m$, we consider the problem of finding a vector $z = \begin{pmatrix} x \\ \lambda \end{pmatrix} \in R^s$ satisfying

$$Mz + \bar{q} \geq 0, \quad z \geq 0, \quad z^T(Mz + \bar{q}) = 0. \qquad (11.1)$$

For a nonempty subset $\alpha \subset \{1, 2, \ldots, s\}$, $M_{\alpha\alpha}$ denotes the corresponding principal submatrix of M. If $p \in R^s$, then the column-vector with the components $(p_i)_{i \in \alpha}$ is denoted by p_α.

Let $z = (z_1, z_2, \ldots, z_s)$ be a nonzero solution of (11.1), and let $J = \{j : z_j = 0\}$, $I = \{i : z_i > 0\}$. Since $z_J = 0$ and $(Mz + \bar{q})_I = 0$, we have $M_{II} z_I = -\bar{q}_I$. Therefore, if $\det M_{II} \neq 0$ then z is defined uniquely via $\bar{q}$ by the formulae

$$z_J = 0, \quad z_I = -M_{II}^{-1}(\bar{q}_I).$$

If $I \neq \emptyset$ and $\det M_{II} = 0$, then

$$Q_I := \{q \in R^s : -q_I = M_{II} z_I \text{ for some } z \in R^s\}$$

is a proper subspace of R^s. By Baire's Lemma (Brezis (1987), p. 15), the union $Q := \cup\{Q_I : I \subset \{1, 2, \ldots, s\}, I \neq \emptyset, \det M_{II} = 0\}$ is nowhere dense. So there exists a sequence $q^k = \begin{pmatrix} c^k \\ -b^k \end{pmatrix}$ converging to $\bar{q} = \begin{pmatrix} c \\ -b \end{pmatrix}$ such that $q^k \notin Q$ for all k.

Fix any $x \in S(D, A, c, b)$ and let $\varepsilon > 0$ be given arbitrarily. Since the multifunction $S(D, A, \cdot, \cdot)$ is lsc at (c, b), there exists $\delta_\varepsilon > 0$ such that

$$x \in S(D, A, c', b') + \varepsilon B_{R^n}$$

for all (c', b') satisfying $\max\{\|c' - c\|, \|b' - b\|\} < \delta_\varepsilon$. Consequently, for each k sufficiently large, there exists $x^k \in S(D, A, c^k, b^k)$ such that

$$\|x - x^k\| \leq \varepsilon. \tag{11.2}$$

Since $x^k \in S(D, A, c^k, b^k)$, there exists λ^k such that $z^k := \begin{pmatrix} x^k \\ \lambda^k \end{pmatrix}$ is a solution of the LCP problem

$$Mz + q^k \geq 0, \quad z \geq 0, \quad z^T(Mz + q^k) = 0.$$

We put $J_k = \{j : z_j^k = 0\}$, $I_k = \{i : z_i^k > 0\}$. If $I_k = \emptyset$ then $z^k = 0$. If $I_k \neq \emptyset$ then $\det M_{I_k I_k} \neq 0$, because $q^k \notin Q$. Hence

$$z_{J_k}^k = 0, \quad z_{I_k}^k = -M_{I_k I_k}^{-1}(q_{I_k}^k). \tag{11.3}$$

Obviously, there exists a subset $I \subset \{1, 2, \ldots, s\}$ and a subsequence $\{k_i\}$ of $\{k\}$ such that $I_{k_i} = I$ for all k_i. Let Z denote the set of all $z \in R^s$ such that there is a nonempty subset $I \subset \{1, \ldots, s\}$ with the property that $\det M_{II} \neq 0$, $z_I = -M_{II}^{-1}(\bar{q}_I)$ and $z_J = 0$, where $J := \{1, \ldots, s\} \setminus I$. It is clear that Z is finite. From (11.3) it follows that the sequence $z_{I_{k_i}}^{(k_i)}$ converges to a point from the finite set $\tilde{Z} :=$ $Z \cup \{0\}$. For every $z = \begin{pmatrix} \xi \\ \lambda \end{pmatrix}$ let $\mathrm{pr}_1(z) := \xi$. Since $\mathrm{pr}_1(z^{(k_i)}) = x^{(k_i)}$ and $\mathrm{pr}_1(\cdot)$ is a continuous function, the sequence $\{x^{(k_i)}\}$ has a limit $\tilde{\xi}$ in the finite set $\tilde{X} := \{\mathrm{pr}_1(z) : z \in \tilde{Z}\}$. By (11.2), $x \in \tilde{X} + \varepsilon B_{R^n}$. As this inclusion holds for every $\varepsilon > 0$, we have $x \in \tilde{X}$. Thus $S(D, A, c, b) \subset \tilde{X}$. We have shown that $S(D, A, c, b)$ is a finite set. □

The following examples show that the finiteness of $S(D, A, c, b)$ may not be sufficient for the multifunction $S(\cdot)$ to be lower semicontinuous at (D, A, c, b).

Example 11.1. Consider the problem (P_ε) of minimizing the function

$$f_\varepsilon(x) = -\frac{1}{2}x_1^2 - x_2^2 + x_1 - \varepsilon x_2$$

on the set $\Delta = \{x \in R^2 : x \geq 0, -x_1 - x_2 \geq -2\}$. Note that Δ is a compact set with nonempty interior. Denote by $S(\varepsilon)$ the KKT point set of (P_ε). A direct computation using (10.2) gives

$$S(0) = \left\{(0,0), (1,0), (2,0), \left(\frac{5}{3}, \frac{1}{3}\right), (0,2)\right\}, \text{ and}$$

$$S(\varepsilon) = \left\{(2,0), \left(\frac{5+\varepsilon}{3}, \frac{1-\varepsilon}{3}\right), (0,2)\right\}$$

for $\varepsilon > 0$ small enough. For $U := \{x \in R^2 : \frac{1}{2} < x_1 < \frac{3}{2}, -1 < x_2 < 1\}$ we have $S(\varepsilon) \cap U = \emptyset$ for every $\varepsilon > 0$ small enough. Meanwhile, $S(0) \cap U = \{(1,0)\}$. Hence the multifunction $\varepsilon \mapsto S(\varepsilon)$ is not lsc at $\varepsilon = 0$.

Example 11.2. Consider the problem $(\tilde{P}_\varepsilon)$ of minimizing the function

$$\tilde{f}_\varepsilon(x) = \frac{1}{2}x_1^2 - x_2^2 - x_1 - \varepsilon x_2$$

on the set $\Delta = \{x \in R^2 : x \geq 0, -x_1 - x_2 \geq -2\}$. Denote by $\tilde{S}(\varepsilon)$ the KKT point set of $(\tilde{P}_\varepsilon)$. Using (10.2) we can show that $\tilde{S}(0) = \{(1,0), (0,2)\}$, and $\tilde{S}(\varepsilon) = \{(0,2)\}$ for every $\varepsilon > 0$. For $U := \{x \in R^2 : \frac{1}{2} < x_1 < \frac{3}{2}, -1 < x_2 < 1\}$ we have $\tilde{S}(0) \cap U =$

$\{(1,0)\}$, but $\widetilde{S}(\varepsilon) \cap U = \emptyset$ for every $\varepsilon > 0$. Hence the multifunction $\varepsilon \mapsto \widetilde{S}(\varepsilon)$ is not lsc at $\varepsilon = 0$.

In the KKT point set $S(D, A, c, b)$ of (10.1) we distinguish three types of elements: (1) Local solutions of $QP(D, A, c, b)$; (2) Local solutions of $QP(-D, A, c, b)$ which are not local solutions of $QP(D, A, c, b)$; (3) Points of $S(D, A, c, b)$ which do not belong to the first two classes. Elements of the first type (of the second type, of the third type) are called, respectively, the *local minima*, the *local maxima*, and the *saddle points* of (10.1).

In Example 11.1, $(1,0) \in S(0)$ is a local maximum of (P_0) which lies on the boundary of Δ. Similarly, in Example 11.2, $(1,0) \in \widetilde{S}(0)$ is a saddle point of $\widetilde{P}_0$ which lies on the boundary of Δ. If such situations do not happen, then the set of the KKT points is lower semicontinuous at the given parameter.

Theorem 11.2. *Assume that the inequality system $Ax \geq b$, $x \geq 0$ is regular. If the set $S(D, A, c, b)$ is nonempty, finite, and in $S(D, A, c, b)$ there exist no local maxima and no saddle points of (10.1) which are on the boundary of $\Delta(A, b)$, then the multifunction $S(\cdot)$ is lower semicontinuous at (D, A, c, b).*

Proof. For proving the lower semicontinuity of $S(\cdot)$ at (D, A, c, b) it suffices to show that: For any $\bar{x} \in S(D, A, c, b)$ and for any neighborhood U of $\bar{x}$ there exists $\delta > 0$ such that $S(D', A', c', b') \cap U \neq \emptyset$ for every (D', A', c', b') satisfying

$$\max\{\|D' - D\|, \|A' - A\|, \|c' - c\|, \|b' - b\|\} < \delta.$$

First, suppose that $\bar{x}$ is a local minimum of (10.1). As $S(D, A, c, b)$ is a finite set, $\bar{x}$ is an isolated local minimum. Using Theorem 3.7 we can verify that, for any Lagrange multiplier $\bar{\lambda}$ of $\bar{x}$, the second-order sufficient condition in the sense of Robinson (1982) is satisfied at $(\bar{x}, \bar{\lambda})$. According to Theorem 3.1 from Robinson (1982), for each neighborhood U of $\bar{x}$ there exists $\delta > 0$ such that for every (D', A', c', b') satisfying

$$\max\{\|D' - D\|, \|A' - A\|, \|c' - c\|, \|b' - b\|\} < \delta$$

there is a local minimum x' of the problem $QP(D', A', c', b')$ belonging to U. Since $x' \in S(D', A', c', b')$, we have $S(D', A', c', b') \cap U \neq \emptyset$, as desired. Now, suppose that $\bar{x}$ is a local maximum or a saddle point of (10.1). By our assumption, $\bar{x}$ belongs to the interior of $\Delta(A, b)$. Hence $\nabla f(\bar{x}) = D\bar{x} + c = 0$, or equivalently,

$$D\bar{x} = -c. \tag{11.4}$$

As $S(D, A, c, b)$ is finite, $\bar{x}$ is an isolated KKT point of (10.1). Then $\bar{x}$ must be the unique solution of the linear system (11.4). Therefore, the matrix D is nonsingular, and

$$\bar{x} = -D^{-1}c. \tag{11.5}$$

Since the system $Ax \geq b, \ x \geq 0$ is regular, using Lemma 3 from Robinson (1977) we can prove that there exist $\delta_0 > 0$ and an open neighborhood U_0 of $\bar{x}$ such that $U_0 \subset \Delta(A', b')$ for every (A', b') satisfying $\max\{\|A' - A\|, \|b' - b\|\} < \delta_0$. For any neighborhood U of $\bar{x}$, by (11.5) there exists $\delta \in (0, \delta_0)$ such that, for every (D', A', c', b') satisfying $\max\{\|D' - D\|, \|A' - A\|, \|c' - c\|, \|b' - b\|\} < \delta$, the matrix D' is nonsingular and $x' := -(D')^{-1}c'$ belongs to $U \cap U_0$. Since x' is an interior point $\Delta(A', b')$, this implies that $x' \in S(D', A', c', b')$. (It is easily seen that $\lambda' := 0$ is a Lagrange multiplier corresponding to x'.) We have thus shown that, for every (D', A', c', b') satisfying $\max\{\|D' - D\|, \|A' - A\|, \|c' - c\|, \|b' - b\|\} < \delta$, $S(D', A', c', b') \cap U \neq \emptyset$. The proof is complete. $\square$

11.2 The Case of Standard QP Problems

In this section we consider the following QP problem

$$\begin{cases} \text{Minimize} & \dfrac{1}{2}x^T Dx + c^T x \\ \text{subject to} & x \in \Delta(A, b) \end{cases} \tag{11.6}$$

where $A \in R^{m \times n}$ and $D \in R_S^{n \times n}$ are given matrices, $b \in R^m$ and $c \in R^n$ are given vectors,

$$\Delta(A, b) = \{x \in R^n : Ax \geq b\}.$$

Recall that $x \in R^n$ is a Karush-Kuhn-Tucker point of (11.6) if there exists $\lambda \in R^m$ such that

$$\begin{cases} Dx - A^T\lambda + c = 0, \\ Ax \geq b, \quad \lambda \geq 0, \\ \lambda^T(Ax - b) = 0. \end{cases}$$

The KKT point set (resp., the local solution set, the solution set) of (11.6) are denoted by $S(D, A, c, b)$, (resp., $\text{loc}(D, A, c, b)$, $\text{Sol}(D, A, c, b)$).

We will study the lower semicontinuity of the multifunctions

$$(D', A', c', b') \mapsto S(D', A', c', b') \tag{11.7}$$

and

$$(c', b') \mapsto S(D, A, c', b'), \tag{11.8}$$

which will be denoted by $S(\cdot)$ and $S(D, A, \cdot, \cdot)$, respectively. It is obvious that if (11.7) is lsc at $(D, A, c, b) \in R_S^{n \times n} \times R^{m \times n} \times R^n \times R^m$ then (11.8) is lsc at $(c, b) \in R^n \times R^m$.

Necessary conditions for the lsc property of the multifunction (11.8) can be stated as follows.

Theorem 11.3. *Let $(D, A, c, b) \in R_S^{n \times n} \times R^{m \times n} \times R^n \times R^m$. If the multifunction $S(D, A, \cdot, \cdot)$ is lower semicontinuous at (c, b), then*

(a) *the set $S(D, A, c, b)$ is finite, nonempty, and*

(b) *the system $Ax \geq b$ is regular.*

Proof. (a) For each index set $I \subset \{1, \cdots, m\}$, we define a matrix $M_I \in R^{(n+|I|) \times (n+|I|)}$, where $|I|$ is the number of elements of I, by setting

$$M_I = \begin{bmatrix} D & -A_I^T \\ A_I & O \end{bmatrix}.$$

(If $I = \emptyset$ then we set $M_I = D$). Let

$$Q_I = \left\{ (u, v) \in R^n \times R^m : \binom{u}{v_I} = M_I \binom{x}{\lambda_I} \right.$$
$$\left. \text{for some } (x, \lambda) \in R^n \times R^m \right\},$$

and

$$Q = \bigcup \{ Q_I : I \subset \{1, \cdots, m\}, \det M_I = 0 \}.$$

If $\det M_I = 0$ then it is clear that Q_I is a proper linear subspace of $R^n \times R^m$. Since the number of the index sets $I \subset \{1, \ldots, m\}$ is finite, the set Q is nowhere dense in $R^n \times R^m$ according to the Baire Lemma (see Brezis (1987), p. 15). So there exists a sequence $\{(c^k, b^k)\}$ converging to the given point $(c, b) \in R^n \times R^m$ such that $(-c^k, b^k) \notin Q$ for all k.

Fix any $\bar{x} \in S(D, A, c, b)$. Since $S(D, A, \cdot, \cdot)$ is lower semicontinuous at (c, b), one can find a subsequence $\{(c^{k_l}, b^{k_l})\}$ of $\{(c^k, b^k)\}$ and a sequence $\{x^{k_l}\}$ converging to $\bar{x}$ in R^n such that

$$x^{k_l} \in S(D, A, c^{k_l}, b^{k_l})$$

for all k_l. As $x^{k_l} \in S(D, A, c^{k_l}, b^{k_l})$, there exists $\lambda^{k_l} \in R^m$ such that

$$\begin{cases} Dx^{k_l} - A^T \lambda^{k_l} + c^{k_l} = 0, \\ Ax^{k_l} \geq b^{k_l}, \quad \lambda^{k_l} \geq 0, \\ (\lambda^{k_l})^T (Ax^{k_l} - b^{k_l}) = 0. \end{cases} \tag{11.9}$$

For every k_l, let $I_{k_l} := \{i \in \{1, \ldots, m\} : \lambda_i^{k_l} > 0\}$. (It may happen that $I_{k_l} = \emptyset$.) Since the number of the index sets $I \subset \{1, \ldots, m\}$ is finite, there must exist an index set $I \subset \{1, \cdots, m\}$ such that $I_{k_l} = I$ for infinitely many k_l. Without loss of generality we can assume that $I_{k_l} = I$ for all k_l. From (11.9) we deduce that

$$Dx^{k_l} - A_I^T \lambda_I^{k_l} + c^{k_l} = 0, \quad A_I x^{k_l} = b_I^{k_l}.$$

or, equivalently,

$$M_I \begin{pmatrix} x^{k_l} \\ \lambda_I^{k_l} \end{pmatrix} = \begin{pmatrix} -c^{k_l} \\ b_I^{k_l} \end{pmatrix}. \tag{11.10}$$

We claim that $\det M_I \neq 0$. Indeed, if $\det M_I = 0$ then, by (11.10) and by the definitions of Q_I and Q, we have

$$(-c^{k_l}, b^{k_l}) \in Q_I \subset Q,$$

contrary to the fact that $(-c^k, b^k) \notin Q$ for all k. We have proved that $\det M_I \neq 0$. By (11.10), we have

$$\begin{pmatrix} x^{k_l} \\ \lambda_I^{k_l} \end{pmatrix} = M_I^{-1} \begin{pmatrix} -c^{k_l} \\ b_I^{k_l} \end{pmatrix}.$$

Therefore

$$\lim_{l \to \infty} \begin{pmatrix} x^{k_l} \\ \lambda_I^{k_l} \end{pmatrix} = M_I^{-1} \begin{pmatrix} -c \\ b_I \end{pmatrix}. \tag{11.11}$$

If $I = \emptyset$ then formula (11.11) has the form

$$\lim_{l \to \infty} x^{k_l} = D^{-1}(-c). \tag{11.12}$$

From (11.11) it follows that the sequence $\{\lambda_I^{k_l}\}$ converges to some $\lambda_I \geq 0$ in $R^{|I|}$. Since the sequence $\{x^{k_l}\}$ converges to $\bar{x}$, from (11.11) and (11.12) it follows that

$$\begin{pmatrix} \bar{x} \\ \lambda_I \end{pmatrix} = M_I^{-1} \begin{pmatrix} -c \\ b_I \end{pmatrix}. \tag{11.13}$$

(Recall that $M_I = D$ if $I = \emptyset$). We set

$$Z = \{(x, \lambda) \in R^n \times R^m : \text{ there exists } J \subset \{1, \cdots, m\}$$
$$\text{such that } \det M_J \neq 0 \text{ and } \left(\begin{smallmatrix} x \\ \lambda_J \end{smallmatrix}\right) = M_J^{-1}\left(\begin{smallmatrix} -c \\ b_J \end{smallmatrix}\right)\},$$

and

$$X = \{x \in R^n : \text{ there exists } \lambda \in R^m \text{ such that } (x, \lambda) \in Z\}.$$

From the definitions of Z and X, we can deduce that X is a finite set (although Z may have infinitely many elements). We observe also that Z and X do not depend on the choice of $\bar{x}$. Actually, these sets depend only on the parameters (D, A, c, b). From (11.13) we have $\bar{x} \in X$. Since $\bar{x} \in S(D, A, c, b)$ can be chosen arbitrarily and since X is finite, we conclude that $S(D, A, c, b)$ is a finite set.

(b) If $Ax \geq b$ is irregular then there exists a sequence $\{b^k\}$ converging in R^n to b such that $\Delta(A, b^k)$ is empty for all k (Robinson (1977), Lemma 3). Clearly, $S(D, A, c, b^k) = \emptyset$ for all k. As $\{b^k\}$ converges to b, this shows that $S(D, A, \cdot, \cdot)$ cannot be lower semicontinuous at (c, b). The proof is complete. $\square$

Examples 11.1 and 11.2 show that finiteness and nonemptiness of $S(D, A, c, b)$ together with the regularity of the system $Ax \geq b$, in general, does not imply that $S(D, A, \cdot, \cdot)$ is lower semicontinuous at (c, b).

Let $(D, A, c, b) \in R_S^{n \times n} \times R^{m \times n} \times R^n \times R^m$. Let $x \in S(D, A, c, b)$ and let $\lambda \in R^m$ be a Lagrange multiplier corresponding to x. We define $I = \{1, 2, \ldots, m\}$,

$$K = \{i \in I : A_i x = b_i, \ \lambda_i > 0\} \tag{11.14}$$

and

$$J = \{i \in I : A_i x = b_i, \ \lambda_i = 0\}. \tag{11.15}$$

It is clear that K and J are two disjoint sets (possibly empty).

We now obtain a sufficient condition for the lsc property of the multifunction $S(D, A, \cdot, \cdot)$ at a given point $(c, b) \in R^n \times R^m$.

Theorem 11.4. *Let* $(D, A, c, b) \in R_S^{n \times n} \times R^{m \times n} \times R^n \times R^m$. *Suppose that*

(i) *the set* $S(D, A, c, b)$ *is finite, nonempty,*

(ii) *the system* $Ax \geq b$ *is regular,*

and suppose that for every $x \in S(D, A, c, b)$ there exists a Lagrange multiplier λ corresponding to x such that at least one of the following conditions holds:

(c1) $x \in \text{loc}(D, A, c, b)$,

(c2) $J = K = \emptyset$,

(c3) $J = \emptyset$, $K \neq \emptyset$, *and the system* $\{A_i : i \in K\}$ *is linearly independent,*

(c4) $J \neq \emptyset$, $K = \emptyset$, D *is nonsingular and* $A_J D^{-1} A_J^T$ *is a positive definite matrix,*

where K and J are defined via (x, λ) by (11.14) and (11.15). Then, the multifunction $S(D, A, \cdot, \cdot)$ is lower semicontinuous at (c, b).

Proof. Since $S(D, A, c, b)$ is nonempty, in order to prove that $S(D, A, \cdot, \cdot)$ is lower semicontinuous at (c, b) we only need to show that, for any $x \in S(D, A, c, b)$ and for any open neighborhood V_x of x, there exists $\delta > 0$ such that

$$S(D, A, c', b') \cap V_x \neq \emptyset \tag{11.16}$$

for every $(c', b') \in R^n \times R^m$ satisfying $\|(c', b') - (c, b)\| < \delta$.

Let $x \in S(D, A, c, b)$ and let V_x be an open neighborhood of x. By our assumptions, there exists a Lagrange multiplier λ corresponding to x such that at least one of the four conditions (c1)-(c4) holds.

We first examine the case where (c1) holds, that is

$$x \in \text{loc}(D, A, c, b).$$

Since $S(D, A, c, b)$ is finite by (i), $\text{loc}(D, A, c, b)$ is finite. So x is an isolated local solution of (11.1). Using Theorem 3.7 we can verify that, for any Lagrange multiplier $\bar{\lambda}$ of $\bar{x}$, the second-order sufficient condition in the sense of Robinson (1982), Definition 2.1, is satisfied at $(\bar{x}, \bar{\lambda})$. By assumption (ii), we can apply Theorem 3.1 from Robinson (1982) to find an $\delta > 0$ such that

$$\text{loc}(D, A, c', b') \cap V_x \neq \emptyset$$

for every $(c', b') \in R^n \times R^n$ with $\|(c', b') - (c, b)\| < \delta$. Since $\text{loc}(D, A, c, b) \subset S(D, A, c', b')$, we conclude that (11.16) is valid for every (c', b') satisfying $\|(c', b') - (c, b)\| < \delta$.

Consider the case where (c2) holds, that is $A_i x > b_i$ for every $i \in I$. Since λ is a Lagrange multiplier corresponding to x, the system

$$Dx - A^T\lambda + c = 0, \quad Ax \geq b, \quad \lambda \geq 0, \quad \lambda^T(Ax - b) = 0$$

is satisfied. As $Ax > b$, from this we deduce that $\lambda = 0$. Hence the first equality in the above system implies that $Dx = -c$. Thus x is a solution of the linear system

$$Dz = -c \quad (z \in R^n). \tag{11.17}$$

Since $S(D, A, c, b)$ is finite, x is a locally unique KKT point of (11.6). Combining this with the fact that x is an interior point of $\Delta(A, b)$, we can assert that x is a unique solution of (11.17). Hence matrix D is nonsingular and we have

$$x = -D^{-1}c. \tag{11.18}$$

Since $Ax > b$, there exist $\delta_1 > 0$ and an open neighborhood $U_x \subset V_x$ of x such that $U_x \subset \Delta(A, b')$ for all $b' \in R^m$ satisfying $\|b' - b\| < \delta_1$. By (11.18), there exists $\delta_2 > 0$ such that if $\|c' - c\| < \delta_2$ and $x' = -D^{-1}c'$ then $x' \in U_x$. Set $\delta = \min\{\delta_1, \delta_2\}$. Let (c', b') be such that $\|(c', b') - (c, b)\| < \delta$. Since $x' := -D^{-1}c'$ belongs to the open set $U_x \subset \Delta(A, b')$, we deduce that

$$Dx' + c' = 0, \quad Ax' > b'.$$

From this it follows that $x' \in S(D, A, c', b')$. (Observe that $\lambda' = 0$ is a Lagrange multiplier corresponding to x'.) We have thus shown that (11.16) is valid for every $(c', b') \in R^n \times R^m$ satisfying $\|(c', b') - (c, b)\| < \delta$.

We now suppose that (c3) holds. First, we establish that the matrix $M_K \in R^{(n+|K|) \times (n+|K|)}$ defined by setting

$$M_K = \begin{bmatrix} D & -A_K^T \\ A_K & 0 \end{bmatrix},$$

where $|K|$ denotes the number of elements in K, is nonsingular. To obtain a contradiction, suppose that M_K is singular. Then there exists a nonzero vector $(v, w) \in R^n \times R^{|K|}$ such that

$$M_K \begin{pmatrix} v \\ w \end{pmatrix} = \begin{bmatrix} D & -A_K^T \\ A_K & 0 \end{bmatrix} \begin{pmatrix} v \\ w \end{pmatrix} = 0.$$

This implies that

$$Dv - A_K^T w = 0, \quad A_K v = 0. \tag{11.19}$$

Since the system $\{A_i : i \in K\}$ is linearly independent by (c3), from (11.19) it follows that $v \neq 0$. As $A_{I \setminus K} x > b_{I \setminus K}$ and $\lambda_K > 0$, there exists $\delta_3 > 0$ such that $A_{I \setminus K}(x + tv) \geq b_{I \setminus K}$ and $\lambda_K + tw \geq 0$ for every $t \in [0, \delta_3]$. By (11.19), we have

$$\begin{cases} D(x + tv) - A_K^T(\lambda_K + tw) + c = 0, \\ A_K(x + tv) = b_K, \quad \lambda_K + tw \geq 0, \\ A_{I \setminus K}(x + tv) \geq b_{I \setminus K}, \quad \lambda_{I \setminus K} = 0 \end{cases} \tag{11.20}$$

for every $t \in [0, \delta_3]$. From (11.20) we deduce that $x + tv \in S(D, A, c, b)$ for all $t \in [0, \delta_3]$. This contradicts the assumption that $S(D, A, c, b)$ is finite. We have thus proved that M_K is nonsingular. From the definition of K it follows that

$$\begin{cases} Dx - A_K^T \lambda_K + c = 0, \\ A_K x = b_K, \quad \lambda_K > 0, \\ A_{I \setminus K} x > b_{I \setminus K}, \quad \lambda_{I \setminus K} = 0. \end{cases}$$

The last system can be rewritten equivalently as follows

$$M_K \begin{pmatrix} x \\ \lambda_K \end{pmatrix} = \begin{pmatrix} -c \\ b_K \end{pmatrix}, \quad \lambda_K > 0, \quad \lambda_{I \setminus K} = 0, \quad A_{I \setminus K} x > b_{I \setminus K}. \tag{11.21}$$

As M_K is nonsingular, (11.21) yields

$$\begin{pmatrix} x \\ \lambda_K \end{pmatrix} = M_K^{-1} \begin{pmatrix} -c \\ b_K \end{pmatrix}, \quad \lambda_K > 0, \quad \lambda_{I \setminus K} = 0, \quad A_{I \setminus K} x > b_{I \setminus K}.$$

So there exists $\delta > 0$ such that if $(c', b') \in R^n \times R^m$ is such that $\|(c', b') - (c, b)\| < \delta$, then the formula

$$\begin{pmatrix} x' \\ \lambda'_K \end{pmatrix} = M_K^{-1} \begin{pmatrix} c' \\ b'_K \end{pmatrix}$$

defines a vector $(x', \lambda'_K) \in R^n \times R^{|K|}$ satisfying the conditions

$$x' \in V_x, \quad \lambda'_K > 0, \quad A_{I \setminus K} x' > b'_{I \setminus K}.$$

We see at once that vector x' defined in this way belongs to the set

$$S(D, A, c', b') \cap V_x$$

and $\lambda' := (\lambda'_K, \lambda'_{I \setminus K})$, where $\lambda'_{I \setminus K} = 0$, is a Lagrange multiplier corresponding to x'. We have shown that (11.16) is valid for every $(c', b') \in R^n \times R^m$ satisfying $\|(c', b') - (c, b)\| < \delta$.

Finally, suppose that (c4) holds. In this case, we have

$$Dx + c = 0, \quad A_J x = b_J, \quad \lambda_J = 0, \quad A_{I \setminus J} x > b_{I \setminus J}, \quad \lambda_{I \setminus J} = 0. \tag{11.22}$$

To prove that there exists $\delta > 0$ such that (11.16) is valid for every $(c', b') \in R^n \times R^m$ satisfying $\|(c', b') - (c, b)\| < \delta$, we consider the following system of equations and inequalities of variables $(z, \mu) \in R^n \times R^m$:

$$\begin{cases} Dz - A_J^T \mu_J + c' = 0, \quad A_J z \geq b'_J, \quad \mu_J \geq 0, \\ A_{I \setminus J} z \geq b'_{I \setminus J}, \quad \mu_{I \setminus J} = 0, \quad \mu_J^T (A_J z - b'_J) = 0. \end{cases} \tag{11.23}$$

Since D is nonsingular, (11.23) is equivalent to the system

$$\begin{cases} z = D^{-1}(-c' + A_J^T \mu_J), \quad A_J z \geq b'_J, \quad \mu_J \geq 0, \\ A_{I \setminus J} z \geq b'_{I \setminus J}, \quad \mu_{I \setminus J} = 0, \quad \mu_J^T (A_J z - b'_J) = 0. \end{cases} \tag{11.24}$$

By (11.22), $A_{I \setminus J} x > b_{I \setminus J}$. Hence there exist $\delta_4 > 0$ and an open neighborhood $U_x \subset V_x$ of x such that $A_{I \setminus J} z \geq b'_{I \setminus J}$ for any $z \in U_x$ and $(c', b') \in R^n \times R^m$ satisfying $\|(c', b') - (c, b)\| < \delta_4$. Consequently, for every (c', b') satisfying $\|(c', b') - (c, b)\| < \delta_4$, the verification of (11.16) is reduced to the problem of finding $z \in U_x$ and $\mu_J \in R^{|J|}$ such that (11.24) holds. Here $|J|$ denotes the number of elements in J. We substitute z from the first equation of (11.24) into the first inequality and the last equation of that system to get

$$\begin{cases} A_J D^{-1} A_J^T \mu_J \geq b'_J + A_J D^{-1} c', \quad \mu_J \geq 0, \\ \mu_J^T (A_J D^{-1} A_J^T \mu_J - b'_J - A_J D^{-1} c') = 0. \end{cases} \tag{11.25}$$

Let $S := A_J D^{-1} A_J^T$ and $q' := -b'_J - A_J D^{-1} c'$. We can rewrite (11.25) as follows

$$S\mu_J + q' \geq 0, \quad \mu_J \geq 0, \quad (\mu_J)^T (S\mu_J + q') = 0. \tag{11.26}$$

Problem of finding $\mu_J \in R^{|J|}$ satisfying (11.26) is the linear complementarity problem defined by the matrix $S \in R^{|J| \times |J|}$ and the vector $q' \in R^{|J|}$. By assumption (c4), S is a positive definite matrix, that is $y^T S y > 0$ for every $y \in R^{|J|} \setminus \{0\}$. Then S is a P-matrix. The latter means that every principal minor of S is positive (see Cottle et al. (1992), Definition 3.3.1). According to Cottle et al. (1992),

Theorem 3.3.7, for each $q' \in R^{|J|}$, problem (11.26) has a unique solution $\mu_J \in R^{|J|}$. Since D is nonsingular, from (11.22) it follows that

$$A_J D^{-1}(-c) - b_J = 0.$$

Setting $q = -b_J - A_J D^{-1} c$ we have $q = 0$. Substituting $q' = q = 0$ into (11.26) we find the unique solution $\bar{\mu}_J = 0 = \lambda_J$. By Theorem 7.2.1 from Cottle et al. (1992), there exist $\ell > 0$ and $\varepsilon > 0$ such that for every $q' \in R^{|J|}$ satisfying $\|q' - q\| < \varepsilon$ we have

$$\|\mu_J - \lambda_J\| \leq \ell \|q' - q\|.$$

Therefore

$$\|\mu_J\| = \|\mu_J - \lambda_J\| \leq \ell \|b'_J - b_J + A_J D^{-1}(c' - c)\|.$$

From this we conclude that there exists $\delta \in (0, \delta_4]$ such that if (c', b') satisfies the condition $\|(c', b') - (c, b)\| < \delta$, then the vector

$$z = D^{-1}(-c' + A_J^T \mu_J),$$

where μ_J is the unique solution of (11.26), belongs to U_x. From the definition of μ_J and z we see that system (11.24), where $\mu_{I \setminus J} := 0$, is satisfied. Then $z \in S(D, A, c', b')$. We have thus shown that, for any (c', b') satisfying $\|(c', b') - (c, b)\| < \delta$, property (11.16) is valid. The proof is complete. $\square$

To verify condition (c1), we can use Theorem 3.5.

We now consider three examples to see how the conditions (c1)–(c4) can be verified for concrete QP problems.

Example 11.3. (See Robinson (1980), p. 56) Let

$$f(x) = \frac{1}{2}x_1^2 - \frac{1}{2}x_2^2 - x_1 \quad \text{for all} \quad x = (x_1, x_2) \in R^2. \tag{11.27}$$

Consider the QP problem

$$\min\{f(x) : x = (x_1, x_2) \in R^2, x_1 - 2x_2 \geq 0, x_1 + 2x_2 \geq 0\}. \tag{11.28}$$

For this problem, we have

$$D = \begin{bmatrix} 1 & 0 \\ 0 & -1 \end{bmatrix}, \quad A = \begin{bmatrix} 1 & -2 \\ 1 & 2 \end{bmatrix}, \quad c = \begin{pmatrix} -1 \\ 0 \end{pmatrix}, \quad b = \begin{pmatrix} 0 \\ 0 \end{pmatrix},$$

$$S(D, A, c, b) = \left\{ (1, 0), \left(\frac{4}{3}, \frac{2}{3} \right), \left(\frac{4}{3}, -\frac{2}{3} \right) \right\},$$

$$\mathrm{loc}(D, A, c, b) = \left\{ \left(43, \frac{2}{3} \right), \left(\frac{4}{3}, -\frac{2}{3} \right) \right\}.$$

For any feasible vector $x = (x_1, x_2)$ of (11.28), we have $x_1 \geq 2|x_2|$. Therefore

$$f(x) + \frac{2}{3} = \frac{1}{2}x_1^2 - \frac{1}{2}x_2^2 - x_1 + \frac{2}{3} \geq \frac{3}{8}x_1^2 - x_1 + \frac{2}{3} \geq 0. \quad (11.29)$$

For $\bar{x} := \left(\frac{4}{3}, \frac{2}{3} \right)$ and $\hat{x} := \left(\frac{4}{3}, -\frac{2}{3} \right)$, we have $f(\bar{x}) = f(\hat{x}) = -\frac{2}{3}$. Hence from (11.29) it follows that $\bar{x}$ and $\hat{x}$ are the solutions of (11.28). Actually,

$$\mathrm{Sol}(D, A, c, b) = \mathrm{loc}(D, A, c, b) = \{\bar{x}, \hat{x}\}.$$

Setting $\tilde{x} = (1, 0)$ we have $\tilde{x} \in S(D, A, c, b) \setminus \mathrm{loc}(D, A, c, b)$. Note that $\tilde{\lambda} := (0, 0)$ is a Lagrange multiplier corresponding to $\tilde{x}$. We check at once that conditions (i) and (ii) in Theorem 11.4 are satisfied and, for each KKT point $x \in S(D, A, c, b)$, either (c1) or (c2) is satisfied. Theorem 11.4 shows that the multifunction $S(D, A, \cdot, \cdot)$ is lower semicontinuous at (c, b).

Example 11.4. Let $f(\cdot)$ be defined by (11.27). Consider the QP problem

$$\min\{f(x) : x = (x_1, x_2) \in R^2, x_1 - 2x_2 \geq 0, x_1 + 2x_2 \geq 0, x_1 \geq 1\}.$$

For this problem, we have

$$D = \begin{bmatrix} 1 & 0 \\ 0 & -1 \end{bmatrix}, \quad A = \begin{bmatrix} 1 & -2 \\ 1 & 2 \\ 1 & 0 \end{bmatrix}, \quad c = \begin{pmatrix} -1 \\ 0 \end{pmatrix}, \quad b = \begin{pmatrix} 0 \\ 0 \\ 1 \end{pmatrix}.$$

Let $\bar{x}, \hat{x}, \tilde{x}$ be the same as in the preceding example. Note that $\tilde{\lambda} := (0, 0, 0)$ is a Lagrange multiplier corresponding to $\tilde{x}$. We have

$$S(D, A, c, b) = \{\tilde{x}, \bar{x}, \hat{x}\}, \quad \mathrm{Sol}(D, A, c, b) = \mathrm{loc}(D, A, c, b) = \{\bar{x}, \hat{x}\}.$$

Clearly, for $x = \bar{x}$ and $x = \hat{x}$, assumption (c1) is satisfied. It is easily seen that, for the pair $(\tilde{x}, \tilde{\lambda})$, we have $K = \emptyset$, $J = \{3\}$. Since $A_J = (1\ 0)$ and $D^{-1} = D$, we get $A_J D^{-1} A_J^T = 1$. Thus (c4) is satisfied. By Theorem 11.4, $S(D, A, \cdot, \cdot)$ is lower semicontinuous at (c, b).

Example 11.5. Let $f(x)$ be as in (11.27). Consider the QP problem

$$\min\{f(x) : x = (x_1, x_2) \in R^2, x_1 - 2x_2 \geq 0, x_1 + 2x_2 \geq 0, x_1 \geq 2\}.$$

For this problem, we have

$$D = \begin{bmatrix} 1 & 0 \\ 0 & -1 \end{bmatrix}, \quad A = \begin{bmatrix} 1 & -2 \\ 1 & 2 \\ 1 & 0 \end{bmatrix}, \quad c = \begin{pmatrix} -1 \\ 0 \end{pmatrix}, \quad b = \begin{pmatrix} 0 \\ 0 \\ 2 \end{pmatrix},$$

$$S(D, A, c, b) = \{(2, 0), (2, 1), (2, -1)\},$$

$$\text{Sol}(D, A, c, b) = \text{loc}(D, A, c, b) = \{(2, 1), (2, -1)\}.$$

Let $\bar{x} = (2, -1)$, $\hat{x} = (2, 1)$, $\tilde{x} = (2, 0)$. Note that $\tilde{\lambda} := (0, 0, 1)$ is a Lagrange multiplier corresponding to $\tilde{x}$. For $x = \bar{x}$ and $x = \hat{x}$, we see at once that (c1) is satisfied. For the pair $(\tilde{x}, \tilde{\lambda})$, we have $K = \{3\}$, $J = \emptyset$. Since

$$\{A_i : i \in K\} = \{A_3\} = \{(1\ 0)\},$$

assumption (c3) is satisfied. According to Theorem 11.4, $S(D, A, \cdot, \cdot)$ is lower semicontinuous at (c, b).

The idea of the proof of Theorem 11.4 is adapted from Robinson (1980), Theorem 4.1, and the proof of Theorem 11.2. In Robinson (1980), some results involving Schur complements were obtained.

Let $(D, A, c, b) \in R_S^{n \times n} \times R^{m \times n} \times R^n \times R^m$. Let $x \in S(D, A, c, b)$ and let $\lambda \in R^m$ be a Lagrange multiplier corresponding to x. We define K and J by (11.14) and (11.15), respectively. Consider the case where both the sets K and J are nonempty. If the matrix

$$M_K = \begin{bmatrix} D & -A_K^T \\ A_K & 0 \end{bmatrix} \in R^{(n+|K|) \times (n+|K|)}$$

is nonsingular, then we denote by S_J the *Schur complement* (see Cottle et al. (1992), p. 75) of M_K in the following matrix

$$\begin{bmatrix} D & -A_K^T & -A_J^T \\ A_K & 0 & 0 \\ A_J & 0 & 0 \end{bmatrix} \in R^{(n+|K|+|J|) \times (n+|K|+|J|)}.$$

This means that

$$S_J = [A_J\ 0] M_K^{-1} [A_J\ 0]^T.$$

Note that S_J is a symmetric matrix (see Robinson (1980), p. 56). Consider the following condition:

(c5) $J \neq \emptyset$, $K \neq \emptyset$, the system $\{A_i : i \in K\}$ is linearly independent, $v^T Dv \neq 0$ for every nonzero vector v satisfying
$A_K v = 0$, and S_J is positive definite.

Modifying some arguments of the proof of Theorem 11.4 we can
show that if $J \neq \emptyset$, $K \neq \emptyset$, the system $\{A_i : i \in K\}$ is linearly
independent, and $v^T Dv \neq 0$ for every nonzero vector v satisfying
$A_K v = 0$, then M_K is nonsingular.

It can be proved that the assertion of Theorem 11.4 remains valid
if instead of (c1)–(c4) we use (c1)–(c3) and (c5). The method of
dealing with (c5) is similar to that of dealing with (c4) in the proof
of Theorem 11.4. Up to now we have not found any example of
QP problems of the form (11.1) for which there exists a pair (x, λ),
$x \in S(D, A, c, b)$ and λ is a Langrange multiplier corresponding
to x, such that (c1)–(c4) are not satisfied, but (c5) is satisfied.
Thus the usefulness of (c5) in characterizing the lsc property of the
multifunction $S(D, A, \cdot, \cdot)$ is to be investigated furthermore. This is
the reason why we omit (c5) in the formulation of Theorem 11.4.

We observe that the sufficient condition in Theorem 11.2 for the
lsc property of the following multifunction

$$(D', A', c', b') \rightarrow S(D', A', c', b'), \qquad (11.30)$$

where $(D', A', c', b') \in R_S^{n \times n} \times R^{m \times n} \times R^n \times R^m$, can be reformulated
equivalently as follows.

Theorem 11.5. Let $(D, A, c, b) \in R_S^{n \times n} \times R^{m \times n} \times R^n \times R^m$. Suppose
that

(i)   the set $S(D, A, c, b)$ is finite, nonempty,

(ii)   the system $Ax \geq b$ is regular,

and suppose that for every $x \in S(D, A, c, b)$ at least one of the
following conditions holds:

(c1) $x \in \mathrm{loc}(D, A, c, b)$,

(c2) $Ax > b$.

Then, multifunction (11.30) is lower semicontinuous at (D, A, c, b).

It is easy to check that (c2) in the above theorem is equivalent
to (c2) in Theorem 11.4.

11.3 Commentaries

The material of this chapter is taken from Tam and Yen (1999) and Lee et al. (2002b, 2002c).

Chapter 12

Continuity of the Solution Map in Quadratic Programming

In this chapter we study the lower semicontinuity and the upper semicontinuity properties of the multifunction $(D, A, c, b) \mapsto \mathrm{Sol}(D, A, c, b)$, where $\mathrm{Sol}(D, A, c, b)$ denotes the solution set of the canonical quadratic programming problem.

12.1 USC Property of the Solution Map

Let $D \in R_S^{n \times n}$, $A \in R^{m \times n}$, $c \in R^n$, and $b \in R^m$. Consider the following QP problem of the canonical form:

$$(P) \qquad \begin{cases} \text{Minimize} \quad f(x) := \dfrac{1}{2} x^T D x + c^T x \\ \text{subject to} \quad Ax \geq b, \quad x \geq 0. \end{cases}$$

Let $\Delta(A, b)$, $\mathrm{Sol}(D, A, c, b)$, and $S(D, A, c, b)$ denote, respectively, the constraint set, the solution set, and the Karush-Kuhn-Tucker point set of (P).

In Chapter 10 we have studied the upper semicontinuity of the set-valued map

$$(D, A, c, b) \mapsto S(D, A, c, b).$$

In this section we will examine in detail the usc property of the solution map

$$(D, A, c, b) \mapsto \mathrm{Sol}(D, A, c, b). \tag{12.1}$$

.

A complete characterization of the lsc property of the map will be given in the next section.

Recall that the inequality system

$$Ax \geq b, \quad x \geq 0 \tag{12.2}$$

is called regular if there exists $x^0 \in R^n$ such that $Ax^0 > b$, $x^0 > 0$.

The next result is due to Nhan (1995), Theorem 3.4.

Theorem 12.1. *Assume that*

(a_1) $\mathrm{Sol}(D, A, 0, 0) = \{0\}$,

(a_2) *the system (12.2) is regular.*

Then, for any $c \in R^n$, the multifunction $\mathrm{Sol}(\cdot)$ is upper semicontinuous at (D, A, c, b).

Proof. To obtain a contradiction, suppose that there is a pair $(c, b) \in R^n \times R^m$ such that $\mathrm{Sol}(D, A, c, b) \neq \emptyset$ and there exist an open set Ω containing $\mathrm{Sol}(D, A, c, b)$, a sequence $\{(D^k, A^k, c^k, b^k)\}$ converging to (D, A, c, b), a sequence $\{x^k\}$ such that

$$x^k \in \mathrm{Sol}(D^k, A^k, c^k, b^k) \setminus \Omega \quad \text{for every } k \in N.$$

If the sequence $\{x^k\}$ is bounded, then there is no loss of generality in assuming that $x^k \to x^0$ for some $x^0 \in R^n$. It is clear that $x^0 \in \Delta(A, b)$. Fix any $x \in \Delta(A, b)$. By assumption (b_2), there exists a sequence $\{\xi^k\}$, $\xi^k \in \Delta(A^k, b^k)$ for all $k \in N$, such that $\lim\limits_{k \to \infty} x^k = x$ (see Lemma 13.1 in Chapter 13). Since $x^k \in \mathrm{Sol}(D^k, A^k, c^k, b^k)$, we have $f(x^k) \leq f(\xi^k)$. Letting $k \to \infty$ we get $f(x^0) \leq f(x)$. This shows that $x^0 \in \mathrm{Sol}(D, A, c, b) \subset \Omega$. We have arrived at a contradiction, because $x^k \notin \Omega$ for all k, and Ω is open.

Now suppose that the sequence $\{x^k\}$ is bounded. Without loss of generality we can assume that $\|x^k\|^{-1} x^k \to \bar{v}$, $\bar{v} \in \Delta(A, 0)$. Fix any $x \in \Delta(A, b)$. By (a_2) there exists a sequence $\{\xi^k\}$, $\xi^k \in \Delta(A^k, b^k)$ for all k and $\xi^k \to x$. Dividing the inequality

$$\frac{1}{2}(x^k)^T D^k x^k + (c^k)^T x^k \leq \frac{1}{2}(\xi^k)^T D^k \xi^k + (c^k)^T \xi^k$$

by $\|x^k\|^2$ and letting $k \to \infty$ we get $\bar{v}^T D \bar{v} \leq 0$. If $\bar{v}^T D \bar{v} < 0$, then $\mathrm{Sol}(D, A, 0, 0) = \emptyset$, contrary to (a_1). If $\bar{v}^T D \bar{v} = 0$, then we have $\bar{v} \in \mathrm{Sol}(D, A, 0, 0)$, which is also impossible.

The proof is complete.   □

Corollary 12.1. *If the system* (12.2) *is regular and if the set* $\Delta(A, b)$ *is bounded, then* Sol(·) *is upper semicontinuous at* (D, A, c, b).

Proof. Since the system (12.2) is regular, $\Delta(A, b)$ is nonempty. The boundedness of $\Delta(A, b)$ implies that $\Delta(A, 0) = \{0\}$. Hence Sol$(D, A, 0, 0) = \{0\}$, and the desired property follows from Theorem 12.1.   □

Condition (a_1) amounts to saying that $x^T D x > 0$ for every $x \in \Delta(A, 0) \setminus \{0\}$, i.e., the quadratic form $x^T D x$ is strictly copositive on the cone $\Delta(A, 0)$.

The next statement is a complement to Theorem 12.1.

Theorem 12.2. *Assume that:*

(b_1) $S(D, A, 0, 0) = \{0\}$,

(b_2) *the system* $Ax \geq 0$, $x \geq 0$ *is regular.*

Then, for any $(c, b) \in R^n \times R^m$, *the multifunction* Sol(·) *is upper semicontinuous at* (D, A, c, b).

Proof. Suppose that the assertion of the theorem is false. Then there is a pair $(c, b) \in R^n \times R^m$ such that Sol$(D, A, c, b) \neq \emptyset$ and there exist an open set Ω containing Sol(D, A, c, b), a sequence $\{(D^k, A^k, c^k, b^k)\}$ converging to (D, A, c, b), a sequence $\{x^k\}$ such that

$$x^k \in \text{Sol}(D^k, A^k, c^k, b^k) \setminus \Omega \quad \text{for every } k \in N.$$

If the sequence $\{x^k\}$ is bounded, then we can assume that $x^k \to x^0$ for some $x^0 \in R^n$. We have $x^0 \in \Delta(A, b)$. Fix any $x \in \Delta(A, b)$. Assumption (b_2) implies that system (12.2) is regular. Then there exists a sequence $\{\xi^k\}$, $\xi^k \in \Delta(A^k, b^k)$ for all $k \in N$, such that $\lim_{k \to \infty} x^k = x$. Since $f(x^k) \leq f(\xi^k)$, letting $k \to \infty$ we obtain $f(x^0) \leq f(x)$. Thus $x^0 \in \text{Sol}(D, A, c, b) \subset \Omega$. This contradicts the fact that $x^k \notin \Omega$ for all k.

Now assume that $\{x^k\}$ is unbounded. By taking a subsequence if necessary, we may assume that $||x^k|| \to \infty$. Since $x^k \in \text{Sol}(D^k, A^k, c^k, b^k)$, for each k there exists $\lambda^k \in R^m$ such that

$$D^k x^k - (A^k)^T \lambda^k + c^k \geq 0, \quad A^k x^k - b^k \geq 0, \qquad (12.3)$$

$$x^k \geq 0, \quad \lambda^k \geq 0, \qquad (12.4)$$

$$(x^k)^T (D^k x^k - (A^k)^T \lambda^k + c^k) + (\lambda^k)^T (A^k x^k - b^k) = 0. \qquad (12.5)$$

Since $||(x^k, \lambda^k)|| \to \infty$, without loss of generality we can assume that $||(x^k, \lambda^k)|| \neq 0$ for all k, and the sequence of vectors

$$\frac{(x^k, \lambda^k)}{||(x^k, \lambda^k)||} = \left(\frac{x^k}{||(x^k, \lambda^k)||}, \frac{\lambda^k}{||(x^k, \lambda^k)||} \right)$$

converges to some $(\bar{x}, \bar{\lambda}) \in R^n \times R^m$ with $||(\bar{x}, \bar{\lambda})|| = 1$. Dividing both sides of (12.3) and of (12.4) by $||(x^k, \lambda^k)||$, dividing both sides of (12.5) by $||(x^k, \lambda^k)||^2$, and taking the limits as $k \to \infty$, we obtain

$$D\bar{x} - A^T\bar{\lambda} \geq 0, \quad A\bar{x} \geq 0, \tag{12.6}$$

$$\bar{x} \geq 0, \quad \bar{\lambda} \geq 0, \tag{12.7}$$

$$\bar{x}^T(D\bar{x} - A^T\bar{\lambda}) + \bar{\lambda}^T A\bar{x} = 0. \tag{12.8}$$

The system (12.6)–(12.8) proves that $\bar{x} \in S(D, A, 0, 0)$. By (b_1), $\bar{x} = 0$. Hence

$$-A^T\bar{\lambda} \geq 0, \quad \bar{\lambda} \geq 0. \tag{12.9}$$

Combining (12.9) and (b_2) yields $\bar{\lambda} = 0$ (see Lemma 10.1), hence $||(\bar{x}, \bar{\lambda})|| = 0$, a contradiction. The proof is complete. $\quad \square$

Remark 12.1. Since $\Delta(A, b) + \Delta(A, 0) \subset \Delta(A, b)$, (b_2) implies (a_2). However, (b_1) does not imply (a_1).

Observe that neither (a_1) nor (a_2) is a necessary condition for the upper semicontinuity of the solution map $\mathrm{Sol}(\cdot)$ at a given point (D, A, c, b).

Example 12.1. Let $n = m = 1$, $D = [0]$, $A = [1]$, $c = 1$, $b = 1$. It is easily verified that $\mathrm{Sol}(D, A, c, b) = \{1\}$ and the multifunction $\mathrm{Sol}(\cdot)$ is usc at (D, A, c, b). Meanwhile, $\mathrm{Sol}(D, A, 0, 0) = \{x \in R : x \geq 0\}$, so (a_1) fails to hold.

Example 12.2. Let $n = m = 1$, $A = [-1]$, $b = 0$. If $A' = [-1 + \alpha]$, $b' = \beta$, where α and β are sufficiently small, then $\Delta(A', b') = \left\{ x \in R : 0 \leq x \leq \dfrac{-\beta}{1 - \alpha} \right\}$. It is easily seen that, for arbitrarily chosen D and c, the multifunction $\mathrm{Sol}(\cdot)$ is usc at (D, A, c, b), while condition (a_2) does not hold.

12.2 LSC Property the Solution Map

By definition, multifunction $\mathrm{Sol}(\cdot)$ is continuous at (D, A, c, b) if it is simultaneously upper semicontinuous and lower semicontinuous at that point.

Our main result in this section can be stated as follows.

Theorem 12.3. *The solution map* $\mathrm{Sol}(\cdot)$ *of* (P) *is lower semicontinuous at* (D, A, c, b) *if and only if the following three conditions are satisfied:*

(a) *the system* $Ax \geq b$, $x \geq 0$ *is regular,*

(b) $\mathrm{Sol}(D, A, 0, 0) = \{0\}$,

(c) $|\mathrm{Sol}(D, A, c, b)| = 1$.

For proving Theorem 12.3 we need some lemmas.

Lemma 12.1. *If* $\mathrm{Sol}(\cdot)$ *is lower semicontinuous at* (D, A, c, b), *then the system* $Ax \geq b$, $x \geq 0$ *is regular.*

Proof. If the system $Ax \geq b$, $x \geq 0$ is irregular then, according to Lemma 3 in Robinson (1977), there exists a sequence $(A^k, b^k) \in R^{m \times n} \times R^m$ tending to (A, b) such that $\Delta(A^k, b^k) = \emptyset$ for each k. Therefore, $\mathrm{Sol}(D, A^k, c, b^k) = \emptyset$ for each k, contrary to the assumed lower semicontinuity of the solution map. $\square$

Lemma 12.2. *If the multifunction* $\mathrm{Sol}(\cdot)$ *is lower semicontinuous at* (D, A, c, b), *then* $\mathrm{Sol}(D, A, 0, 0) = \{0\}$.

Proof. On the contrary, suppose that $\mathrm{Sol}(D, A, 0, 0) \neq \{0\}$. Then there is a nonzero vector $\bar{x} \in R^n$ such that

$$A\bar{x} \geq 0, \quad \bar{x} \geq 0, \quad \bar{x}^T D \bar{x} \leq 0. \tag{12.10}$$

Since $\Delta(A, b) \neq \emptyset$, from (12.10) it follows that $\Delta(A, b)$ is unbounded. For every $\varepsilon > 0$, we get from (12.10) that $\bar{x}^T(D - \varepsilon E)\bar{x} < 0$. Hence, for any $x \in \Delta(A, b)$,

$$f(x + t\bar{x}) = \frac{1}{2}(x + t\bar{x})^T(D - \varepsilon E)(x + t\bar{x}) + c^T(x + t\bar{x}) \to -\infty$$

as $t \to \infty$. Thus, $\mathrm{Sol}(D - \varepsilon E, A, c, b) = \emptyset$. This contradicts our assumption that $\mathrm{Sol}(\cdot)$ is lower semicontinuous at (D, A, c, b). $\square$

Lemma 12.3. (i) *If* $\mathrm{Sol}(D, A, 0, 0) = \{0\}$ *then, for any* $(c, b) \in R^n \times R^m$, $\mathrm{Sol}(D, A, c, b)$ *is a compact set.*

(ii) *If* $\mathrm{Sol}(D, A, 0, 0) = \{0\}$ *and if* $\Delta(A, b)$ *is nonempty, then* $\mathrm{Sol}(D, A, c, b)$ *is nonempty for every* $c \in R^n$.

Proof. (i) Suppose that $\mathrm{Sol}(D, A, 0, 0) = \{0\}$, but $\mathrm{Sol}(D, A, c, b)$ is unbounded for some (c, b). Then there is a sequence $\{x^k\} \subset$

$\text{Sol}(D, A, c, b)$ such that $||x^k|| \to \infty$ as $k \to \infty$. Fixing any $x \in \Delta(A, b)$, one has

$$\frac{1}{2}(x^k)^T D x^k + c^T x^k \leq \frac{1}{2} x^T D x + c^T x, \qquad (12.11)$$

$$A x^k \geq b, \quad x^k \geq 0. \qquad (12.12)$$

We can assume that the sequence $||x^k||^{-1} x^k$ converges to some $\bar{x}$ with $||\bar{x}|| = 1$. Using (12.11) and (12.12) it is easy to show that $\bar{x}^T D \bar{x} \leq 0$, $A \bar{x} \geq 0$, $\bar{x} \geq 0$. This contradicts the fact that $\text{Sol}(D, A, 0, 0) = \{0\}$. We have thus proved that $\text{Sol}(D, A, c, b)$ is a bounded set. Then $\text{Sol}(D, A, c, b)$, being closed, is a compact set.

(ii) Let $\text{Sol}(D, A, 0, 0) = \{0\}$, $\Delta(A, b) \neq \emptyset$, and let $c \in R^n$ be given arbitrarily. If the quadratic form $f(x) = \frac{1}{2} x^T D x + c^T x$ is bounded below on the polyhedron $\Delta(A, b)$ then, by the Frank-Wolfe Theorem (see Theorem 2.1), the solution set $\text{Sol}(D, A, c, b)$ is nonempty. Now assume that there exists a sequence $x^k \in \Delta(A, b)$ such that $f(x^k) \to -\infty$ as $k \to \infty$. By taking a subsequence, if necessary, we can assume that

$$\frac{1}{2}(x^k)^T D x^k + (c^k)^T x^k \leq 0 \qquad (12.13)$$

for all k, $||x^k|| \to \infty$, and $||x^k||^{-1} x^k$ converges to some $\bar{x}$ as $k \to \infty$. It is a simple matter to show that $\bar{x} \in \Delta(A, 0)$. Dividing both sides of (12.13) by $||x^k||^2$ and letting $k \to \infty$ one has $\bar{x}^T D \bar{x} \leq 0$. As $||\bar{x}|| = 1$, one gets $\text{Sol}(D, A, 0, 0) \neq \{0\}$, which is impossible. $\square$

We omit the proof of the next lemma, because it is similar to the proof of Theorem 11.1.

Lemma 12.4. *If the multifunction* $\text{Sol}(\cdot)$ *is lower semicontinuous at* (D, A, c, b), *then the set* $\text{Sol}(D, A, c, b)$ *is finite.*

Lemma 12.5. *The set* $G := \{(D, A) : \text{Sol}(D, A, 0, 0) = \{0\}\}$ *is open in* $R_S^{n \times n} \times R^{m \times n}$.

Proof. On the contrary, suppose that there is a sequence $\{(D^k, A^k)\}$ converging to $(D, A) \in G$ such that $\text{Sol}(D^k, A^k, 0, 0) \neq \{0\}$ for all k. Then for each k there exists a vector x^k such that $||x^k|| = 1$ and

$$A^k x^k \geq 0, \quad x^k \geq 0, \quad (x^k)^T D^k x^k \leq 0. \qquad (12.14)$$

Without loss of generality, we can assume that $\{x^k\}$ converges to some x^0 with $||x^0|| = 1$. Taking the limits in (12.14) as $k \to \infty$, we obtain

$$A x^0 \geq 0, \quad x^0 \geq 0, \quad (x^0)^T D x^0 \leq 0.$$

This contradicts the assumption that $\text{Sol}(D, A, 0, 0) = \{0\}$. The proof of is complete. □

For each subset $\alpha \subset \{1, \ldots, m\}$ with the complement $\bar{\alpha}$, and for each subset $\beta \subset \{1, \ldots, n\}$ with the complement $\bar{\beta}$, let

$$F(\alpha, \beta) := \{x \in R^n : (Ax)_\alpha > b_\alpha, (Ax)_{\bar{\alpha}} = b_{\bar{\alpha}}, x_\beta > 0, x_{\bar{\beta}} = 0\}.$$

Note that $F(\alpha, \beta)$ is a pseudo-face of $\Delta(A, b)$. Obviously,

$$\Delta(A, b) = \bigcup_{(\alpha, \beta)} F(\alpha, \beta).$$

Besides, for every $x \in \Delta(A, b)$ there exists a unique pair (α, β) such that $x \in F(\alpha, \beta)$. In addition, if $(\alpha, \beta) \neq (\alpha', \beta')$ then $F(\alpha, \beta) \cap F(\alpha', \beta') = \emptyset$.

The following lemma is immediate from Theorem 4.5.

Lemma 12.6. *If the solution set* $\text{Sol}(D, A, c, b)$ *is finite then, for any* $\alpha \subset \{1, \ldots, m\}$ *and for any* $\beta \subset \{1, \ldots, n\}$, *we have*

$$|\text{Sol}(D, A, c, b) \cap F(\alpha, \beta)| \leq 1.$$

Lemma 12.7. *If the multifunction* $\text{Sol}(\cdot)$ *is lower semicontinuous at* (D, A, c, b), *then*

$$|\text{Sol}(D, A, c, b)| = 1.$$

Proof. On the contrary, suppose that in $\text{Sol}(D, A, c, b)$ we can find two distinct vectors $\bar{x}$, $\bar{y}$. Let $J(\bar{x}) = \{j : \bar{x}_j = 0\}$, $J(\bar{y}) = \{j : \bar{y}_j = 0\}$.

If $J(\bar{x}) \neq J(\bar{y})$, then there exists j_0 such that $\bar{x}_{j_0} = 0$ and $\bar{y}_{j_0} > 0$, or there exists j_1 such that $\bar{x}_{j_1} > 0$ and $\bar{y}_{j_1} = 0$. By symmetry, it is enough to consider the first case. As $\bar{y} \in \text{Sol}(D, A, c, b)$ and $y_{j_0} > 0$, there is an open neighborhood U of $\bar{y}$ such that $f(y) \geq f(\bar{y})$ and $y_{j_0} > 0$ for every $y \in U$. Fix any $\varepsilon > 0$ and put $c(\varepsilon) = (c_i(\varepsilon))$ where

$$c_i(\varepsilon) = \begin{cases} c_i & \text{if } i \neq j_0 \\ c_i + \varepsilon & \text{if } i = j_0. \end{cases}$$

Let $f_\varepsilon(x) = f(x) + \varepsilon x_{j_0}$, where, as before, $f(x) = \frac{1}{2}x^T D x + c^T x$. Consider the quadratic program

$$\text{Minimize } f_\varepsilon(x) \quad \text{subject to } x \in \Delta(A, b),$$

whose solution set is $\mathrm{Sol}(D, A, c(\varepsilon), b)$. For every $y \in U$, we have

$$f_\varepsilon(y) = f(y) + \varepsilon y_{j_0} > f(y) \geq f(\bar{y})$$
$$= f(\bar{x}) = f_\varepsilon(\bar{x}).$$

Hence $y \notin \mathrm{Sol}(D, A, c(\varepsilon), b)$. So

$$\mathrm{Sol}(D, A, c(\varepsilon), b) \cap U = \emptyset. \tag{12.15}$$

Since $\varepsilon > 0$ can be arbitrarily small, (12.15) contradicts our assumption that $\mathrm{Sol}(\cdot)$ is lower semicontinuous at (D, A, c, b).

We now suppose that $J(\bar{x}) = J(\bar{y})$. Let α and α' be the index sets such that

$$\bar{x} \in F(\alpha, \beta), \quad \bar{y} \in F(\alpha', \beta),$$

where β is the complement of $J(\bar{x}) = J(\bar{y})$ in $\{1, \ldots, n\}$. By Lemma 12.4, $\mathrm{Sol}(D, A, c, b)$ is a finite set. Then, by Lemma 12.6, $\alpha \neq \alpha'$. Hence at least one of the sets $\alpha \setminus \alpha'$ and $\alpha' \setminus \alpha$ must be nonempty. By symmetry, it suffices to consider the first case. Let $i_0 \in \alpha \setminus \alpha'$. Then we have

$$(A\bar{x})_{i_0} > b_{i_0}, \quad (A\bar{y})_{i_0} = b_{i_0}. \tag{12.16}$$

As $\mathrm{Sol}(D, A, c, b)$ is finite, one can find a neighborhood W of $\bar{y}$ such that

$$\mathrm{Sol}(D, A, c, b) \cap W = \{\bar{y}\}. \tag{12.17}$$

Fix any $\varepsilon > 0$ and put $b(\varepsilon) = (b_i(\varepsilon))$, where

$$b_i(\varepsilon) = \begin{cases} b_i & \text{if } i \neq i_0 \\ b_i + \varepsilon & \text{if } i = i_0. \end{cases}$$

By (12.16), there exists $\delta > 0$ such that $\bar{x} \in \Delta(A, b(\varepsilon))$ for every $\varepsilon \in (0, \delta)$. Since $\Delta(A, b(\varepsilon)) \subset \Delta(A, b)$, we have

$$\inf_{x \in \Delta(A, b(\varepsilon))} f(x) \geq \inf_{x \in \Delta(A, b)} f(x) = f(\bar{x}).$$

Therefore, for every $\varepsilon \in (0, \delta)$, $\bar{x} \in \mathrm{Sol}(D, A, c, b(\varepsilon))$. Moreover,

$$\mathrm{Sol}(D, A, c, b(\varepsilon)) \subset \mathrm{Sol}(D, A, c, b).$$

It is clear that $\bar{y} \notin \Delta(A, b(\varepsilon))$. Then we have $\mathrm{Sol}(D, A, c, b(\varepsilon)) \subset \mathrm{Sol}(D, A, c, b) \setminus \{\bar{y}\}$. Hence, by (12.17), $\mathrm{Sol}(D, A, c, b(\varepsilon)) \cap W = \emptyset$ for every $\varepsilon \in (0, \delta)$. This contradicts the lower semicontinuity of $\mathrm{Sol}(\cdot)$ at (D, A, c, b). Lemma 12.7 is proved. $\square$

Proof of Theorem 12.3.

If $\text{Sol}(\cdot)$ is lower semicontinuous at (D, A, c, b) then from Lemmas 12.1, 12.2, and 12.7, we get (a), (b), and (c).

Conversely, assume that the conditions (a), (b) and (c) are fulfilled. Let Ω be an open set containing the unique solution $\bar{x} \in \text{Sol}(D, A, c, b)$. By (a), there exists $\delta_1 > 0$ such that $\Delta(A', b') \neq \emptyset$ for every pair (A', b') satisfying $\max\{\|A' - A\|, \|b' - b\|\} < \delta_1$ (see Lemma 13.1 in Chapter 13). By (b) and by Lemma 12.5, there exists $\delta_2 > 0$ such that $\text{Sol}(D', A', 0, 0) = \{0\}$ for every pair (D', A') satisfying $\max\{\|D' - D\|, \|A' - A\|\} \leq \delta_2$. For $\delta := \min\{\delta_1, \delta_2\}$, by the second assertion of Lemma 12.3 we have $\text{Sol}(D', A', c', b') \neq \emptyset$ for every (D', A', c', b') satisfying

$$\max\{\|D' - D\|, \|A' - A\|, \|c' - c\|, \|b' - b\|\} < \delta. \qquad (12.18)$$

By (a) and (b), it follows from Theorem 2.1 that $\text{Sol}(\cdot)$ is upper semicontinuous at (D, A, c, b). Hence $\text{Sol}(D', A', c', b') \subset \Omega$ for every (D', A', c', b') satisfying (12.18), if $\delta > 0$ is small enough. For such a δ, from what has been said it follows that $\text{Sol}(D', b', c', b') \cap \Omega \neq \emptyset$ for every (D', A', c', b') satisfying (12.18). This shows that $\text{Sol}(\cdot)$ is lower semicontinuous at (D, A, c, b). The proof is complete. $\square$

The following fact follows directly from Theorems 12.3 and 12.1.

Corollary 12.2. *If the multifunction* $\text{Sol}(\cdot)$ *is lower semicontinuous at* (D, A, c, b) *then it is upper semicontinuous at* (D, A, c, b), *hence it is continuous at the point.*

Let us mention two other interesting consequences of Theorem 12.3.

Corollary 12.3. *If* D *is a negative semidefinite matrix, then the multifunction* $\text{Sol}(\cdot)$ *is continuous at* (D, A, c, b) *if and only if the following conditions are satisfied*

(i) *the system* $Ax \geq b$, $x \geq 0$ *is regular,*

(ii) $\Delta(A, b)$ *is a compact set, and*

(iii) $|\text{Sol}(D, A, c, b)| = 1$.

Proof. Assume that $\text{Sol}(\cdot)$ is lower semicontinuous at (D, A, c, b). By Theorem 12.3, conditions (i) and (iii) are satisfied. Moreover,

$$\text{Sol}(D, A, 0, 0) = \{0\}. \qquad (12.19)$$

We claim that $\Delta(A, 0) = \{0\}$. Indeed, by assumption, $x^T D x \leq 0$ for every $x \in \Delta(A, 0)$. If there exists no $\bar{x} \in \Delta(A, 0)$ with the property that $\bar{x}^T D \bar{x} < 0$ then $\mathrm{Sol}(D, A, 0, 0) = \Delta(A, 0)$, and (12.19) forces $\Delta(A, 0) = \{0\}$. If $\bar{x}^T D \bar{x} < 0$ for some $\bar{x} \in \Delta(A, 0)$ then it is obvious that $\mathrm{Sol}(D, A, 0, 0) = \emptyset$, which is impossible. Our claim is proved. Property (ii) follows directly from the equality $\Delta(A, 0) = \{0\}$.

Conversely, suppose that (i), (ii) and (iii) are satisfied. As $\Delta(A, b) \neq \emptyset$ by (i), assumption (ii) implies that $\Delta(A, 0) = \{0\}$. Therefore, $\mathrm{Sol}(D, A, 0, 0) = \{0\}$. Since the conditions (a), (b) and (c) in Theorem 12.3 are satisfied, $\mathrm{Sol}(\cdot)$ is lower semicontinuous at (D, A, c, b). The proof is complete. $\square$

Corollary 12.4. *If D is a positive definite matrix, then the multi-function $\mathrm{Sol}(\cdot)$ is continuous at (D, A, c, b) if and only if the system $Ax \geq b$, $x \geq 0$ is regular.*

The proof of this corollary is simple, so it is omitted.

12.3   Commentaries

The material of this chapter is adapted from Tam (1999). The proof of the 'necessity' part of Theorem 12.3 can be shortened greatly by using an argument in Phu and Yen (2001).

Chapter 13

Continuity of the Optimal Value Function in Quadratic Programming

In this chapter we will characterize the continuity property of the optimal value function in a general parametric QP problem. The lower semicontinuity and upper semicontinuity properties of the optimal value function are studied as well. Directional differentiability of the optimal value function in QP problems will be addressed in the next chapter.

13.1   Continuity of the Optimal Value Function

Consider the following general quadratic programming problem with linear constraints, which will be denoted by $QP(D, A, c, b)$,

$$\begin{cases} \text{Minimize } f(x, c, D) := \dfrac{1}{2}x^T D x + c^T x \\ \text{subject to } x \in \Delta(A, b) := \{x \in R^n : Ax \geq b\} \end{cases} \tag{13.1}$$

depending on the parameter $\omega = (D, A, c, b) \in \Omega$, where

$$\Omega := R_S^{n \times n} \times R^{m \times n} \times R^n \times R^m.$$

The solution set of (13.1) will be denoted by $\text{Sol}(D, A, c, b)$. The function $\varphi : \Omega \longrightarrow \overline{R}$ defined by

$$\varphi(\omega) = \inf\{f(x, c, D) : x \in \Delta(A, b)\}.$$

is the optimal value function of the parametric problem (13.1).

If $v^T D v \geq 0$ (resp., $v^T D v \leq 0$) for all $v \in R^n$ then $f(\cdot, c, D)$ is a convex (resp., concave) function and (13.1) is a convex (resp., concave) QP problem. If such conditions are not required then (13.1) is an *indefinite* QP problem (see Section 1.5).

In this section, complete characterizations of the continuity of the function φ at a given point are obtained. In Section 13.2, sufficient conditions for the upper and lower semicontinuity of φ at a given point will be established. For proving the results, we rely on some results due to Robinson (1975, 1977) on stability of the feasible region $\Delta(A, b)$ and the Frank-Wolfe Theorem.

Before obtaining the desired characterizations, we state some lemmas.

Lemma 13.1. *Let $A \in R^{m \times n}$, $b \in R^m$. The system $Ax \geq b$ is regular if and only if the multifunction $\Delta(\cdot) : R^{m \times n} \times R^m \longrightarrow 2^{R^n}$, defined by $\Delta(A', b') = \{x \in R^n : A'x \geq b'\}$, is lower semicontinuous at (A, b).*

Proof. Suppose that $Ax \geq b$ is a regular system and $x^0 \in R^n$ is such that $Ax^0 > b$. Obviously, $\Delta(A, b)$ is nonempty. Let V be an open subset in R^n satisfying $\Delta(A, b) \cap V \neq \emptyset$. Take $x \in \Delta(A, b) \cap V$. For every $t \in [0, 1]$, we set

$$x_t := (1 - t)x + tx^0.$$

Since $x_t \to x$ as $t \to 0$, there is $t_0 > 0$ such that $x_{t_0} \in V$. Since

$$Ax_{t_0} = (1 - t_0)Ax + t_0 Ax^0 > (1 - t_0)b + t_0 b = b,$$

there exists $\delta_{t_0} > 0$ such that

$$A'x_{t_0} > b'$$

for all $(A', b') \in R^{m \times n} \times R^m$ satisfying

$$\|(A', b') - (A, b)\| < \delta_{t_0}. \tag{13.2}$$

Thus $x_t \in \Delta(A', b')$ for every (A', b') fulfilling (13.2). Therefore $\Delta(\cdot)$ is lower semicontinuous at (A, b).

Conversely, if $\Delta(\cdot)$ is lower semicontinuous at (A, b) then there exists $\delta > 0$ such that $Ax \geq b'$ is solvable for every $b' \in R^m$ satisfying $b' > b$ and $\|b' - b\| < \delta$. This implies that $Ax > b$ is solvable. Thus $Ax \geq b$ is a regular system. $\square$

Remark 13.1. If the inequality system $Ax \geq b$ is irregular then there exists a sequence $\{(A^k, b^k)\}$ in $R^{m \times n} \times R^m$ converging to (A, b) such that, for every k, the system $A^k x \geq b^k$ has no solutions. This fact follows from the results of Robinson (1977).

Lemma 13.2 (cf. Robinson (1977), Lemma 3). *Let $A \in R^{m \times n}$. If the system $Ax \geq 0$ is regular then, for every $b \in R^m$, the system $Ax \geq b$ is regular.*

Proof. Assume that $Ax \geq 0$ is a regular and $\bar{x} \in R^n$ is such that $A\bar{x} > 0$. Setting $\bar{b} = A\bar{x}$, we have $\bar{b} > 0$. Let $b \in R^m$ be given arbitrarily. Then there exists $t > 0$ such that $t\bar{b} > b$. We have $A(t\bar{x}) = tA\bar{x} = t\bar{b}$. Therefore $A(t\bar{x}) > b$, hence the system $Ax \geq b$ is regular. $\square$

The set

$$G := \{(D, A) \in R_S^{n \times n} \times R^{m \times n} : \text{Sol}(D, A, 0, 0) = \{0\}\} \quad (13.3)$$

is open in $R_S^{n \times n} \times R^{m \times n}$. This fact can be proved similarly as Lemma 12.5. It is worthy to stress that Lemma 12.5 is applicable only to canonical QP problems while, in this chapter, the standard QP problems are considered.

Lemma 13.3. *If $\Delta(A, b)$ is nonempty and if $\text{Sol}(D, A, 0, 0) = \{0\}$ then, for every $c \in R^n$, $\text{Sol}(D, A, c, b)$ is a nonempty compact set.*

Proof. Let $\Delta(A, b)$ be nonempty and $\text{Sol}(D, A, 0, 0) = \{0\}$. Suppose that $\text{Sol}(D, A, c, b) = \emptyset$ for some $c \in R^n$. By the Frank-Wolfe Theorem, there exists a sequence $\{x^k\}$ such that $Ax^k \geq b$ for every k and

$$f(x^k, c, D) = \frac{1}{2}(x^k)^T Dx^k + c^T x^k \to -\infty \quad \text{as} \quad k \to \infty.$$

It is clear that $\|x^k\| \to +\infty$ as $k \to \infty$. By taking a subsequence if necessary, we can assume that $\|x^k\|^{-1} x^k \to \bar{x} \in R^n$ and

$$f(x^k, c, D) = \frac{1}{2}(x^k)^T Dx^k + c^T x^k < 0 \quad \text{for every } k. \quad (13.4)$$

We have

$$A \frac{x^k}{\|x^k\|} \geq \frac{b}{\|x^k\|}.$$

Letting $k \to \infty$, we obtain $\bar{x} \in \Delta(A, 0)$. Dividing both sides of the inequality in (13.4) by $\|x^k\|^2$ and letting $k \to \infty$, we get $\bar{x}^T D\bar{x} \leq 0$. Since $\|\bar{x}\| = 1$, we have $\text{Sol}(D, A, 0, 0) \neq \{0\}$. This contradicts the assumption $\text{Sol}(D, A, 0, 0) = \{0\}$. Thus $\text{Sol}(D, A, c, b)$ is nonempty for each $c \in R^n$.

Suppose, contrary to our claim, that $\text{Sol}(D, A, c, b)$ is unbounded for some $c \in R^n$. Then there exists a sequence $\{x^k\} \subset \text{Sol}(D, A, c, b)$ such that $\|x^k\| \to \infty$ as $k \to \infty$ and $\{\|x^k\|^{-1} x^k\}$ converges to a certain $\bar{x} \in R^n$. Taking any $x \in \Delta(A, b)$, we have

$$\frac{1}{2}(x^k)^T D x^k + c^T x^k \leq \frac{1}{2} x^T D x + c^T x, \qquad (13.5)$$

$$A x^k \geq b. \qquad (13.6)$$

Dividing both sides of (13.5) by $\|x^k\|^2$, both sides of (13.6) by $\|x^k\|$, and letting $k \to \infty$, we obtain

$$\bar{x}^T D \bar{x} \leq 0, \quad A \bar{x} \geq 0.$$

Thus $\text{Sol}(D, A, 0, 0) \neq \{0\}$, a contradiction. We have proved that, for every $c \in R^n$, the solution set $\text{Sol}(D, A, c, b)$ is bounded. Fixing any $\bar{x} \in \text{Sol}(D, A, c, b)$ one has

$$\text{Sol}(D, A, c, b) = \{x \in \Delta(A, b) : f(x, c, D) = f(\bar{x}, c, D)\}.$$

Hence $\text{Sol}(D, A, c, b)$ is a closed set and, therefore, $\text{Sol}(D, A, c, b)$ is a compact set. $\square$

We are now in a position to state our first theorem on the continuity of the optimal value function φ. This theorem gives a set of conditions which is necessary and sufficient for the continuity of φ at a point $\omega = (D, A, c, b)$ where φ has a finite value.

Theorem 13.1. *Let $(D, A, c, b) \in \Omega$. Assume that $\varphi(D, A, c, b) \neq \pm\infty$. Then, the optimal value function $\varphi(\cdot)$ is continuous at (D, A, c, b) if and only if the following two conditions are satisfied:*

(a) *the system $Ax \geq b$ is regular,*

(b) $\text{Sol}(D, A, 0, 0) = \{0\}$.

Proof. *Necessity:* First, suppose that $\varphi(\cdot)$ is continuous at $\omega := (D, A, c, b)$ and $\varphi(\omega) \neq \pm\infty$. If (a) is violated then, by Remark 13.1, there exists a sequence $\{(A^k, b^k)\}$ in $R^{m \times n} \times R^m$ converging to (A, b) such that, for every k, the system $A^k x \geq b^k$ has no solutions. Consider the sequence $\{(D, A^k, c, b^k)\}$. Since $\Delta(A^k, b^k) = \emptyset$,

$\varphi(D, A^k, c, b^k) = +\infty$ for every k. As $\varphi(\cdot)$ is continuous at ω and $\{(D, A^k, c, b^k)\}$ converges to ω, we have

$$\lim_{k \to \infty} \varphi(D, A^k, c, b^k) = \varphi(D, A, c, b) \neq \pm\infty.$$

We have arrived at a contradiction. Thus (a) is fulfilled.

Now we suppose that (b) fails to hold. Then there is a nonzero vector $\bar{x} \in R^n$ such that

$$A\bar{x} \geq 0, \quad \bar{x}^T D\bar{x} \leq 0. \tag{13.7}$$

Consider the sequence $\{(D^k, A, c, b)\}$, where $D^k := D - \dfrac{1}{k}E$, E is the unit matrix in $R^{n \times n}$. From the assumption $\varphi(\omega) \neq \pm\infty$ it follows that $\Delta(A, b)$ is nonempty. Then from (13.7) we can deduce that $\Delta(A, b)$ is unbounded. For every k, by (13.7) we have

$$\bar{x}^T D^k \bar{x} = \bar{x}^T (D - \frac{1}{k}E)\bar{x} < 0.$$

Hence, for any x belonging to $\Delta(A, b)$ and for any $t > 0$, we have $x + t\bar{x} \in \Delta(A, b)$ and

$$f(x + t\bar{x}, c, D^k) = \frac{1}{2}(x + t\bar{x})^T D^k (x + t\bar{x}) + c^T(x + t\bar{x}) \to -\infty$$

as $t \to \infty$. This implies that, for all k, $\text{Sol}(D^k, A, c, b) = \emptyset$ and $\varphi(D^k, A, c, b) = -\infty$. We have arrived at a contradiction, because $\varphi(\cdot)$ is continuous at ω and $\varphi(\omega) \neq \pm\infty$. We have proved that (b) holds true.

Sufficiency: Suppose that (a), (b) are satisfied and

$$\{(D^k, A^k, c^k, b^k)\} \subset \Omega$$

is a sequence converging to ω. By Lemma 13.1, assumption (a) implies the existence of a positive integer k_0 such that $\Delta(A^k, b^k) \neq \emptyset$ for every $k \geq k_0$. From assumption (b) it follows that the set G defined by (13.3) is open. Hence there exists a positive integer $k_1 \geq k_0$ such that $\text{Sol}(D^k, A^k, 0, 0) = \{0\}$ for every $k \geq k_1$. By Lemma 13.3, $\text{Sol}(D^k, A^k, c^k, b^k) \neq \emptyset$ for every $k \geq k_1$. Therefore, for every $k \geq k_1$ there exists $x^k \in R^n$ satisfying

$$\varphi(D^k, A^k, c^k, b^k) = \frac{1}{2}(x^k)^T Dx^k + (c^k)^T x^k, \tag{13.8}$$

$$A^k x^k \geq b^k. \tag{13.9}$$

Since $\varphi(\omega) \neq \pm\infty$, the Frank-Wolfe Theorem shows that

$$\mathrm{Sol}(D, A, c, b) \neq \emptyset.$$

Taking any $x^0 \in \mathrm{Sol}(D, A, c, b)$, we have

$$\varphi(D, A, c, b) = \frac{1}{2}(x^0)^T D x^0 + c^T x^0, \tag{13.10}$$

$$A x^0 \geq b. \tag{13.11}$$

By Lemma 13.1, there exists a sequence $\{y^k\}$ in R^n converging to x^0 and

$$A^k y^k \geq b^k \quad \text{for every } k \geq k_1. \tag{13.12}$$

From (13.12) it follows that $y^k \in \Delta(A^k, b^k)$ for $k \geq k_1$. So

$$\varphi(D^k, A^k, c^k, b^k) \leq \frac{1}{2}(y^k)^T D^k y^k + (c^k)^T y^k. \tag{13.13}$$

From (13.13) it follows that

$$\limsup_{k \to \infty} \varphi(D^k, A^k, c^k, b^k)$$

$$\leq \limsup_{k \to \infty} \left[\frac{1}{2}(y^k)^T D^k y^k + (c^k)^T y^k \right]$$

$$= \lim_{k \to \infty} \left[\frac{1}{2}(y^k)^T D^k y^k + (c^k)^T y^k \right].$$

Therefore, taking account of (13.10) and (13.11), we get

$$\limsup_{k \to \infty} \varphi(D^k, A^k, c^k, b^k) \leq \varphi(D, A, c, b). \tag{13.14}$$

We now claim that the sequence $\{x^k\}$, $k \geq k_1$, is bounded. Indeed, if it is unbounded then, by taking a subsequence if necessary, we can assume that $\|x^k\| \to \infty$ as $k \to \infty$ and $\|x^k\| \neq 0$ for all $k \geq k_1$. Then the sequence $\{\|x^k\|^{-1} x^k\}$, $k \geq k_1$, has a convergent subsequence. Without loss of generality we can assume that $\|x^k\|^{-1} x^k \to \hat{x}$, $\|\hat{x}\| = 1$. From (13.9) we have

$$A^k \frac{x^k}{\|x^k\|} \geq \frac{b^k}{\|x^k\|}.$$

Letting $k \to \infty$, we obtain

$$A\hat{x} \geq 0. \tag{13.15}$$

By (13.8) and (13.13),

$$\frac{1}{2}(x^k)^T D^k x^k + (c^k)^T x^k \le \frac{1}{2}(y^k)^T D^k y^k + (c^k)^T y^k. \qquad (13.16)$$

Dividing both sides of (13.16) by $\|x^k\|^2$ and taking limits as $k \to \infty$, we get

$$\hat{x}^T D\hat{x} \le 0. \qquad (13.17)$$

By (13.15) and (13.17), we have $\mathrm{Sol}(D, A, 0, 0) \ne \{0\}$. This contradicts (b). We have thus shown that the sequence $\{x^k\}$, $k \ge k_1$, is bounded; hence it has a convergent sequence. There is no loss of generality in assuming that $x^k \to \tilde{x} \in R^n$. By (13.8) and (13.9),

$$\lim_{k \to \infty} \varphi(D^k, A^k, c^k, b^k) = \frac{1}{2}\tilde{x}^T D\tilde{x} + c^T\tilde{x} = f(\tilde{x}, c, D), \qquad (13.18)$$

$$A\tilde{x} \ge b. \qquad (13.19)$$

From (13.19) it follows that $\tilde{x} \in \Delta(A, b)$. Hence

$$f(\tilde{x}, c, D) \ge \varphi(D, A, c, b).$$

Therefore, by (13.18),

$$\lim_{k \to \infty} \varphi(D^k, A^k, c^k, b^k) \ge \varphi(D, A, c, b). \qquad (13.20)$$

Combining (13.14) with (13.20) gives

$$\lim_{k \to \infty} \varphi(D^k, A^k, c^k, b^k) = \varphi(D, A, c, b).$$

This shows that φ is continuous at (D, A, c, b). The proof is complete. $\square$

Example 13.1. Consider the problem $QP(D, A, c, b)$ where $m = 3$, $n = 2$,

$$D = \begin{bmatrix} 1 & 0 \\ 0 & -1 \end{bmatrix}, \quad A = \begin{bmatrix} 1 & 0 \\ 0 & 1 \\ 1 & -2 \end{bmatrix}, \quad c = \begin{pmatrix} 1 \\ 1 \end{pmatrix}, \quad b = \begin{pmatrix} 0 \\ 0 \\ 0 \end{pmatrix}.$$

It can be verified that $\varphi(D, A, c, b) = 0$, $\mathrm{Sol}(D, A, 0, 0) = \{0\}$, and the system $Ax \ge b$ is regular. By Theorem 13.1, φ is continuous at (D, A, c, b).

Example 13.2. Consider the problem $QP(D, A, c, b)$ where $m = n = 1$, $D = [1]$, $A = [0]$, $c = (1)$, $b = (0)$. It can be shown that

$\varphi(D, A, c, b) = 0$, and the system $Ax \geq b$ is irregular. By Theorem 13.1, φ is not continuous at (D, A, c, b).

Remark 13.2. If $\Delta(A, b)$ is nonempty then $\Delta(A, 0)$ is the recession cone of $\Delta(A, b)$. By definition, $\text{Sol}(D, A, 0, 0)$ is the solution set of the problem $QP(D, A, 0, 0)$. So, verifying the assumption $\text{Sol}(D, A, 0, 0) = \{0\}$ is equivalent to solving one special QP problem.

Now we study the continuity of the optimal value function $\varphi(\cdot)$ at a point where its value is infinity. Let $\alpha \in \{+\infty, -\infty\}$ and $\varphi(D, A, c, b) = \alpha$. We say that $\varphi(\cdot)$ is continuous at (D, A, c, b) if, for every sequence $\{(D^k, A^k, c^k, b^k)\} \subset \Omega$ converging to (D, A, c, b),

$$\lim_{k \to \infty} \varphi(D^k, A^k, c^k, b^k) = \alpha.$$

The next theorem characterizes the continuity of φ at a point $\omega = (D, A, c, b)$ where φ has the value $-\infty$.

Theorem 13.2. *Let $(D, A, c, b) \in \Omega$ and $\varphi(D, A, c, b) = -\infty$. Then, the optimal value function φ is continuous at (D, A, c, b) if and only if the system $Ax \geq b$ is regular.*

Proof. Suppose that $\varphi(D, A, c, d) = -\infty$ and φ is continuous at (D, A, c, b) but the system $Ax \geq b$ is irregular. By Remark 13.1, there exists a sequence $\{(A^k, b^k)\}$ in $R^{m \times n} \times R^m$ converging to (A, b) such that, for every k, the system $A^k x \geq b^k$ has no solutions. Since $\Delta(A^k, b^k) = \emptyset$, $\varphi(D, A^k, c, b^k) = +\infty$ for every k. Therefore, $\lim_{k \to \infty} \varphi(D, A^k, c, b^k) = +\infty$. On the other hand, since φ is continuous at (D, A, c, b) and since

$$(D, A^k, c, b^k) \longrightarrow (D, A, c, b) \quad \text{as } k \to \infty,$$

we obtain

$$+\infty = \lim_{k \to \infty} \varphi(D, A^k, c, b^k) = \varphi(D, A, c, b) = -\infty,$$

a contradiction. Thus $Ax \geq b$ must be a regular system.

Conversely, suppose that $\varphi(D, A, c, d) = -\infty$ and the system $Ax \geq b$ is regular. Let $\{(D^k, A^k, c^k, b^k)\} \subset \Omega$ be a sequence converging to (D, A, c, b). By the assumption, $\varphi(D, A, c, b) = -\infty$, hence there is a sequence $\{x^i\}$ in R^n such that $Ax^i \geq b$ and

$$f(x^i, c, D) \longrightarrow -\infty \quad \text{as } i \to \infty. \tag{13.21}$$

By Lemma 13.1, for every i, there exists a sequence $\{y^{ik}\}$ in R^n with the property that

$$A^k y^{ik} \geq b^k, \tag{13.22}$$

$$\lim_{k \to \infty} y^{ik} = x^i. \tag{13.23}$$

By (13.22),

$$\varphi(D^k, A^k, c^k, b^k) \leq \frac{1}{2}(y^{ik})^T D^k y^{ik} + (c^k)^T y^{ik}. \tag{13.24}$$

From (13.23) and (13.24) it follows that

$$\limsup_{k \to \infty} \varphi(D^k, A^k, c^k, b^k) \leq \frac{1}{2}(x^i)^T D x^i + c^T x^i. \tag{13.25}$$

Combining (13.25) with (13.21), we obtain

$$\limsup_{k \to \infty} \varphi(D^k, A^k, c, b^k) = -\infty.$$

This implies that

$$\lim_{k \to \infty} \varphi(D^k, A^k, c^k, b^k) = -\infty = \varphi(D, A, c, b).$$

Thus φ is continuous at (D, A, c, b). The proof is complete. $\square$

The following theorem characterizes the continuity of φ at a point $\omega = (D, A, c, b)$ where φ has the value $+\infty$.

Theorem 13.3. *Let $(D, A, c, b) \in \Omega$ and $\varphi(D, A, c, b) = +\infty$. Then, the optimal value function φ is continuous at (D, A, c, b) if and only if $\mathrm{Sol}(D, A, 0, 0) = \{0\}$.*

Proof. Suppose that $\varphi(D, A, c, b) = +\infty$ and that φ is continuous at (D, A, c, b) but $\mathrm{Sol}(D, A, 0, 0) \neq \{0\}$. Then there exists a nonzero vector $\bar{x} \in R^n$ such that

$$A\bar{x} \geq 0, \quad \bar{x}^T D \bar{x} \leq 0.$$

Let $\bar{x} = (\bar{x}_1, \ldots, \bar{x}_n)$. We define a matrix $M \in R^{m \times n}$ by setting $M = [m_{ij}]$, where

$$m_{ij} = \bar{x}_j \quad \text{for } 1 \leq i \leq m, \ 1 \leq j \leq n.$$

Let

$$D^k = D - \frac{1}{k}E, \quad A^k = A + \frac{1}{k}M,$$

where E is the unit matrix in $R^{n \times n}$. Consider the sequence

$$\{(D^k, A^k, c, b)\}.$$

A simple computation shows that

$$A^k \bar{x} > 0 \quad \text{for every } k.$$

By Lemma 13.2, for every k the system $A^k x \geq b$ is regular. Let z be a solution of the system $A^k x \geq b$. Since $A^k \bar{x} > 0$ and

$$\bar{x}^T D^k \bar{x} = \bar{x}^T D \bar{x} - \frac{\bar{x}^T \bar{x}}{k} < 0$$

for every k, we have

$$f(z + t\bar{x}, c, D^k) = \frac{1}{2}(z + t\bar{x})^T D^k(z + t\bar{x}) + c^T(z + t\bar{x}) \to -\infty$$

as $t \to \infty$. Since $z + t\bar{x} \in \Delta(A^k, b)$ for every k and for every $t > 0$, $\mathrm{Sol}(D^k, A^k, c, b) = \emptyset$. We have arrived at a contradiction, because φ is continuous at (D, A, c, b) and

$$-\infty = \lim_{k \to \infty} \varphi(D^k, A^k, c, b) = \varphi(D, A, c, b) = +\infty.$$

Conversely, assume that $\mathrm{Sol}(D, A, 0, 0) = \{0\}$ and

$$\{(D^k, A^k, c^k, b^k)\} \subset \Omega$$

is a sequence converging to (D, A, c, b). We shall show that

$$\liminf_{k \to \infty} \varphi(D^k, A^k, c^k, b^k) = +\infty.$$

Suppose that $\liminf_{k \to \infty} \varphi(D^k, A^k, c^k, b^k) < +\infty$. Without loss of generality we can assume that

$$\liminf_{k \to \infty} \varphi(D^k, A^k, c^k, b^k) = \lim_{k \to \infty} \varphi(D^k, A^k, c^k, b^k) < +\infty.$$

Then, there exist a positive integer k_1 and a constant $\gamma \geq 0$ such that

$$\varphi(D^k, A^k, c^k, b^k) \leq \gamma$$

for every $k \geq k_1$. As $\mathrm{Sol}(D, A, 0, 0) = \{0\}$, we can assume that there is an positive integer k_2 such that $\mathrm{Sol}(D^k, A^k, 0, 0) = \{0\}$ for every $k \geq k_2$. By Lemma 13.3 we can assume that

$$\mathrm{Sol}(D^k, A^k, c^k, b^k) \neq \emptyset$$

for every $k \geq k_2$. Hence there exists a sequence $\{x^k\}$ in R^n such that, for every $k \geq k_2$, we have

$$\varphi(D^k, A^k, c^k, b^k) = \frac{1}{2}(x^k)^T D^k x^k + (c^k)^T x^k \leq \gamma, \qquad (13.26)$$

$$A^k x^k \geq b^k. \qquad (13.27)$$

We now prove that $\{x^k\}$ is a bounded sequence. Suppose, contrary to our claim, that the sequence $\{x^k\}$ is unbounded. Without loss of generality we can assume that $\|x^k\| \neq 0$ for every k and that $\|x^k\| \to \infty$ as $k \to \infty$. Then the sequence $\{\|x^k\|^{-1}x^k\}$ has a convergent subsequence. We can assume that the sequence itself converges to a point $x^0 \in R^n$ with $\|x^0\| = 1$. By (13.27) we have

$$A^k \frac{x^k}{\|x^k\|} \geq \frac{b^k}{\|x^k\|},$$

hence

$$Ax^0 \geq 0. \qquad (13.28)$$

By dividing both sides of the inequality in (13.26) by $\|x^k\|^2$ and taking the limits as $k \to \infty$, we get

$$(x^0)^T Dx^0 \leq 0. \qquad (13.29)$$

From (13.28) and (13.29) we deduce that $\mathrm{Sol}(D, A, 0, 0) \neq \{0\}$. This contradicts our assumption. Thus the sequence $\{x^k\}$ is bounded, and it has a convergent subsequence. Without loss of generality we can assume that $\{x^k\}$ converges to $\bar{x} \in R^n$. Letting $k \to \infty$, from (13.27) we obtain

$$A\bar{x} \geq b.$$

This means that $\Delta(A, b) \neq \emptyset$. We have arrived at a contradiction because $\varphi(D, A, c, b) = +\infty$. The proof is complete. $\square$

From Theorems 13.1–13.3 it follows that conditions (a), (b) in Theorem 13.1 are sufficient for the function $\varphi(\cdot)$ to be continuous at the given parameter value (D, A, c, b).

13.2 Semicontinuity of the Optimal Value Function

As it has been shown in the preceding section, continuity of the optimal value function holds under a special set of conditions. In some situations, only the upper semicontinuity or the lower semicontinuity of that function is required. So we wish to have simple sufficient conditions for the upper semicontinuity and the lower semicontinuity of φ at a given point. Such conditions are given in this section.

A sufficient condition for the upper semicontinuity of the function $\varphi(\cdot)$ at a given parameter value is given in the following theorem.

Theorem 13.4. *Let $(D, A, c, b) \in \Omega$. If the system $Ax \geq b$ is regular then $\varphi(\cdot)$ is upper semicontinuous at (D, A, c, b).*

Proof. As $Ax \geq b$ is regular, we have $\Delta(A, b) \neq \emptyset$. Hence

$$\varphi(D, A, c, b) < +\infty.$$

Let $\{(D^k, A^k, c^k, b^k)\} \subset \Omega$ be a sequence converging to (D, A, c, b). Since $\varphi(D, A, c, b) < +\infty$, there is a sequence $\{x^i\}$ in R^n such that $Ax^i \geq b$ and

$$f\left(x^i, c, D\right) = \frac{1}{2}\left(x^i\right)^T Dx^i + c^T x^i \longrightarrow \varphi(D, A, c, b) \quad \text{as} \quad i \to \infty.$$

By Lemma 13.1 and by the regularity of the system $Ax \geq b$, for each i one can find a sequence $\{y^{ik}\}$ in R^n such that $A^k y^{ik} \geq b^k$ and

$$\lim_{k \to \infty} y^{ik} = x^i.$$

Since $y^{ik} \in \Delta(A^k, b^k)$,

$$\varphi(D^k, A^k, c^k, b^k) \leq f(y^i, c^k, D^k).$$

This implies that

$$\limsup_{k \to \infty} \varphi(D^k, A^k, c^k, b^k) \leq f(x^i, c, D).$$

Taking limits in the last inequality as $i \to \infty$, we obtain

$$\limsup_{k \to \infty} \varphi(D^k, A^k, c^k, b^k) \leq \varphi(D, A, c, b).$$

We have proved that $\varphi(\cdot)$ is upper semicontinuous at (D, A, c, b). $\square$

The next example shows that the regularity condition in Theorem 13.4 does not guarantee the lower semicontinuity of φ at (D, A, c, b).

Example 13.3. Consider the problem $QP(D, A, c, b)$ where $m = n = 1$, $D = [0]$, $A = [1]$, $c = (0), b = (0)$. It is clear that $Ax \geq 0$ is regular, $\text{Sol}(D, A, c, b) = \Delta(A, b) = \{x : x \geq 0\}$, and $\varphi(D, A, c, b) = 0$. Consider the sequence $\{(D^k, A, c, b)\}$, where $D^k = D - \begin{bmatrix} \frac{1}{k} \end{bmatrix}$. We have $\varphi(D^k, A, c, b) = -\infty$ for every k, so

$$\liminf_{k \to \infty} \varphi(D^k, A, c, b) < \varphi(D, A, c, b).$$

Thus φ is not lower semicontinuous at (D, A, c, b).

The following example is designed to show that the regularity condition in Theorem 13.4 is sufficient but not necessary for the upper semicontinuity of φ at (D, A, c, b).

Example 13.4. Choose a matrix $A \in R^{m \times n}$ and a vector $b \in R^m$ such that $\Delta(A, b) = \emptyset$ (then the system $Ax \geq b$ is irregular). Fix an arbitrary matrix $D \in R_S^{n \times n}$ and an arbitrary vector $c \in R^n$. Since $\varphi(D, A, c, b) = +\infty$, for any sequence $\{(D^k, A^k, c^k, b^k)\}$ converging to (D, A, c, b), we have

$$\limsup_{k \to \infty} \varphi(D^k, A^k, c^k, b^k) \leq \varphi(D, A, c, b).$$

Thus φ is upper semicontinuous at (D, A, c, b).

A sufficient condition for the lower semicontinuity of the function $\varphi(\cdot)$ is given in the following theorem.

Theorem 13.5. Let $(D, A, c, b) \in \Omega$. If $\text{Sol}(D, A, 0, 0) = \{0\}$ then $\varphi(\cdot)$ is lower semicontinuous at (D, A, c, b).

Proof. Assume that $\text{Sol}(D, A, 0, 0) = \{0\}$. Let

$$\{(D^k, A^k, c^k, b^k)\} \subset \Omega$$

be a sequence converging to (D, A, c, b). We claim that

$$\liminf_{k \to \infty} \varphi(D^k, A^k, c^k, b^k) \geq \varphi(D, A, c, b).$$

Indeed, suppose that

$$\liminf_{k \to \infty} \varphi(D^k, A^k, c^k, b^k) < \varphi(D, A, c, b).$$

Without loss of generality we can assume that

$$\liminf_{k \to \infty} \varphi(D^k, A^k, c^k, b^k) = \lim_{k \to \infty} \varphi(D^k, A^k, c^k, b^k).$$

Then there exist an index k_1 and a real number γ such that $\gamma < \varphi(D, A, c, b)$ and

$$\varphi(D^k, A^k, c^k, b^k) \leq \gamma \quad \text{for every} \ k \geq k_1.$$

Since $\varphi(D^k, A^k, c^k, b^k) < +\infty$, we must have $\Delta(A^k, b^k) \neq \emptyset$ for every $k \geq k_1$. Since $\text{Sol}(D, A, 0, 0) = \{0\}$, there exists an integer $k_2 \geq k_1$ such that
$$\text{Sol}(D^k, A^k, 0, 0) = \{0\}$$
for every $k \geq k_2$. As $\Delta(A^k, b^k) \neq \emptyset$, by Lemma 13.3 we have $\text{Sol}(D^k, A^k, c^k, b^k) \neq \emptyset$ for every $k \geq k_2$. Hence there exists a sequence $\{x^k\}$ such that we have $A^k x^k \geq b^k$ for every $k \geq k_2$, and

$$\frac{1}{2}(x^k)^T D^k x^k + (c^k)^T x^k = \varphi(D^k, A^k, c^k, b^k) \leq \gamma.$$

The sequence $\{x^k\}$ must be bounded. Indeed, if $\{x^k\}$ is unbounded then, without loss of generality, we can assume that $\|x^k\| \neq 0$ for every k and $\|x^k\| \to \infty$ as $k \to \infty$. Then the sequence $\{\|x^k\|^{-1}x^k\}$ has a convergent subsequence. We can assume that this sequence itself converges to a vector $v \in R^n$ with $\|v\| = 1$. Since

$$A^k \frac{x^k}{\|x^k\|} \geq \frac{b^k}{\|x^k\|} \quad \text{for every} \ k \geq k_2,$$

we have $Av \geq 0$. On the other hand, since for each $k \geq k_2$ it holds

$$\frac{1}{2}\frac{(x^k)^T}{\|x^k\|}D^k\frac{x^k}{\|x^k\|} + (c^k)^T\frac{x^k}{\|x^k\|} \leq \frac{\gamma}{\|x^k\|^2},$$

we deduce that
$$v^T Dv \leq 0.$$

Combining all the above we get $v \in \text{Sol}(D, A, 0, 0) \setminus \{0\}$, a contradiction. We have thus proved that the sequence $\{x^k\}$ is bounded. Without loss of generality we can assume that $x^k \to \bar{x} \in R^n$. Since $A^k x^k \geq b^k$ for every k, we get $A\bar{x} \geq b$. Since

$$\frac{1}{2}(x^k)^T D^k x^k + (c^k)^T x^k \leq \gamma,$$

we have
$$f(\bar{x}, c, D) = \frac{1}{2}\bar{x}^T D\bar{x} + c^T\bar{x} \leq \gamma.$$

As $\gamma < \varphi(D, A, c, b)$, we see that $f(\bar{x}, c, D) < \varphi(D, A, c, b)$. This is an absurd because $\bar{x} \in \Delta(A, b)$. We have thus proved that $\varphi(\cdot)$ is lower semicontinuous at (D, A, c, b). $\square$

The next example shows that the condition $\text{Sol}(D, A, 0, 0) = \{0\}$ in Theorem 13.5 does not guarantee the upper semicontinuity of φ at (D, A, c, b).

Example 13.5. Consider the problem $QP(D, A, c, b)$ where $m = n = 1$, $D = [1]$, $A = [0]$, $c = (0), b = (0)$. It is clear that $\text{Sol}(D, A, 0, 0) = \{0\}$. Consider the sequence $\{(D, A, c, b^k)\}$, where $b^k = (\frac{1}{k})$. We have $\varphi(D, A, c, b) = 0$ and $\varphi(D, A, c, b^k) = +\infty$ for all k (because $\Delta(A, b^k) = \emptyset$ for all k). Therefore

$$\limsup_{k \to \infty} \varphi(D, A, c, b^k) = +\infty > 0 = \varphi(D, A, c, b).$$

Thus φ is not upper semicontinuous at (D, A, c, b).

The condition $\text{Sol}(D, A, 0, 0) = \{0\}$ in Theorem 13.5 is sufficient but not necessary for the lower semicontinuity of φ at (D, A, c, b).

Example 13.6. Consider the problem $QP(D, A, c, b)$ where $m = n = 1$, $D = [-1]$, $A = [1]$, $c = (1), b = (0)$. It is clear that $\text{Sol}(D, A, 0, 0) = \emptyset$. Since $\varphi(D, A, c, b) = -\infty$, for any sequence $\{(D^k, A^k, c^k, b^k)\}$ converging to (D, A, c, b), we have

$$\liminf_{k \to \infty} \varphi(D^k, A^k, c^k, b^k) \geq \varphi(D, A, c, b).$$

Thus φ is lower semicontinuous at (D, A, c, b).

13.3 Commentaries

The results presented in this chapter are due to Tam (2002).

Lemma 13.1 is a well-known fact (see, for example, Robinson (1975), Theorem 1, and Bank et al. (1982), Theorem 3.1.5).

In Best and Chakravarti (1990) and Best and Ding (1995) the authors have considered convex quadratic programming problems and obtained some results on the continuity and differentiability of the optimal value function of the problem as a function of a parameter specifying the magnitude of the perturbation. In Auslender and Coutat (1996), similar questions for the case of linear-quadratic programming problems were investigated. Continuity and Lipschitzian properties of the function $\varphi(D, A, \cdot, \cdot)$ (the matrices D and A are fixed) were studied in Bank et al. (1982), Bank and Hansel (1984), Klatte (1985), Rockafellar and Wets (1998).

We have considered indefinite QP problems and obtained several results on the continuity, the upper and lower semicontinuity of the optimal value function φ at a given point ω. In comparison with the preceding results of Best and Chakravarti (1990), Best and Ding (1995), the advantage here is that the quadratic objective function is allowed to be indefinite.

The obtained results can be used for analyzing algorithms for solving the indefinite QP problems.

Chapter 14

Directional Differentiability of the Optimal Value Function

In this chapter we establish an explicit formula for computing the directional derivative of the optimal value function in a general parametric QP programming problem. We will consider one illustrative example to see how the formula works for concrete QP problems.

In Section 14.1 we prove several lemmas. In Section 14.2 we introduce *condition* (G) and describe a general situation where (G) holds. Section 14.3 is devoted to proving the above-mentioned formula for computing the directional derivative of the optimal value function in indefinite QP problems. In the same section, the obtained result is compared with the corresponding results on differential stability in nonlinear programming of Auslender and Cominetti (1990), and Minchenko and Sakolchik (1996).

14.1   Lemmas

Consider the general quadratic programming problem (13.1) which is abbreviated to $QP(D, A, c, b)$. The problem depends on the parameter $\omega = (D, A, c, b) \in \Omega$, where

$$\Omega := R_S^{n \times n} \times R^{m \times n} \times R^n \times R^m.$$

As in Chapter 13, the solution set of this problem will be denoted by $\mathrm{Sol}(D, A, c, b)$, and its optimal value function $\varphi : \Omega \longrightarrow \overline{R}$ is given by

$$\varphi(\omega) = \inf\{f(x, c, D) : x \in \Delta(A, b)\}.$$

The proofs of Theorem 14.1 and Theorem 14.2, the main results in this chapter, are based on some lemmas established in the present section.

Let $\omega = (D, A, c, b)$ and $\omega^0 = (D^0, A^0, c^0, b^0)$ be two elements of Ω. Denote

$$\omega + t\omega^0 = (D + tD^0, A + tA^0, c + tc^0, b + tb^0),$$
$$\varphi^+(\omega; \omega^0) = \limsup_{t \downarrow 0} \frac{\varphi(\omega + t\omega^0) - \varphi(\omega)}{t},$$
$$\varphi^-(\omega; \omega^0) = \liminf_{t \downarrow 0} \frac{\varphi(\omega + t\omega^0) - \varphi(\omega)}{t}.$$

If $\varphi^+(\omega; \omega^0) = \varphi^-(\omega; \omega^0)$ then the optimal value function $\varphi(\cdot)$ is directionally differentiable at ω in direction ω^0 (see Definition 1.8). The common value is denoted by $\varphi'(\omega; \omega^0)$ which is the directional derivative of φ at ω in direction ω^0. We have

$$\varphi'(\omega; \omega^0) = \lim_{t \downarrow 0} \frac{\varphi(\omega + t\omega^0) - \varphi(\omega)}{t}.$$

For every $\bar{x} \in \Delta(A, b)$, we set

$$I = \alpha(\bar{x}) = \{i : (A\bar{x})_i = b_i\},$$

and define

$$F(\bar{x}, \omega, \omega^0) = \{v \in R^n : \exists \varepsilon > 0 \text{ such that } \bar{x} + tv \in \Delta(A + tA^0, b + tb^0) \text{ for every } t \in [0, \varepsilon]\},$$

$$R(\bar{x}, \omega, \omega^0) = \begin{cases} R^n & \text{if } I = \emptyset \\ \{v \in R^n : A_I v + A_I^0 \bar{x} - b_I^0 \geq 0\} & \text{if } I \neq \emptyset. \end{cases}$$

The following lemma is originated from Seeger (1988), Auslender and Cominetti (1990).

Lemma 14.1. *If the system $Ax \geq b$ is regular, then*

$$\emptyset \neq \text{int}\, R(\bar{x}, \omega, \omega^0) \subset F(\bar{x}, \omega, \omega^0) \subset R(\bar{x}, \omega, \omega^0) \tag{14.1}$$

for every $\bar{x} \in \Delta(A, b)$.

Proof. Let $\bar{x} \in \Delta(A, b)$, $I = \alpha(\bar{x}) = \{i : (A\bar{x})_i = b_i\}$. If $I = \emptyset$ then $A\bar{x} > b$. Thus, for every $v \in R^n$ there is an $\varepsilon = \varepsilon(v) > 0$ such that for each $t \in [0, \varepsilon]$ we have

$$A\bar{x} + t(Av + A^0\bar{x} - b^0 + tA^0v) \geq b.$$

The above inequality is equivalent to the following one

$$(A + tA^0)(\bar{x} + tv) \geq b + tb^0.$$

Hence $\bar{x} + tv \in \Delta(A + tA^0, b + tb^0)$ for each $t \in [0, \varepsilon]$. This implies that $F(\bar{x}, \omega, \omega^0) = R^n$. By definition, in this case we also have $R(\bar{x}, \omega, \omega^0) = R^n$. Therefore

$$F(\bar{x}, \omega, \omega^0) = R^n = R(\bar{x}, \omega, \omega^0),$$

and we have (14.1). Consider the case where $I \neq \emptyset$. We first show that

$$\text{int} R(\bar{x}, \omega, \omega^0) \neq \emptyset.$$

Since $Ax \geq b$ is a regular system, there exists $x^0 \in R^n$ such that $Ax^0 > b$. Then we have

$$A_I x^0 > b_I.$$

As $A_I\bar{x} = b_I$ and $A_I x^0 > b_I$, we have

$$A_I(x^0 - \bar{x}) > 0.$$

Putting $\hat{v} = x^0 - \bar{x}$, we get

$$A_I \hat{v} > 0.$$

By Lemma 13.2, the inequality system (of the unknown v)

$$A_I v \geq b_I^0 - A_I^0 \bar{x}$$

is regular, hence there exists $\bar{v} \in R^n$ such that

$$A_I \bar{v} > b_I^0 - A_I^0 \bar{x}.$$

This proves that $\bar{v} \in \text{int} R(\bar{x}, \omega, \omega^0)$, therefore $\text{int} R(\bar{x}, \omega, \omega^0) \neq \emptyset$. We now prove that

$$\text{int} R(\bar{x}, \omega, \omega^0) \subset F(\bar{x}, \omega, \omega^0).$$

Suppose that $v \in \operatorname{int} R(\bar{x}, \omega, \omega^0)$. We have

$$A_I v + A_I^0 \bar{x} - b_I^0 > 0.$$

Hence there is $\varepsilon_1 > 0$ such that for each $t \in [0, \varepsilon_1]$ one has

$$A_I v + A_I^0 \bar{x} - b_I^0 + t A_I^0 v > 0.$$

Then, for each $t \in [0, \varepsilon_1]$,

$$t(A_I v + A_I^0 \bar{x} - b_I^0 + t A_I^0 v) \geq 0. \qquad (14.2)$$

As $A_i \bar{x} > b_i$ for every $i \in \{1, \ldots, m\} \setminus I$, one can find $\varepsilon_2 > 0$ such that for each $t \in [0, \varepsilon_2]$ it holds

$$A_i \bar{x} + t(A_i v + A_i^0 \bar{x} - b_i^0 + t A_i^0 v) \geq b_i \qquad (14.3)$$

for every $i \in \{1, \ldots, m\} \setminus I$. Let $\varepsilon := \min\{\varepsilon_1, \varepsilon_2\}$. It follows from (14.2) and (14.3) that

$$A\bar{x} + t(Av + A^0 \bar{x} - b^0 + t A^0 v) \geq b \qquad (14.4)$$

for every $t \in [0, \varepsilon]$. This implies that

$$\bar{x} + tv \in \Delta(A + t A^0, b + t b^0)$$

for every $t \in [0, \varepsilon]$. Hence $v \in F(\bar{x}, \omega, \omega^0)$, and we have

$$\operatorname{int} R(\bar{x}, \omega, \omega^0) \subset F(\bar{x}, \omega, \omega^0).$$

Finally, we shall prove that

$$F(\bar{x}, \omega, \omega^0) \subset R(\bar{x}, \omega, \omega^0).$$

Take any $v \in F(\bar{x}, \omega, \omega^0)$. By definition, there is an $\varepsilon > 0$ such that for each $t \in [0, \varepsilon]$ we have

$$(A_I + t A_I^0)(\bar{x} + tv) \geq b + t b^0.$$

Consequently,

$$A_I \bar{x} + t(A_I v + A_I^0 \bar{x} - b_I^0 + t A_I^0 v) \geq b_I$$

for every $t \in [0, \varepsilon]$. As $A_I \bar{x} = b_I$, we have

$$t(A_I v + A_I^0 \bar{x} - b_I^0 + t A_I^0 v) \geq 0$$

for each $t \in [0, \varepsilon]$. Hence, for every $t \in (0, \varepsilon]$,

$$A_I v + A_I^0 \bar{x} - b_I^0 + t A_I^0 v \geq 0.$$

Letting $t \to 0$, we obtain

$$A_I v + A_I^0 \bar{x} - b_I^0 \geq 0.$$

This shows that $v \in R(\bar{x}, \omega, \omega^0)$, hence $F(\bar{x}, \omega, \omega^0) \subset R(\bar{x}, \omega, \omega^0)$. We have thus shown that the inclusions in (14.1) are valid. The proof is complete. □

If $\bar{x} \in \text{Sol}(D, A, c, b)$, then there exists a Lagrange multiplier $\lambda \in R^m$ such that

$$D\bar{x} - A^T \lambda + c = 0,$$
$$A\bar{x} \geq b, \quad \lambda \geq 0,$$
$$\lambda^T (A\bar{x} - b) = 0$$

The set of all the Lagrange multipliers corresponding to $\bar{x}$ is denoted by $\Lambda(\bar{x}, \omega)$, where $\omega = (D, A, c, b)$.

The forthcoming result is well known in nonlinear programming (see, for instance, Gauvin (1977) and Dien (1985)). For the sake of completeness, we give a proof for the case of QP problems.

Lemma 14.2. *If the system $Ax \geq b$ is regular, then for every $\bar{x} \in \text{Sol}(D, A, c, b)$ the set $\Lambda(\bar{x}, \omega)$ is compact.*

Proof. Let $\omega = (D, A, c, b)$. Suppose that there is $\bar{x} \in \text{Sol}(D, A, c, b)$ such that $\Lambda(\bar{x}, \omega)$ is noncompact. Then there exists a sequence $\{\lambda^k\}$ in R^m such that $\|\lambda^k\| \neq 0$,

$$D\bar{x} - A^T \lambda^k + c = 0, \tag{14.5}$$

$$\lambda^k \geq 0, \tag{14.6}$$

$$(\lambda^k)^T (A\bar{x} - b) = 0, \tag{14.7}$$

for every k, and $\|\lambda^k\| \to \infty$ as $k \to \infty$. Without loss of generality we can assume that $\{\|\lambda^k\|^{-1} \lambda^k\}$ converges to $\bar{\lambda}$ with $\|\bar{\lambda}\| = 1$. Dividing each expression in (14.5)–(14.7) by $\|\lambda^k\|$ and taking the limits as $k \to \infty$, we get

$$A^T \bar{\lambda} = 0, \quad \bar{\lambda} \geq 0, \quad \bar{\lambda}^T (A\bar{x} - b) = 0. \tag{14.8}$$

Since $\bar{\lambda}^T A\bar{x} = \bar{x}^T (A^T \bar{\lambda}) = 0$, from (14.8) it follows that

$$A^T \bar{\lambda} = 0, \quad \bar{\lambda} \geq 0, \quad \bar{\lambda}^T b = 0.$$

For every $t > 0$, we set $b_t = b + t\bar{\lambda}$. Since $\bar{\lambda}^T \bar{\lambda} = \|\bar{\lambda}\|^2 = 1$,

$$\bar{\lambda}^T b_t = \bar{\lambda}^T b + t\bar{\lambda}^T \bar{\lambda} = \bar{\lambda}^T b + t = t.$$

Consequently, for every $t > 0$, $\bar{\lambda}$ is a solution of the following system

$$A^T \lambda = 0, \quad \lambda \geq 0, \quad \lambda^T b_t > 0.$$

Hence, for every $t > 0$, the system $Ax \geq b_t$ has no solutions (see Cottle et al. (1992), Theorem 2.7.8). Since $\Delta(A, b) \neq \emptyset$ and $\|b_t - b\| = t \to 0$ as $t \to 0$, the system $Ax \geq b$ is irregular (see Mangasarian (1980), Lemma 2.1), contrary to our assumption. The proof is complete. $\square$

Lemma 14.3 (cf. Auslender and Cominetti (1990), Lemma 2). *If the system $Ax \geq b$ is regular and $\bar{x} \in \mathrm{Sol}(D, A, c, b)$ then*

$$\inf_{v \in R(\bar{x}, \omega, \omega^0)} (D\bar{x} + c)^T v = \max_{\lambda \in \Lambda(\bar{x}, \omega)} (b^0 - A^0 \bar{x})^T \lambda, \qquad (14.9)$$

where $\Lambda(\bar{x}, \omega)$ stands for the Lagrange multiplier set corresponding to $\bar{x}$.

Proof. Let $\bar{x} \in \mathrm{Sol}(D, A, c, b)$. If $I = \alpha(\bar{x}) = \{i : (A\bar{x})_i = b_i\}$ is empty then, by definition, $R(\bar{x}, \omega, \omega^0) = R^n$. As $\bar{x} \in \mathrm{Sol}(D, A, c, b)$ and $A\bar{x} > b$, Theorem 3.3 applied to $\bar{x}$ shows that $(D\bar{x} + c)^T v = 0$ for every $v \in R^n$. Then we have

$$\inf_{v \in R(\bar{x}, \omega, \omega^0)} (D\bar{x} + c)^T v = 0.$$

Again, by the just cited first-order necessary optimality condition, for every $\bar{x}$ we have $\Lambda(\bar{x}, \omega) \neq \emptyset$. Since $A\bar{x} > b$, $\Lambda(\bar{x}, \omega) = \{0\}$. Therefore

$$\max_{\lambda \in \Lambda(\bar{x}, \omega)} (b^0 - A^0 \bar{x})^T \lambda = 0.$$

Thus, in the case $I = \emptyset$ the assertion of the lemma is valid. We now consider the case where $I = \alpha(\bar{x}) = \{i : (A\bar{x})_i = b_i\} \neq \emptyset$. We have

$$\inf_{v \in R(\bar{x}, \omega, \omega^0)} (D\bar{x} + c)^T v = \inf\{(D\bar{x} + c)^T v : v \in R^n, A_I v \geq b_I^0 - A_I^0 \bar{x}\}.$$

Consider a pair of dual linear programs

$$(P) \qquad \begin{cases} (D\bar{x} + c)^T v \longrightarrow \min \\ v \in R^n, \quad A_I v \geq b_I^0 - A_I^0 \bar{x} \end{cases}$$

and

(P')
$$\begin{cases} (b_I^0 - A_I^0\bar{x})^T\lambda_I \longrightarrow \max \\ \lambda_I \in R^{|I|}, \quad A_I^T\lambda_I = D\bar{x} + c, \quad \lambda_I \geq 0 \end{cases}$$

where $|I|$ is the number of the elements of I. From the definition of $\Lambda(\bar{x}, \omega)$ it follows that if λ_I is a feasible point of (P') then $(\lambda_I, 0_J) \in \Lambda(\bar{x}, \omega)$, where $J = \{1, \cdots, m\} \setminus I$. Conversely, if $\lambda = (\lambda_I, \lambda_J) \in \Lambda(\bar{x}, \omega)$ then $\lambda_J = 0_J$. The regularity of the system $Ax \geq b$ and Lemma 14.2 imply that $\Lambda(\bar{x}, \omega)$ is nonempty and compact. Therefore, by the above observation, the feasible domain of (P') is nonempty and compact. By the duality theorem in linear programming (see Theorem 1.10(iv)), the optimal values of (P) and (P') are both finite and equal to each other. Therefore

$$\begin{aligned} &\inf\{(D\bar{x} + c)^T v \ : \ v \in R(\bar{v}, \omega, \omega^0)\} \\ &= \inf\{(D\bar{x} + c)^T v \ : \ v \in R^n, A_I v \geq b_I^0 - A_I^0\bar{x}\} \\ &= \max\{(b_I^0 - A_I^0\bar{x})^T\lambda_I \ : \ \lambda_I \in R^{|I|}, \ \lambda_I \geq 0, \ A_I^T\lambda_I = D\bar{x} + c\} \\ &= \max\{(b^0 - A^0\bar{x})^T\lambda \ : \ \lambda \in R^m, \ \lambda = (\lambda_I, 0_J) \geq 0, \\ &\qquad A^T\lambda = D\bar{x} + c\} \\ &= \max_{\lambda \in \Lambda(\bar{x}, \omega)} (b^0 - A_0\bar{x})^T\lambda. \end{aligned}$$

Formula (14.9) is proved. $\square$

Lemma 14.4. *Suppose that* $\omega^k = (D^k, A^k, c^k, b^k)$, $k \in N$, *is a sequence in* Ω *converging to* $\omega = (D, A, c, b)$, $\{x^k\}$ *is a sequence in* R^n *such that* $x^k \in \text{Sol}(D^k, A^k, c^k, b^k)$ *for every* k. *If the system* $Ax \geq b$ *is regular and* $\text{Sol}(D, A, 0, 0) = \{0\}$ *then there exists a subsequence* $\{x^{k_i}\}$ *of* $\{x^k\}$ *such that* $\{x^{k_i}\}$ *converges to* $\bar{x} \in \text{Sol}(D, A, c, b)$ *as* $i \to \infty$.

Proof. Suppose that $Ax \geq b$ is a regular system and $\text{Sol}(D, A, 0, 0) = \{0\}$. We have

$$A^k x^k \geq b^k. \tag{14.10}$$

Take $x \in \Delta(A, b)$. Then there exists a sequence $\{y^k\}$ in R^n tending to x such that

$$A^k y^k \geq b^k \quad \text{for every } k \tag{14.11}$$

(see Lemma 13.1). The inequality in (14.11) shows that $y^k \in \Delta(A^k, b^k)$. Since $x^k \in \text{Sol}(D^k, A^k, c^k, b^k)$,

$$\frac{1}{2}(x^k)^T D^k x^k + (c^k)^T x^k \leq \frac{1}{2}(y^k)^T D^k y^k + (c^k)^T y^k. \tag{14.12}$$

We claim that the sequence $\{x^k\}$ is bounded. Suppose for a while that $\{x^k\}$ is unbounded. Then, without loss of generality, we may assume that $\|x^k\| \to \infty$ as $k \to \infty$ and $\|x^k\| \neq 0$ for every k. So the sequence $\{\|x^k\|^{-1}x^k\}$ has a convergent subsequence. We may assume that the sequence $\{\|x^k\|^{-1}x^k\}$ itself converges to $\hat{x} \in R^n$ with $\|\hat{x}\| = 1$. From (14.10) we have

$$A^k \frac{x^k}{\|x^k\|} \geq \frac{b^k}{\|x^k\|}.$$

Letting $k \to \infty$, we obtain

$$A\hat{x} \geq 0. \tag{14.12}$$

Dividing both sides of (14.12) by $\|x^k\|^2$ and taking limit as $k \to \infty$, we obtain

$$\hat{x}^T D\hat{x} \leq 0. \tag{14.14}$$

Combining (14.13) and (14.14), we have $\text{Sol}(D, A, 0, 0) \neq \{0\}$, contrary to our assumptions. Thus the sequence $\{x^k\}$ is bounded and it has a convergent subsequence, say, $\{x^{k_i}\}$. Suppose that $\{x^{k_i}\}$ converges to $\bar{x}$. From (14.12) we have

$$\frac{1}{2}(x^{k_i})^T D^{k_i} x^{k_i} + (c^{k_i})^T x^{k_i} \leq \frac{1}{2}(y^{k_i})^T D^{k_i} y^{k_i} + (c^{k_i})^T y^{k_i}. \tag{14.15}$$

From (14.10) we have

$$A^{k_i} x^{k_i} \geq b^{k_i}. \tag{14.16}$$

Taking limits in (14.15) and (14.16) as $i \to \infty$, we obtain

$$\frac{1}{2}\bar{x}^T D\bar{x} + c^T \bar{x} \leq \frac{1}{2}x^T Dx + c^T x, \tag{14.17}$$

$$A\bar{x} \geq b. \tag{14.18}$$

As $x \in \Delta(A, b)$ is arbitrarily chosen, (14.17) and (14.18) yield $\bar{x} \in \text{Sol}(D, A, c, b)$. The lemma is proved. $\quad\square$

14.2 Condition (G)

Let $\omega = (D, A, c, b) \in \Omega$ be a given parameter value and $\omega^0 = (D^0, A^0, c^0, b^0) \in \Omega$ be a given direction. Consider the following condition which we call *condition* (G):

For every sequence $\{t^k\}$, $t^k \downarrow 0$, *for every sequence* $\{x^k\}$,

$$x^k \longrightarrow \bar{x} \in \text{Sol}(D, A, c, b),$$

where $x^k \in \text{Sol}(\omega + t^k \omega^0)$ for each k, the following inequality is satisfied

$$\liminf_{k \to \infty} \frac{(x^k - \bar{x})^T D(x^k - \bar{x})}{t^k} \geq 0.$$

Remark 14.1. If D is a positive semidefinite matrix, then condition (G) holds. Indeed, if D is positive semidefinite then $(x^k - \bar{x})^T D(x^k - \bar{x}) \geq 0$, hence the inequality in (G) is satisfied.

Remark 14.2. If the system $Ax \geq b$ is regular then (G) is weaker than the condition saying that the $(SOSC)_u$ property introduced in Auslender and Cominetti (1990) (applied to QP problems) holds at every $\bar{x} \in \text{Sol}(D, A, c, b)$. It is interesting to note that if the system $Ax \geq b$ is regular then (G) is also weaker than the condition (H3) introduced by Minchenko and Sakolchik (1996) (applied to QP problems). There exist many QP problems where the conditions $(SOSC)_u$ and (H3) do not hold but condition (G) is satisfied. A detailed comparison of our results with the ones in Auslender and Cominetti (1990) and Minchenko and Sakolchik (1996) will be given in Section 14.3.

Now we describe a general situation where (G) is fulfilled.

Theorem 14.1. *If $Ax \geq b$ is a regular system and every solution $\bar{x} \in \text{Sol}(D, A, c, b)$ is a locally unique solution of problem (13.1), then condition (G) is satisfied.*

Proof. From the statement of (G) it is obvious that the condition is satisfied if $\text{Sol}(D, A, c, b) = \emptyset$. Consider the case where $\text{Sol}(D, A, c, b) \neq \emptyset$. For any given $\bar{x} \in \text{Sol}(D, A, c, b)$ we set $I = \alpha(\bar{x}) = \{i : (A\bar{x})_i = b_i\}$ and

$$F_{\bar{x}} = \{v \in R^n : (Av)_i \geq 0 \text{ for every } i \in I\}.$$

For every $\bar{x} \in \text{Sol}(D, A, c, b)$, Theorem 3.7 shows that the following conditions are equivalent:

(a) $\bar{x}$ is a locally unique solution of problem (13.1),

(b) for every $v \in F_{\bar{x}} \setminus \{0\}$, if $(D\bar{x} + c)^T v = 0$ then $v^T D v > 0$.

We shall use the above equivalence to prove our theorem. Suppose, contrary to our claim, that (G) does not hold. Then there exist a sequence $\{t^k\}$, $t^k \downarrow 0$, and a sequence $\{x^k\}$, $x^k \to \bar{x} \in$

$Sol(D, A, c, b)$, $x^k \in Sol(D + t^k D^0, A + t^k A^0, c + t^k c^0, b + t^k b^0)$ for every k, such that

$$\lim_{k \to \infty} \frac{(x^k - \bar{x})^T D(x^k - \bar{x})}{t^k} < 0. \qquad (14.19)$$

By taking a subsequence if necessary, we can assume that

$$(x^k - \bar{x})^T D(x^k - \bar{x}) < 0, \quad \|x^k - \bar{x}\| \neq 0 \quad \text{for every } k, \qquad (14.20)$$

and

$$\lim_{k \to \infty} \frac{\|x^k - \bar{x}\|}{t^k} = +\infty. \qquad (14.21)$$

Then the sequence $\{\|x^k - \bar{x}\|^{-1}(x^k - \bar{x})\}$ has a convergent subsequence. Without loss of generality, we may assume that $\{\|x^k - \bar{x}\|^{-1}(x^k - \bar{x})\}$ converges to some $v \in R^n$ with $\|v\| = 1$. Dividing both sides of the inequality in (14.20) by $\|x^k - \bar{x}\|^2$ and letting $k \to \infty$, we get

$$v^T D v \leq 0. \qquad (14.22)$$

Since $x^k \in Sol(D + t^k D^0, A + t^k A^0, c + t^k c^0, b + t^k b^0)$, we have

$$(A_I + t^k A_I^0)x^k \geq b_I + t^k b_I^0,$$

where $I = \{i : (A\bar{x})_i = b_i\}$. Since $b_I = A_I \bar{x}$,

$$A_I(x^k - \bar{x}) \geq t^k(b_I^0 - A_I^0 x^k).$$

Dividing both sides of the inequality above by $\|x^k - \bar{x}\|$, taking account of (14.21) and letting $k \to \infty$, we obtain

$$A_I v \geq 0.$$

Then

$$v \in F_{\bar{x}} \setminus \{0\}. \qquad (14.23)$$

Now we are going to show that $(D\bar{x} + c)^T v = 0$. We have

$$\varphi(\omega + t^k \omega^0) - \varphi(\omega)$$
$$= \frac{1}{2}(x^k)^T(D + t^k D^0)x^k + (c + t^k c^0)^T x^k - \frac{1}{2}\bar{x}^T D\bar{x} - c^T \bar{x}$$
$$= \frac{1}{2}(x^k - \bar{x})^T D(x^k - \bar{x}) + (D\bar{x} + c)^T(x^k - \bar{x})$$
$$+ t^k \left(\frac{1}{2}(x^k)^T D^0 x^k + (c^0)^T x^k \right).$$

$$(14.24)$$

Since $Ax \geq b$ is a regular system, by Lemma 14.1 we have

$$F(\bar{x}, \omega, \omega^0) \neq \emptyset.$$

Take $\bar{v} \in F(\bar{x}, \omega, \omega^0)$. Then, for every small enough positive number t^k, we have

$$\bar{x} + t^k \bar{v} \in \Delta(A + t^k A^0, b + t^k b^0).$$

Hence, for small enough t^k, we have

$$
\begin{aligned}
\varphi(\omega + t^k \omega^0) - \varphi(\omega) &= \frac{1}{2}(x^k)^T(D + t^k D^0)x^k \\
&+ (c + t^k c^0)^T x^k + \left(-\frac{1}{2}\bar{x}^T D\bar{x} - c^T \bar{x}\right) \\
&\leq \frac{1}{2}(\bar{x} + t^k \bar{v})^T(D + t^k D^0)(\bar{x} + t^k \bar{v}) \\
&+ (c + t^k c^0)^T(\bar{x} + t^k \bar{v}) + \left(-\frac{1}{2}\bar{x}^T D\bar{x} - c^T \bar{x}\right).
\end{aligned}
\tag{14.25}
$$

From (14.24) and (14.25), for k large enough, we have

$$
\begin{aligned}
&(D\bar{x} + c)^T(x^k - \bar{x}) + \frac{1}{2}(x^k - \bar{x})^T D(x^k - \bar{x}) \\
&+ t^k \left(\frac{1}{2}(x^k)^T D^0 x^k + (c^0)^T x^k\right) \\
&\leq t^k(c^0)^T(\bar{x} + t^k \bar{v}) + \frac{1}{2}t^k(\bar{x} + t^k \bar{v})^T D^0(\bar{x} + t^k \bar{v}) + \\
&+ t^k \left(c^T \bar{v} + \bar{v}^T D\bar{v} + \frac{1}{2}t^k \bar{v}^T D\bar{v}\right).
\end{aligned}
\tag{14.26}
$$

Dividing both sides of (14.26) by $\|x^k - \bar{x}\|$, letting $k \to \infty$ and taking account of (14.21), we get

$$(D\bar{x} + c)^T v \leq 0. \tag{14.27}$$

As $\bar{x}$ is a solution of (13.1) and (14.23) is valid, we have $(D\bar{x}+c)^T v \geq 0$ (see Theorem 3.5). Combining this with (14.27), we conclude that

$$(D\bar{x} + c)^T v = 0. \tag{14.28}$$

Properties (14.22), (14.23) and (14.28) show that (b) does not hold. Thus $\bar{x}$ cannot be a locally unique solution of (13.1), a contrary to our assumptions. The proof is complete. $\square$

14.3   Directional Differentiability of $\varphi(\cdot)$

The following theorem describes a sufficient condition for $\varphi(\cdot)$ to be directionally differentiable and gives an explicit formula for computing the directional derivative of $\varphi(\cdot)$.

Theorem 14.2. *Let $\omega = (D, A, c, b) \in \Omega$ be a given point and $\omega^0 = (D^0, A^0, c^0, b^0) \in \Omega$ be a given direction. If (G) and the following two conditions*

(i) *the system $Ax \geq b$ is regular,*

(ii) $\text{Sol}(D, A, 0, 0) = \{0\}$

are satisfied, then the optimal value function φ is directionally differentiable at $\omega = (D, A, c, b)$ in direction $\omega^0 = (D^0, A^0, c^0, b^0)$, and

$$\varphi'(\omega; \omega^0) = \inf_{\bar{x} \in \text{Sol}(D,A,c,b)} \max_{\lambda \in \Lambda(\bar{x},\omega)} \left[\frac{1}{2}\bar{x}^T D^0 \bar{x} + (c^0)^T \bar{x} + (b^0 - A^0\bar{x})^T \lambda \right],$$

$$\tag{14.29}$$

where $\Lambda(\bar{x}, \omega)$ is the Lagrange multipliers set corresponding to the solution $\bar{x} \in \text{Sol}(D, A, c, b)$.

Proof.

1) Suppose that the conditions (i) and (ii) are satisfied. According to Lemma 13.3, $\text{Sol}(D, A, c, b)$ is a nonempty compact set. Take any $\bar{x} \in \text{Sol}(D, A, c, b)$. By (i) and Lemma 14.1, $F(\bar{x}, \omega, \omega^0) \neq \emptyset$. Take any $v \in F(\bar{x}, \omega, \omega^0)$. For $t > 0$ small enough, we have

$$\bar{x} + tv \in \Delta(A + tA^0, b + tb^0),$$

hence

$$\varphi(\omega + t\omega^0) - \varphi(\omega) \leq \frac{1}{2}(\bar{x} + tv)^T(D + tD^0)(\bar{x} + tv)$$

$$+ (c + tc^0)^T(\bar{x} + tv) - \left(\frac{1}{2}\bar{x}^T D\bar{x} + c^T\bar{x} \right)$$

$$= t(D\bar{x} + c)^T v + t\left(\frac{1}{2}\bar{x}^T D^0 \bar{x} + (c^0)^T\bar{x} \right)$$

$$+ \frac{1}{2}t^2 v^T Dv + t^2 v^T D\bar{x} + \frac{1}{2}t^3 v^T D^0 v.$$

Multiplying the above double inequality by t^{-1} and taking lim sup as $t \to 0^+$, we obtain

$$\varphi^+(\omega; \omega^0) \leq \frac{1}{2}\bar{x}^T D^0 \bar{x} + (D\bar{x} + c)^T v + + (c^0)^T\bar{x}.$$

This inequality is valid for any $v \in F(\bar{x}, \omega, \omega^0)$ and any

$$\bar{x} \in \mathrm{Sol}(D, A, c, b).$$

Consequently,

$$\varphi^+(\omega; \omega^0) \leq \inf_{\bar{x} \in \mathrm{Sol}(D,A,c,b)} \inf_{v \in F(\bar{x},\omega,\omega^0)} \left[\frac{1}{2}\bar{x}D^0\bar{x} + (c^0)^T\bar{x} + (D\bar{x} + c)^T v \right].$$

By Lemmas 14.2 and 14.3,

$$\inf_{v \in F(\bar{x},\omega,\omega^0)}(D\bar{x} + c)^T v = \inf_{v \in R(\bar{x},\omega,\omega^0)}(D\bar{x} + c)^T v$$
$$= \max_{\lambda \in \Lambda(\bar{x},\omega)}(b^0 - A^0\bar{x})^T\lambda.$$

Hence

$$\varphi^+(\omega; \omega^0) \leq \inf_{\bar{x} \in \mathrm{Sol}(D,A,c,b)} \max_{\lambda \in \Lambda(\bar{x},\omega)} \left[\frac{1}{2}\bar{x}^T D^0\bar{x} + (c^0)^T\bar{x} + (b^0 - A^0\bar{x})^T\lambda \right].$$

$$(14.30)$$

2) Let $\{t^k\}$ be a sequence of real numbers such that $t^k \downarrow 0$ and

$$\varphi^-(\omega; \omega^0) = \lim_{k \to \infty} \frac{\varphi(\omega + t^k\omega^0) - \varphi(\omega)}{t^k}.$$

Due to the assumptions (i) and (ii), taking account of Lemmas 13.1 and 13.3 and the openness of the set G defined in (13.3) we can assume that

$$\mathrm{Sol}(\omega + t^k\omega^0) \neq \emptyset \quad \text{for every } k.$$

Let $\{x^k\}$ be a sequence in R^n such that $x^k \in \mathrm{Sol}(\omega + t^k\omega^0)$ for every k. By Lemma 14.4, without loss of generality we can assume that $x^k \to \hat{x} \in \mathrm{Sol}(D, A, c, b)$ as $k \to \infty$. We have

$$\varphi(\omega + t^k\omega^0) - \varphi(\omega) = \frac{1}{2}(x^k)^T(D + t^k D^0)x^k$$
$$+ (c + t^k c^0)^T x^k + \left(-\frac{1}{2}\hat{x}^T D\hat{x} - c^T\hat{x} \right). \quad (14.31)$$

Take $\lambda \in \Lambda(\bar{x}, \omega)$. As

$$\lambda^T(A\hat{x} - b) = 0, \quad \lambda \geq 0,$$

and

$$(A + t^k A^0)x^k \geq b + t^k b^0,$$

from (14.31) we get

$$\varphi(\omega + t^k \omega^0) - \varphi(\omega) \geq \frac{1}{2}(x^k)^T(D + t^k D^0)x^k$$
$$+(c + t^k c^0)^T x^k - \frac{1}{2}\hat{x}^T D\hat{x} - c^T\hat{x}$$
$$+\lambda^T(A\hat{x} - b) - [(A + t^k A_0)x^k - b - t^k b^0]^T \lambda$$
$$= (D\hat{x} - A^T\lambda + c)^T(x^k - \hat{x}) + \frac{1}{2}(x^k - \hat{x})^T D(x^k - \hat{x})$$
$$+t^k \left[\frac{1}{2}(x^k)^T D^0 x^k + (c^0)^T x^k + (b^0 - A^0 x^k)^T \lambda\right].$$

Since $\lambda \in \Lambda(\hat{x}, \omega)$, $D\hat{x} - A^T\lambda + c = 0$. Then we have

$$\varphi(\omega + t^k \omega^0) - \varphi(\omega) \geq \frac{1}{2}(x^k - \hat{x})^T D(x^k - \hat{x})$$
$$+t^k \left[\frac{1}{2}(x^k)^T D^0 x^k + (c^0)^T x^k + (b^0 - A^0 x^k)^T \lambda\right].$$

Multiplying both sides of this inequality by $(t^k)^{-1}$, taking lim inf as $k \to \infty$ and using condition (G), we obtain

$$\varphi^-(\omega; \omega^0) \geq \left(\frac{1}{2}\hat{x}^T D^0 \hat{x} + c^0\right)^T \hat{x} + (b^0 - A^0 \hat{x})^T \lambda.$$

As $\lambda \in \Lambda(\hat{x}, \omega)$ can be chosen arbitrarily, we conclude that

$$\varphi^-(\omega; \omega^0) \geq \max_{\lambda \in \Lambda(\hat{x}, \omega)} \left[\frac{1}{2}\hat{x}^T D^0 \hat{x} + (c^0)^T \hat{x} + (b^0 - A^0 \hat{x})^T \lambda\right]$$
$$\geq \inf_{\bar{x} \in \mathrm{Sol}(D, A, c, b)} \max_{\lambda \in \Lambda(\bar{x}, \omega)} [\frac{1}{2}\bar{x}^T D^0 \bar{x} + (c^0)^T \bar{x} + (b^0 - A^0 \bar{x})^T \lambda].$$

Combining this with (14.30), we have

$$\varphi^-(\omega; \omega^0) = \varphi^+(\omega; \omega^0)$$

and, therefore,

$$\varphi'(\omega; \omega^0) = \inf_{\bar{x} \in \mathrm{Sol}(\omega)} \max_{\lambda \in \Lambda(\bar{x}, \omega)} [\frac{1}{2}\bar{x}^T D^0 \bar{x} + (c^0)^T \bar{x} + (b^0 - A^0 \bar{x})^T \lambda].$$

The proof is complete. $\square$

We now apply Theorem 14.2 to a concrete example.

Example 14.1. Let $n = 2$, $m = 3$,

$$D = \begin{bmatrix} 1 & 0 \\ 0 & -1 \end{bmatrix}, \quad A^T = \begin{bmatrix} 1 & -1 & 0 \\ -1 & 0 & 1 \end{bmatrix},$$

$$b^T = (0, -1, 0), \quad c = \begin{pmatrix} 0 \\ 0 \end{pmatrix},$$

$$D^0 = \begin{bmatrix} 1 & 0 \\ 0 & -1 \end{bmatrix}, \quad (A^0)^T = \begin{bmatrix} 0 & 0 & 0 \\ 0 & 0 & 0 \end{bmatrix},$$

$$(b^0)^T = (0, -1, 0), \quad c^0 = \begin{pmatrix} 0 \\ 0 \end{pmatrix},$$

$$\omega = (D, A, c, b), \quad \omega^0 = (D^0, A^0, c^0, b^0).$$

It is easy to verify that $Ax \geq b$ is a regular system, $\mathrm{Sol}(D, A, 0, 0) = \{0\}$ and

$$\mathrm{Sol}(D, A, c, b) = \mathrm{Sol}(\omega) = \{(x_1, x_2)^T \in R^2 : x_1 = x_2, \; 0 \leq x_1 \leq 1\}$$
$$\mathrm{Sol}(\omega + t\omega^0) = \{(x_1, x_2)^T \in R^2 : x_1 = x_2, \quad 0 \leq x_1 \leq 1+t\}$$

for every $t \geq 0$. For $\bar{x} = (\bar{x}_1, \bar{x}_2) \in \mathrm{Sol}(\omega)$, we have

$$\Lambda(\bar{x}, \omega) = \{(\lambda_1, \lambda_2, \lambda_3)^T \in R^3 : \lambda_1 = \bar{x}_1, \lambda_2 = \lambda_3 = 0\}.$$

Suppose that $x^k = (x_1^k, x_2^k) \in \mathrm{Sol}(\omega + t^k \omega^0)$ and the sequence $\{x^k\}$ converges to $\bar{x} = (\bar{x}_1, \bar{x}_2) \in \mathrm{Sol}(\omega)$. We have $x_1^k = x_2^k$ and $\bar{x}_1 = \bar{x}_2$. Then

$$\frac{(x^k - \bar{x})^T D (x^k - \bar{x})}{t^k} = \frac{(x_1^k - \bar{x}_1)^2 - (x_2^k - \bar{x}_2)^2}{t^k} = 0,$$

hence condition (G) is satisfied. By Theorem 14.2,

$$\varphi'(\omega; \omega^0) = \inf_{\bar{x} \in \mathrm{Sol}(\omega)} \max_{\lambda \in \Lambda(\bar{x}, \omega)} \left(\left(\frac{1}{2} \bar{x}^T D^0 \bar{x} + (c^0)^T \bar{x} \right) + b^0 \right)^T \lambda$$
$$= \inf_{\bar{x} \in \mathrm{Sol}(\omega)} 0 = 0.$$

Observe that, in Example 14.1, $x^T Dx$ is an indefinite quadratic form (the sign of the expression $x^T Dx$ depends on the choice of x) and the solutions of the QP problem are not locally unique, so the assumptions of Theorem 14.1 are not satisfied.

Consider problem (13.1) and assume that $\bar{x} \in \mathrm{Sol}(D, A, c, b)$ is one of its solutions. Let $u = \omega^0 = (D^0, A^0, c^0, b^0) \in \Omega$ be a given direction. Applied to the solution $\bar{x}$ of problem (13.1), condition $(SOSC)_u$ in Auslender and Cominetti (1990) is stated as follows:

$$(SOSC)_u \quad \begin{cases} \textit{For every vector } v \in F_{\bar{x}} \setminus \{0\}, \quad \textit{if } (D\bar{x} + c)^T v = 0 \\ \textit{then} \quad v^T Dv > 0, \end{cases}$$

where $F_{\bar{x}}$ is the cone of the feasible directions of $\Delta(A, b)$ at $\bar{x}$. That is

$$F_{\bar{x}} = \{v \in R^n : (Av)_i \geq 0 \text{ for every } i \text{ satisfying } (A\bar{x})_i = b_i\}.$$

Observe that, in the case of QP problems, condition $(SOSC)_u$ is equivalent to the requirement saying that $\bar{x}$ is a locally unique solution of (13.1) (see Theorem 3.7). This remark allows us to deduce from Theorem 1 in Auslender and Coutat (1990) the following result.

Proposition 14.1. *Let $\omega = (D, A, c, b) \in \Omega$ be a given point and $u = \omega^0 = (D^0, A^0, c^0, b^0) \in \Omega$ be a given direction. If all the solutions of problem (13.1) are locally unique and the two conditions*

(i) *the system $Ax \geq b$ is regular,*

(ii) $\mathrm{Sol}(D, A, 0, 0) = \{0\}$

are satisfied, then the optimal value function φ is directionally differentiable at $\omega = (D, A, c, b)$ in direction $u = \omega^0 = (D^0, A^0, c^0, b^0)$, and formula (14.29) is valid.

Proof. By Theorem 12.1, from the assumptions (i) and (ii) it follows that the map $\mathrm{Sol}(\cdot)$ is upper semicontinuous at (D, A, c, b). Besides, by Lemma 13.3, $\mathrm{Sol}(D, A, c, b)$ is a nonempty compact set. Then there exists a compact set $B \subset R^n$ and a constant $\varepsilon > 0$ such that

$$\emptyset \neq \mathrm{Sol}(\omega + t\omega^0) \subset B \quad \text{for every } t \in [0, \varepsilon].$$

Under the conditions of our proposition, all the assumptions of Theorem 1 in Auslender and Coutat (1990) are fulfilled. So the desired conclusion follows from applying Theorem 1 in Auslender and Coutat (1990). □

Observe that Proposition 14.1 is a direct corollary of our Theorems 14.1 and 14.2. It is worth noting that the result stated in Proposition 14.1 cannot be applied to the problem described in Example 14.1 (because condition $(SOSC)_u$, where $u := \omega^0$, does not hold at any solution $\bar{x} \in \mathrm{Sol}(\omega)$). That result cannot be applied also to convex QP problems whose solution sets have more than one element. This is because, for such a problem, the solution set is a convex set consisting of more than one element. Using Remark 14.1 we can conclude that Theorem 14.2 is applicable to convex QP problems.

Consider problem (13.1) and denote $\omega = (D, A, c, b)$. Suppose that $\omega^0 = (D^0, A^0, c^0, b^0) \in \Omega$ is a given direction. In this case, condition (H3) in Minchenko and Sakolchik (1996) is stated as follows:

(H3) *For every sequence* $\{t^k\}$, $t^k \downarrow 0$, *and every sequence* $\{x^k\}$, $x^k \to \bar{x} \in \mathrm{Sol}(D, A, c, b)$, $x^k \in \mathrm{Sol}(\omega + t^k \omega^0)$ *for each k, the following inequality is satisfied*

$$\limsup_{k \to \infty} \frac{\|x^k - \bar{x}\|^2}{t^k} < +\infty.$$

Applying Theorem 4.1 in Minchenko and Sakolchik (1996) to problem (13.1) we get the following result.

Proposition 14.2. *Let* $\omega = (D, A, c, b)$ *and* $\omega^0 = (D^0, A^0, c^0, b^0)$ *be given as in Proposition 14.1. If* (H3) *and the two conditions*

(i) *the system $Ax \geq b$ is regular,*

(ii) *there exist a compact set $B \subset R^n$ and a neighborhood U of $(A, b) \in R^{m \times n} \times R^m$ such that $\Delta(A', b') \subset B$ for every $(A', b') \in U$*

are satisfied, then the optimal value function φ is directionally differentiable at $\omega = (D, A, c, b)$ in direction $u = \omega^0 = (D^0, A^0, c^0, b^0)$, and formula (14.29) is valid.

Consider the problem described in Example 14.1. Choose $\bar{x} = (0, 0) \in \mathrm{Sol}(\omega)$, $t^k = k^{-1}$,

$$x^k = (k^{-\frac{1}{4}}, k^{-\frac{1}{4}}) \in \mathrm{Sol}(\omega + t^k \omega^0).$$

We have $x^k \to \bar{x}$ as $k \to \infty$ and

$$\limsup_{k \to \infty} \frac{\|x^k - \bar{x}\|^2}{t^k} = \limsup_{k \to \infty} \frac{k^{-\frac{1}{2}} + k^{-\frac{1}{2}}}{k^{-1}} = +\infty,$$

so (H3) does not hold and Proposition 14.2 cannot be applied to this QP problem.

We have shown that Theorem 14.2 can be applied even to some kinds of QP problems where the existing results on differential stability in nonlinear programming cannot be used.

Now we want to show that, for problem (13.1), if the system $Ax \geq b$ is regular then (H3) implies (G).

Proposition 14.3. *Let* $\omega = (D, A, c, b)$ *and* $\omega^0 = (D^0, A^0, c^0, b^0)$ *be given as in Proposition 14.1. If the system* $Ax \geq b$ *is regular, then condition* (H3) *implies condition* (G).

Proof. Suppose that (H3) holds. Let $\{t^k\}$, $t^k \downarrow 0$, and $\{x^k\}$, where $x^k \in \text{Sol}(\omega + t^k \omega^0)$ for each k, be arbitrary sequences. If

$$x^k \to \bar{x} \in \text{Sol}(D, A, c, b)$$

then, by (H3), we have

$$\limsup_{k \to \infty} \frac{\|x^k - \bar{x}\|^2}{t^k} < +\infty. \tag{14.32}$$

We have to prove that the inequality written in condition (G) is satisfied. Let $\{(t^{k'})^{-1}(x^{k'} - \bar{x})^T D(x^{k'} - \bar{x})\}$ be a subsequence of $\{(t^k)^{-1}(x^k - \bar{x})^T D(x^k - \bar{x})\}$ satisfying

$$\liminf_{k \to \infty}(t^k)^{-1}(x^k - \bar{x})^T D(x^k - \bar{x})$$
$$= \lim_{k' \to \infty}(t^{k'})^{-1}(x^{k'} - \bar{x})^T D(x^{k'} - \bar{x}). \tag{14.33}$$

From (14.32) it follows that the sequence $\{(t^k)^{-1}\|x^k - \bar{x}\|^2\}$ is bounded. Then the sequence $\{(t^k)^{-1/2}\|x^k - \bar{x}\|\}$ is bounded. Without loss of generality, we may assume that

$$(t^k)^{-1/2}\|x^k - \bar{x}\| \to v \in R^n. \tag{14.34}$$

As $x^k \in \text{Sol}(D + t^k D^0, A + t^k A^0, c + t^k c^0, b + t^k b^0)$, we have

$$(A_I + t^k A_I^0)x^k \geq b_I + t^k b_I^0,$$

where $I = \{i : (A\bar{x})_i = b_i\}$. Since $b_I = A_I \bar{x}$,

$$A_I(x^k - \bar{x}) \geq t^k(b_I^0 - A_I^0 x^k).$$

Multiplying both sides of this inequality by $(t^k)^{-1/2}$ and letting $k \to \infty$, due to (14.34) we can conclude that $A_I v \geq 0$. Hence $v \in F_{\bar{x}}$, where $F_{\bar{x}}$ is defined as in the formulation of condition $(SOSC)_u$. Furthermore, note that the expression (14.24) holds. As $Ax \geq b$ is a regular system, by Lemma 14.1 we have $F(\bar{x}, \omega, \omega^0) \neq \emptyset$. Take any $\bar{v} \in F(\bar{x}, \omega, \omega^0)$. Then, for k large enough,

$$\bar{x} + t^k \bar{v} \in \Delta(A + t^k A^0, b + t^k b^0).$$

Therefore, for k large enough, we have (14.25). From (14.24) and (14.25) we have (14.26). Multiplying both sides of (14.26) by

$$(t^k)^{-1/2},$$

letting $k \to \infty$ and taking account of (14.34), we get (14.27). As $\bar{x}$ is a solution of problem (13.1) and $v \in F_{\bar{x}}$, the situation $(D\bar{x}+c)^T v < 0$ cannot happen. Hence $(D\bar{x} + c)^T v = 0$. Since $\bar{x} \in \text{Sol}(\omega)$, we must have $v^T D v \geq 0$ (see Theorem 3.5). By (14.33) and (14.34),

$$\liminf_{k \to \infty} (t^k)^{-1}(x^k - \bar{x})^T D(x^k - \bar{x})$$
$$= \lim_{k' \to \infty} \left((t^{k'})^{-1/2}(x^{k'} - \bar{x}) \right)^T D \left((t^{k'})^{-1/2}(x^{k'} - \bar{x}) \right)$$
$$= v^T D v \geq 0.$$

Thus (G) is satisfied. $\square$

14.4 Commentaries

The results presented in this chapter are due to Tam (2001b).

Best and Chakravarti (1990) considered parametric convex quadratic programming problems and obtained some results on the directional differentiability of the optimal value function. Auslender and Coutat (1996) investigated similar questions for the case of generalized linear-quadratic programs. A survey of some results on stability and sensitivity of nonlinear mathematical programming problems can be found in Bonnans and Shapiro (1998). A comprehensive theory on perturbation analysis of optimization problems was given by Bonnans and Shapiro (2000).

Chapter 15

Quadratic Programming under Linear Perturbations: I. Continuity of the Solution Maps

Continuity of the local solution map and the solution map of QP problems under linear perturbations is studied in this chapter.

Since it is impossible to give a satisfactory characterization for the usc property of the local solution map and since the usc property of the solution map can be derived from a result of Klatte (1985), Theorem 3, we will concentrate mainly on characterizing the lsc property of the local solution map and the solution map.

Consider the QP problem

$$
\begin{cases}
\text{Minimize } f(x) := \frac{1}{2}x^T D x + c^T x \\
\text{subject to } x \in \Delta(A, b) := \{x \in R^n \, : \, Ax \geq b\}
\end{cases}
\tag{15.1}
$$

depending on the parameter $w = (c, b) \in R^n \times R^m$, where the matrices $D \in R_S^{n \times n}$ and $A \in R^{m \times n}$ are not subject to change. The solution set, the local solution set and the KKT point set of this problem are denoted, respectively, by $\text{Sol}(c, b)$, $\text{loc}(c, b)$ and $S(D, A, c, b)$.

15.1   Lower Semicontinuity of the Local Solution Map

In this section we investigate the lsc property of the local solution map

$$\text{loc}(\cdot) : R^n \times R^m \to 2^{R^m}, \quad (c', b') \in R^n \times R^m \mapsto \text{loc}(c', b'). \quad (15.2)$$

Theorem 15.1. *The multifunction (15.2) is lower semicontinuous at $(c, b) \in R^n \times R^m$ if and only if the system $Ax \geq b$ is regular and the set $\text{loc}(c, b)$ is nonempty and finite.*

Proof. *Necessity:* Since the multifunction (15.2) is lower semicontinuous at (c, b), $\text{loc}(c, b)$ is nonempty and the regularity condition is satisfied. We now prove that $\text{loc}(c, b)$ is finite. Define the sets Q_I ($I \subset \{1, \ldots, m\}$) and Q as in the proof of Theorem 11.3. Since Q is nowhere dense, there exists a sequence $\{(c^k, b^k)\}$ converging to (c, b) in $R^n \times R^m$ such that $(-c^k, b^k) \notin Q$ for all $k \in N$. Fix a point $\bar{x} \in \text{loc}(c, b)$. Since $\text{loc}(\cdot)$ is lower semicontinuous at (c, b), there exist a subsequence $\{(c^{k_l}, b^{k_l})\}$ of $\{(c^k, b^k)\}$ and a sequence $\{x^{k_l}\}$ converging to $\bar{x}$ in R^n such that

$$x^{k_l} \in \text{loc}(c^{k_l}, b^{k_l})$$

for all k_l. For any k_l, since $\text{loc}(c^{k_l}, b^{k_l}) \subset S(D, A, c^{k_l}, b^{k_l})$, there exists $\lambda^{k_l} \in R^m$ such that (11.9) is satisfied. For every k_l, define

$$I_{k_l} = \{i \in \{1, \ldots, m\} : \lambda_i^{k_l} > 0\}.$$

By the same arguments as those used in the proof of Theorem 11.3, we obtain a subset $I \subset \{1, \ldots, m\}$ such that (11.10)–(11.13) hold. Next, let Z and X be defined as in the proof of Theorem 11.3. As before, X is a finite set and we have $\bar{x} \in X$. Since $\bar{x} \in \text{loc}(c, b)$ can be chosen arbitrarily, we have $\text{loc}(c, b) \subset X$. Hence $\text{loc}(c, b)$ is a finite set.

Sufficiency: If the regularity condition is satisfied and the set $\text{loc}(c, b)$ is finite, then from Theorem 5 in Phu and Yen (2001) it follows that the multifunction (15.2) is lower semicontinuous at $(c, b) \in R^n \times R^m$.   □

Since $\text{loc}(c, b) \subset S(D, A, c, b)$, from Theorem 15.1 we obtain the following corollary.

Corollary 15.1. *Let $(D, A, c, b) \in R_S^{n \times n} \times R^{m \times n} \times R^n \times R^m$. Suppose that the system $Ax \geq b$ is regular and the following conditions are satisfied:*

(i) *the set $S(D, A, c, b)$ is finite,*

(ii) *the set $\mathrm{loc}(c, b)$ is nonempty.*

Then, the multifunction (15.2) is lower semicontinuous at (c, b).

15.2 Lower Semicontinuity of the Solution Map

In this section, a complete characterization for the lower semicontinuity of the solution map

$$\mathrm{Sol}(\cdot) : R^n \times R^m \to 2^{R^n}, \quad (c', b') \mapsto \mathrm{Sol}(c', b') \tag{15.3}$$

of the QP problem (15.1) will be given. Before proving the result, we state some lemmas.

Let $(D, A, c, b) \in R_S^{n \times n} \times R^{m \times n} \times R^n \times R^m$.

Lemma 15.1. *If the multifunction (15.3) is lower semicontinuous at (c, b), then the system $Ax \geq b$ is regular and the set $\mathrm{Sol}(c, b)$ is nonempty and finite.*

Proof. (This proof is very similar to the first part of proof of Theorem 15.1.) It is clear that if the multifunction (15.3) is lower semicontinuous at (c, b) then regularity condition is satisfied and the set $\mathrm{Sol}(c, b)$ is nonempty. In order to prove the finiteness of $\mathrm{Sol}(c, b)$, we define Q_I ($I \subset \{1, \ldots, m\}$) and Q as in the proof of Theorem 13.3. Then, there exists a sequence $\{(c^k, b^k)\}$ converging to (c, b) such that $(-c^k, b^k) \notin Q$ for all k. Fix any $\bar{x} \in \mathrm{Sol}(c, b)$. As $\mathrm{Sol}(\cdot)$ is lower semicontinuous at (c, b), there exist a subsequence $\{(c^{k_l}, b^{k_l})\}$ of $\{(c^k, b^k)\}$ and a sequence $\{x^{k_l}\}$ converging to $\bar{x}$ in R^n such that $x^{k_l} \in \mathrm{Sol}(c^{k_l}, b^{k_l})$ for all k_l. Since $\mathrm{Sol}(c^{k_l}, b^{k_l}) \subset S(D, A, c^{k_l}, b^{k_l})$, there exists $\lambda^{k_l} \in R^m$ such that (11.9) is satisfied. Constructing I and defining Z and X as in the proof of Theorem 11.3, we have that X is a finite set and $\bar{x} \in X$. Hence the solution set $\mathrm{Sol}(c, b)$ is finite. $\square$

Lemma 15.2. *If the multifunction (15.3) is lower semicontinuous at (c, b), then the set $\mathrm{Sol}(c, b)$ is a singleton.*

Proof. On the contrary, suppose that $\mathrm{Sol}(\cdot)$ is lower semicontinuous at (c, b) but there exist $\bar{x}, \bar{y} \in \mathrm{Sol}(c, b)$ such that $\bar{x} \neq \bar{y}$. Choose $c_0 \in R^n$ such that

$$c_0^T(\bar{y} - \bar{x}) = 1. \tag{15.4}$$

By Lemma 15.1, $\mathrm{Sol}(c, b)$ is a finite set. Combining this fact with (15.4) we see that there exists a open set $U \subset R^n$ containing $\bar{y}$ such that

$$\mathrm{Sol}(c, b) \cap U = \{\bar{y}\}$$

and

$$c_0^T(y - \bar{x}) > 0 \quad \text{for all } y \in U. \tag{15.5}$$

Let $\delta > 0$ be given arbitrarily. Choose $\varepsilon > 0$ so that

$$0 < \varepsilon < \frac{\delta}{\|c_0\|}.$$

Let $b' = b$, $c' = c + \varepsilon c_0$. We have

$$\|(c', b') - (c, b)\| = \varepsilon \|c_0\| < \delta.$$

We now show that

$$\mathrm{Sol}(c', b') \cap U = \emptyset.$$

For any $y \in \Delta(A, b') = \Delta(A, b)$, since $\bar{x}, \bar{y} \in \mathrm{Sol}(c, b)$, using (15.5) we have

$$
\begin{aligned}
\frac{1}{2} y^T D y + (c')^T y &= \frac{1}{2} y^T D y + (c + \varepsilon c_0)^T y \\
&= (\frac{1}{2} y^T D y + c^T y) + \varepsilon c_0^T y \\
&\geq (\frac{1}{2} \bar{y}^T D \bar{y} + c^T \bar{y}) + \varepsilon c_0^T y \\
&> \left(\frac{1}{2} \bar{y}^T D \bar{y} + c^T \bar{y}\right) + \varepsilon c_0^T \bar{x} \\
&= (\frac{1}{2} \bar{x}^T D \bar{x} + c^T \bar{x}) + \varepsilon c_0^T \bar{x} \\
&= \frac{1}{2} \bar{x}^T D \bar{x} + (c + \varepsilon c_0)^T \bar{x}.
\end{aligned}
\tag{15.6}
$$

Since $\bar{x} \in \Delta(A, b')$, by (15.6) we have $y \notin \mathrm{Sol}(c', b')$. Consequently, for the chosen neighborhood U of $\bar{y} \in \mathrm{Sol}(c, b)$, for every $\delta > 0$ there exists $(c', b') \in R^n \times R^m$ satisfying $\|(c', b') - (c, b)\| < \delta$ and $\mathrm{Sol}(c', b') \cap U = \emptyset$. This contradicts our assumption that $\mathrm{Sol}(\cdot)$ is lower semicontinuous at (c, b). We have shown that $\mathrm{Sol}(c, b)$ cannot have more than one element. Since $\mathrm{Sol}(\cdot)$ is lower semicontinuous at (c, b), we must have $\mathrm{Sol}(c, b) \neq \emptyset$. From what has been proved, we conclude that $\mathrm{Sol}(c, b)$ is a singleton. $\square$

Lemma 15.3. *If $\Delta(A, b)$ is nonempty and if we have*

$$\inf\{v^T(Dx + c) : x \in R^n, \ Ax \geq b\} > 0 \tag{15.7}$$

for some $v \in R^n$, then there exists $\delta > 0$ such that

$$\inf\{v^T(Dx' + c') : x' \in R^n, \; Ax' \geq b'\} \geq 0 \qquad (15.8)$$

for every $(c', b') \in R^n \times R^m$ satisfying $\|(c', b') - (c, b)\| < \delta$. (By convention, $\inf \emptyset = +\infty$.)

Proof. By Corollary 7.2, for the given matrix A there exists a constant $\gamma(A) \geq 0$ such that

$$\Delta(A, b') \subset \Delta(A, b'') + \gamma(A)\|b'' - b'\|B_{R^n} \qquad (15.9)$$

for all $b', b'' \in R^n$ satisfying $\Delta(A, b') \neq \emptyset$ and $\Delta(A, b'') \neq \emptyset$. Define

$$\mu = \inf\{v^T(Dx + c) : x \in R^n, \; Ax \geq b\}. \qquad (15.10)$$

By (15.7), $\mu > 0$. Choose $\delta > 0$ such that

$$\gamma(A)\|v\|\|D\|\delta < \frac{\mu}{4}, \quad \|v\|\delta < \frac{\mu}{4}. \qquad (15.11)$$

Here, as usual, $\|D\| = \max\{\|Dx\| : x \in R^n, \; \|x\| \leq 1\}$. Let $(c', b') \in R^n \times R^m$ be such that $\|(c', b') - (c, b)\| < \delta$. If $\Delta(A, b') = \emptyset$ inequality (15.8) is valid because the infimum in its left-hand-side equal $+\infty$. Now consider the case where $\Delta(A, b') \neq \emptyset$. By our assumption, $\Delta(A, b) \neq \emptyset$. Hence, for every $x' \in \Delta(A, b')$, from (15.9) it follows that there exists $x \in \Delta(A, b)$ such that

$$x' = x + \gamma(A)\|b' - b\|u \qquad (15.12)$$

for some $u \in B_{R^n}$. By (15.11) and (15.12), we have

$$\begin{aligned}
&|v^T(Dx' + c') - v^T(Dx + c)| \\
&\leq |v^T D(x' - x)| + |v^T(c' - c)| \\
&\leq \|v\|\|D\|\|x' - x\| + \|v\|\|c' - c\| \\
&\leq \|v\|\|D\|\gamma(A)\|b' - b\|\|u\| + \|v\|\|c' - c\| \\
&\leq \|v\|\|D\|\gamma(A)\delta + \|v\|\|c' - c\| \\
&< \frac{\mu}{4} + \frac{\mu}{4} = \frac{\mu}{2}.
\end{aligned}$$

Combining this with (15.10), we obtain

$$\begin{aligned}
v^T(Dx' + c') &\geq v^T(Dx + c) - \frac{\mu}{2} \\
&\geq \mu - \frac{\mu}{2} \\
&= \frac{\mu}{2}.
\end{aligned}$$

From this we deduce that (15.8) holds for every $(c', b') \in R^n \times R^m$ with the property that $\|(c', b') - (c, b)\| < \delta$.   □

Lemma 15.4. *Let K denote the cone $\{v \in R^n : Av \geq 0, \ v^T Dv = 0\}$. Assume that the system $Ax \geq b$ is regular, the set $\mathrm{Sol}(c, b)$ is nonempty, and*

$$\inf\{v^T(Dx + c) : x \in R^n, \ Ax \geq b\} > 0$$

for every nonzero $v \in K$. Then there exists $\rho > 0$ such that $\mathrm{Sol}(c', b')$ is nonempty for all $(c', b') \in R^n \times R^m$ satisfying $\|(c', b') - (c, b)\| < \rho$.

Proof. Since the regularity condition is satisfied, by Lemma 13.1 there exists $\rho_0 > 0$ such that for every $b' \in R^m$ satisfying $\|b' - b\| < \rho_0$ we have

$$\Delta(A, b') \neq \emptyset. \tag{15.13}$$

Since $\mathrm{Sol}(c, b) \neq \emptyset$, by Theorem 2.2 we have

$$v^T Dv \geq 0 \tag{15.14}$$

for all $v \in R^n$ satisfying $Av \geq 0$.

We now distinguish two cases.

Case 1. $K = \{v \in R^n : Av \geq 0, \ v^T Dv = 0\} = \{0\}$. In this case, we have

$$v^T(Dx + c') = 0 \tag{15.15}$$

for all $v \in K$, $x \in R^n$ and $c' \in R^n$. On account of Theorem 2.2 and the properties (15.13)–(15.15), we have $\mathrm{Sol}(c', b') \neq \emptyset$ for every $(c', b') \in R^n \times R^m$ satisfying $\|(c', b') - (c, b)\| < \rho_0$.

Case 2. $K = \{v \in R^n : Av \geq 0, \ v^T Dv = 0\} \neq \{0\}$. In this case, from (15.14) it follows that K can be represented as the union of finitely many polyhedral convex cones (see Bank et al. (1982), Lemma 4.5.1). Suppose that

$$K = \bigcup_{j=1}^{s} K_j, \tag{15.16}$$

where K_j $(j = 1, 2, \ldots, s)$ are polyhedral convex cones. Fix an index $j \in \{1, 2, \ldots, s\}$. Let $v^1, v^2, \ldots, v^{k_j}$ be the generators (see Rockafellar (1970), p. 170) of the cone K_j. By our assumption, for every v^i, $1 \leq i \leq k_j$, we have

$$\inf\{(v^i)^T(Dx + c) : x \in R^n, \ Ax \geq b\} > 0.$$

By Lemma 15.3, there exists $\delta_i > 0$ such that

$$\inf\{(v^i)^T(Dx + c') : x \in R^n, \ Ax \geq b'\} \geq 0 \tag{15.17}$$

for all $(c', b') \in R^n \times R^m$ satisfying $\|(c', b') - (c, b)\| < \delta_i$. Define

$$\rho_j = \min\{\delta_i : i = 1, \cdots, k_j\}.$$

From (15.17) it follows that

$$(v^i)^T(Dx + c') \geq 0 \tag{15.18}$$

for all $i = 1, \cdots, k_j$ and for all $x \in R^n$ satisfying $Ax \geq b'$, provided that $\|(c', b') - (c, b)\| < \rho_j$. Define $\bar{\rho}_j = \min\{\rho_0, \rho_j\}$. Let $(c', b') \in R^n \times R^m$ be such that $\|(c', b') - (c, b)\| < \bar{\rho}_j$, $x \in \Delta(A, b')$ and $v \in K_j$. As $v^1, \cdots, v^{k_j}$ are the generators of K_j, there exist nonnegative real numbers $\alpha_1, \cdots, \alpha_{k_j}$ such that $v = \alpha_1 v^1 + \cdots + \alpha_{k_j} v^{k_j}$. By (15.18), we have

$$v^T(Dx + c') = \sum_{i=1}^{k_j} \alpha_i(v^i)^T(Dx + c') \geq 0. \tag{15.19}$$

Set $\rho = \min\{\bar{\rho}_j : j = 1, 2, \ldots, s\}$. Let $(c', b') \in R^n \times R^m$ be such that

$$\|(c', b') - (c, b)\| < \rho, \tag{15.20}$$

and let $x \in \Delta(A, b')$ and $v \in K$ be given arbitrarily. By (15.16), there exists $j \in \{1, 2, \ldots, s\}$ such that $v \in K_j$. Let $v = \alpha_1 v^1 + \cdots + \alpha_{k_j} v^{k_j}$, where $\alpha_i \geq 0$ for $i = 1, 2, \ldots, k_j$. By virtue of (15.19), we have $v^T(Dx + c') \geq 0$. Hence, taking account of (15.13), (15.14), and applying Theorem 2.2 we have $\text{Sol}(c', b') \neq \emptyset$ for all $(c', b') \in R^n \times R^m$ satisfying (15.20). The lemma is proved. $\square$

We can now state the main result of this section.

Theorem 15.2. *The multifunction (15.3) is lower semicontinuous at (c, b) if and only if the system $Ax \geq b$ is regular and the following conditions are satisfied:*

(i) *for every nonzero vector*

$$v \in K := \{v \in R^n : Av \geq 0, \ v^T Dv = 0\}$$

it holds $\inf\{v^T(Dx + c) : x \in R^n, \ Ax \geq b\} > 0$,

(ii) *the set* $\text{Sol}(c, b)$ *is a singleton.*

Proof. *Necessity:* If Sol($\cdot$) is lower semicontinuous at (c, b) then by Lemmas 15.1 and 15.2, the system $Ax \geq b$ is regular and condition (ii) is satisfied. Suppose that the property (i) were false. Then we could find a nonzero vector $\bar{v} \in K = \{v \in R^n : Av \geq 0, v^T Dv = 0\}$ such that

$$\inf\{\bar{v}^T(Dx + c) : x \in R^n, Ax \geq b\} \leq 0. \qquad (15.21)$$

If the infimum in the left-hand-side of (15.21) is $-\infty$ then it is obvious that there exists $\bar{x} \in \Delta(A, b)$ satisfying $\bar{v}^T(D\bar{x} + c) < 0$. If that infimum is finite then, applying the Frank-Wolfe Theorem to the linear programming problem

Minimize $\bar{v}^T(Dx + c)$ subject to $x \in R^n$, $Ax \geq b$,

we find $\bar{x} \in \Delta(A, b)$ such that

$$\bar{v}^T(D\bar{x} + c) = \inf\{\bar{v}^T(Dx + c) : x \in R^n, Ax \geq b\} \leq 0.$$

So, in both cases, we can find $\bar{x} \in \Delta(A, b)$ such that

$$\bar{v}^T(D\bar{x} + c) \leq 0. \qquad (15.22)$$

For every positive integer k, let $c^k := c - \frac{1}{k}\bar{v}$. By (15.22),

$$\bar{v}^T(D\bar{x} + c^k) = \bar{v}^T(D\bar{x} + c) - \frac{1}{k}\bar{v}^T\bar{v} < 0. \qquad (15.23)$$

From (15.23) we see that condition (ii) in Theorem 2.2, where $(D, A, c, b) := (D, A, c^k, b)$, is violated. Hence, by Theorem 2.2 we have Sol(c^k, b) $= \emptyset$ for all k. Since $c^k \to c$ as $k \to \infty$, the latter fact shows that the multifunction (15.3) is not lower semicontinuous at (c, b), a contradiction.

Sufficiency: Suppose that the system $Ax \geq b$ is regular and the conditions (i), (ii) are satisfied. By (ii), we can assume that Sol(c, b) $= \{\bar{x}\}$ for some $\bar{x} \in R^n$. Let U be any open set containing $\bar{x}$. By the regularity assumption, by (i) and Lemma 15.4, there exists $\rho > 0$ such that Sol(c', b') $\neq \emptyset$ for all $(c', b') \in R^n \times R^m$ satisfying $\|(c', b') - (c, b)\| < \rho$. By (ii) and Theorem 3 in Klatte (1985), there exists $\rho_1 > 0$ such that Sol(c', b') $\subset U$ for all $(c', b') \in R^n \times R^m$ satisfying $\|(c', b') - (c, b)\| < \rho_1$. Hence, for $\rho_2 := \min\{\rho, \rho_1\}$, we have Sol($c', b'$) $\cap U \neq \emptyset$ for all $(c', b') \in R^n \times R^m$ satisfying $\|(c', b') - (c, b)\| < \rho_2$. From what already been proved, it may be concluded

that Sol$(\cdot)$ is lower semicontinuous at (c, b). The proof is complete.
$\square$

Let us mention two direct corollaries of Theorem 15.2.

Corollary 15.2. *For $(D, A, c, b) \in R_S^{n \times n} \times R^{m \times n} \times R^n \times R^m$, if $K := \{v \in R^n : Av \geq 0, v^T D v = 0\} = \{0\}$ then the multifunction (15.3) is lower semicontinuous at (c, b) if and only if the system $Ax \geq b$ is regular and the set Sol(c, b) is a singleton.*

Corollary 15.3. *Let $(D, A, c, b) \in R_S^{n \times n} \times R^{m \times n} \times R^n \times R^m$. If D is a positive definite matrix then the multifunction (15.3) is lower semicontinuous at (c, b) if and only if condition the system $Ax \geq b$ is regular.*

In Theorem 15.2 we have established a complete characterization for the lower semicontinuity property of the solution map (15.3). Let us consider an example.

Example 15.1. Let $n = 2$, $m = 4$, and

$$D = \begin{bmatrix} 1 & 0 \\ 0 & -1 \end{bmatrix}, \quad A = \begin{bmatrix} 1 & -1 \\ 1 & 0 \\ 1 & 0 \\ 0 & 1 \end{bmatrix}, \quad c = \begin{pmatrix} 1 \\ 0 \end{pmatrix}, \quad b = \begin{pmatrix} 0 \\ 1 \\ 0 \\ 0 \end{pmatrix}.$$

Then we have the following QP problem

$$\begin{cases} \text{Minimize} & \frac{1}{2}(x_1^2 - x_2^2) + x_1 \\ \text{subject to} & x_1 - x_2 \geq 0, \ x_1 \geq 1, \ x_1 \geq 0, \ x_2 \geq 0. \end{cases}$$

It is easily seen that the system $Ax \geq b$ is regular. For any $x = (x_1, x_2) \in \Delta(A, b)$, we have

$$\frac{1}{2}(x_1^2 - x_2^2) + x_1 \geq x_1 \geq 1.$$

The last two inequalities become equalities if and only if $x_1 = x_2 = 1$. These observations allow us to conclude that Sol$(c, b) = \{(1, 1)\}$. Clearly,

$$\text{Sol}(0, 0) = \{(x_1, x_2) \in R^2 : x_1 \geq 0, \ x_2 \geq 0, \ x_1 - x_2\}$$

and

$$\begin{aligned} K &= \{(v_1, v_2) \in R^2 : v_1 \geq 0, \ v_2 \geq 0, v_1^2 - v_2^2 = 0\} \\ &= \{(v_1, v_2) \in R^2 : v_1 \geq 0, \ v_2 \geq 0, v_1 = v_2\}. \end{aligned}$$

For any $v = (\mu, \mu) \in K \setminus \{0\}$ ($\mu > 0$) and for any $x = (x_1, x_2) \in \Delta(A, b)$, we have

$$
\begin{aligned}
v^T(Dx + c) &= (\mu, \mu)\binom{x_1+1}{-x_2} \\
&= \mu(x_1 + 1 - x_2) \geq \mu.
\end{aligned}
$$

So

$$\inf\{v^T(Dx + c) : x \in R^2, \ Ax \geq b\} \geq \mu > 0.$$

Since the system $Ax \geq b$ is regular and conditions (i), (ii) in Theorem 15.2 are satisfied, we conclude that the multifunction (15.3) is lower semicontinuous at (c, b). Meanwhile, since $\mathrm{Sol}(0,0) \neq \{0\}$, from Theorem 12.3 it follows that the multifunction $(D', A', c', b') \mapsto \mathrm{Sol}(D', A', c', b')$, where $\mathrm{Sol}(D', A', c', b')$ denotes the solution set of the canonical QP problem

$$
\begin{cases}
\text{Minimize} & f(x) := \dfrac{1}{2}x^T D' x + (c')^T x \\
\text{subject to} & A'x \geq b', \quad x \geq 0,
\end{cases}
$$

is not lower semicontinuous at $(D, \widetilde{A}, c, \widetilde{b}) \in R_S^{2\times 2} \times R^{2\times 2} \times R^2 \times R^2$. Here

$$\widetilde{A} = \begin{bmatrix} 1 & -1 \\ 1 & 0 \end{bmatrix}, \quad \widetilde{b} = \binom{0}{1}.$$

15.3 Commentaries

The results presented in this chapter are taken from Lee et al. (2002b, 2002c).

Theorem 15.2 is the main result of this chapter. Example 15.1 shows clearly the difference between the characterization given by Theorem 15.2 and the one provided by Theorem 12.3.

Chapter 16

Quadratic Programming under Linear Perturbations: II. Properties of the Optimal Value Function

In this chapter, we will consider the optimal value function $(c, b) \mapsto \varphi(c, b)$ of the parametric QP problem (15.1). It is proved that φ is directionally differentiable at any point $\bar{w} = (\bar{c}, \bar{b})$ in its effective domain $W := \{w = (c, b) \in R^n \times R^m : -\infty < \varphi(c, b) < +\infty\}$. Formulae for computing the directional derivative $\varphi'(\bar{w}; z)$ of φ at $\bar{w}$ in a direction $z = (u, v) \in R^n \times R^m$ are also obtained.

If D is positive semidefinite, then φ is piecewise linear-quadratic on the set W (which is a polyhedral convex cone). If D is not assumed to be positive semidefinite then W may be nonconvex, but it can be represented as the union of finitely many polyhedral convex cones. We present an example showing that, in general, φ is not piecewise linear-quadratic on W.

16.1 Auxiliary Results

Consider the standard QP problem (15.1) depending on the parameter $w = (c, b) \in R^n \times R^m$, where $D \in R_S^{n \times n}$ and $A \in R^{m \times n}$ are given matrices. Denote by $S(c, b)$, $\text{Sol}(c, b)$, $\text{loc}(c, b)$ and $\varphi(c, b)$, respectively, the set of the Karush-Kuhn-Tucker points, the set of the solutions, the set of the local solutions, and the optimal value of (15.1). Klatte (1985) established several fundamental facts on the

Lipschitzian continuity of the map $(c, b) \mapsto \text{Sol}(c, b)$ and the function $(c, b) \mapsto \varphi(c, b)$. Among other results, he proved that $\varphi(\cdot, \cdot)$ is Lipschitzian on every bounded subset of its effective domain

$$W := \{(c, b) \in R^n \times R^m : -\infty < \varphi(c, b) < +\infty\}. \tag{16.1}$$

Following Klatte (1985), we consider the next auxiliary problem

$$\begin{cases} \text{Minimize} \quad \dfrac{1}{2}(c^T x + b^T \lambda) \\ \text{subject to } (x, \lambda) \in P_{KKT}(c, b) \end{cases} \tag{16.2}$$

where

$$P_{KKT}(c, b) = \{(x, \lambda) \in R^n \times R^m : \quad Dx - A^T \lambda + c = 0, \\ Ax \geq b, \quad \lambda \geq 0, \tag{16.3} \\ \lambda^T(Ax - b) = 0\}.$$

Elements of $P_{KKT}(c, b)$ are the Karush-Kuhn-Tucker pairs of (15.1).

Let

$$\varphi_{KKT}(c, b) = \inf\left\{\dfrac{1}{2}(c^T x + b^T \lambda) : (x, \lambda) \in P_{KKT}(c, b)\right\} \tag{16.4}$$

be the optimal value of the auxiliary problem (16.2). By definition,

$$\varphi(c, b) = \inf\left\{\dfrac{1}{2}x^T Dx + c^T x : Ax \geq b, \ x \in R^n\right\}. \tag{16.5}$$

Denote by $\text{Sol}_{KKT}(c, b)$ the solution set of (16.2).

Lemma 16.1. (See Klatte (1985), p. 820) *If* $\text{Sol}(c, b)$ *is nonempty then* $\text{Sol}_{KKT}(c, b)$ *is nonempty, and*

$$\text{Sol}(c, b) = \pi_{R^n}(\text{Sol}_{KKT}(c, b)), \tag{16.6}$$

$$\varphi(c, b) = \varphi_{KKT}(c, b), \tag{16.7}$$

where, by definition, $\pi_{R^n}(x, \lambda) = x$ *for every* $(x, \lambda) \in R^n \times R^m$.

Proof. Since $\text{Sol}(c, b) \neq \emptyset$, we can select a point $\bar{x} \in \text{Sol}(c, b)$. By Theorem 3.3, there exists $\bar{\lambda} \in R^m$ such that

$$D\bar{x} - A^T \bar{\lambda} + c = 0, \quad A\bar{x} \geq b, \quad \bar{\lambda} \geq 0, \quad \bar{\lambda}^T(A\bar{x} - b) = 0.$$

Let (x, λ) be a feasible point for (16.2), that is

$$Dx - A^T \lambda + c = 0, \quad Ax \geq b, \quad \lambda \geq 0, \quad \lambda^T(Ax - b) = 0. \tag{16.8}$$

By (16.8), $x \in \Delta(A, b)$. Hence $f(x, c) \geq f(\bar{x}, c)$, where $f(x, c) := \frac{1}{2}x^T D x + c^T x$. From (16.8) it follows that

$$\begin{aligned}\frac{1}{2}(c^T x + b^T \lambda) &= \frac{1}{2}(c^T x + x^T A^T \lambda) \\ &= \frac{1}{2}(c^T x + x^T D x + x^T c) \\ &= f(x, c).\end{aligned}$$

Similarly, one has

$$\frac{1}{2}(c^T \bar{x} + b^T \bar{\lambda}) = f(\bar{x}, c). \tag{16.9}$$

Consequently,

$$\frac{1}{2}(c^T x + b^T \lambda) \geq \frac{1}{2}(c^T \bar{x} + b^T \bar{\lambda}).$$

Since this inequality holds for every $(x, \lambda) \in P_{KKT}(c, b)$, we conclude that $(\bar{x}, \bar{\lambda})$ is a solution of (16.2). Combining this with (16.4), (16.5) and (16.9), we obtain (16.7). In order to prove (16.6), we fix any $x \in \mathrm{Sol}(c, b)$. Let λ be a Lagrange multiplier corresponding to that x. The above arguments show that (x, λ) is a solution of (16.2). From this it follows that $x \in \pi_{R^n}(\mathrm{Sol}_{KKT}(c, b))$. Now, let (x, λ) be a solution of (16.2). Since (x, λ) satisfies the inequality system described in (16.8), we have

$$\frac{1}{2}(c^T x + b^T \lambda) = f(x, c).$$

Since $(\bar{x}, \bar{\lambda})$ and (x, λ) are from the solution set of (16.2), it holds

$$\frac{1}{2}(c^T x + b^T \lambda) = \frac{1}{2}(c^T \bar{x} + b^T \bar{\lambda}) = f(\bar{x}, c).$$

Consequently, $f(x, c) = f(\bar{x}, c)$. Since $x \in \Delta(A, b)$, from the last equality we deduce that $x \in \mathrm{Sol}(c, b)$. The equality (16.6) has been proved. $\square$

Note that the set W defined by (16.1) coincides with the effective domain of the multifunction $\mathrm{Sol}(\cdot, \cdot)$, that is

$$\begin{aligned} W = \ &\{(c, b) \in R^n \times R^m : -\infty < \varphi(c, b) < +\infty\} \\ = \ &\{(c, b) \in R^n \times R^m : \mathrm{Sol}(c, b) \neq \emptyset\}. \end{aligned} \tag{16.10}$$

Indeed, for any pair $(c, b) \in R^n \times R^m$, if $\text{Sol}(c, b) \neq \emptyset$ then $-\infty < \varphi(c, b) < +\infty$. Conversely, if $-\infty < \varphi(c, b) < +\infty$ then $\Delta(A, b)$ is nonempty and the function $f(\cdot, c)$ is bounded below on $\Delta(A, b)$. By Theorem 2.1, $\text{Sol}(c, b) \neq \emptyset$.

Taking account of (16.10), we can formulate the results from Klatte (1985) concerning the optimal value function $\varphi(c, b)$ as follows.

Lemma 16.2. (See Klatte (1985), Theorem 2) *The effective domain W of φ is the union of a finitely many polyhedral convex cones, i.e. there exists a finite number of polyhedral convex cones $W_i \subset R^n \times R^m$ $(i = 1, 2, \ldots, s)$ such that*

$$W = \bigcup_{i=1}^{s} W_i. \tag{16.11}$$

Lemma 16.3. (See Klatte (1985), Theorem 3) *The function φ is Lipschitzian on every bounded subset $\Omega_0 \subset W$, i.e., for each bounded subset $\Omega_0 \subset W$ there exists a constant $k_{\Omega_0} > 0$ such that*

$$\|\varphi(c', b') - \varphi(c, b)\| \leq k_{\Omega_0}(\|c' - c\| + \|b' - b\|)$$

for any (c, b), $(c', b') \in \Omega_0$.

For each subset $I \subset \{1, 2, \ldots, m\}$, we define

$$P_{KKT}^I(c, b) = \{(x, \lambda) \in R^n \times R^m : \quad Dx - A^T \lambda + c = 0,$$
$$A_i x \geq b_i, \; ; \lambda_i = 0 \; (\forall i \in I),$$
$$A_j x = b_j, \; \lambda_j \geq 0 \; (\forall j \notin I)\}, \tag{16.12}$$

where A_i $(i \in \{1, \ldots, m\})$ is the i-th row of the matrix A and b_i is the i-th component of b. It is clear that

$$P_{KKT}(c, b) = \bigcup_{I \subset \{1, \ldots, m\}} P_{KKT}^I(c, b). \tag{16.13}$$

Note that $P_{KKT}^I(c, b)$ is the solution set of the following system of linear equalities and inequalities:

$$\begin{cases} Dx - A^T \lambda + c = 0, \\ A_I x \geq b_I, \quad \lambda_I = 0 \\ A_J x = b_J, \quad \lambda_J \geq 0, \\ x \in R^n, \quad \lambda \in R^m, \end{cases} \tag{16.14}$$

where $J = \{1, 2, \ldots, m\} \setminus I$ and, as usual, A_J denotes the matrix composed by the rows A_j $(j \in J)$ of A, and λ_I is the vector with the components λ_i $(i \in I)$.

Let

$$\varphi_{KKT}^I(c, b) = \inf \left\{ \frac{1}{2}(c^T x + b^T \lambda) : (x, \lambda) \in P_{KKT}^I(c, b) \right\}. \quad (16.15)$$

Thus $\varphi_{KKT}^I(c, b)$ is the optimal value of the *linear programming problem* whose objective function is $\frac{1}{2}(c^T x + b^T \lambda)$ and whose constraints are described by (16.14). Note that the pair (c, b) represents the right-hand-side perturbations of the linear system (16.14).

It turns out that, for any $I \subset \{1, 2, \ldots, m\}$, the effective domain of $\varphi_{KKT}^I(\cdot)$ is a polyhedral convex cone on which the function admits a linear-quadratic representation. Namely, using the concept of pseudo-matrix one can establish the following result.

Lemma 16.4. (See Bank et al. (1982), Theorem 5.5.2) *The effective domain*

$$\mathrm{dom}\varphi_{KKT}^I = \{(c, b) \in R^n \times R^m : -\infty < \varphi_{KKT}^I(c, b) < +\infty\}$$

is a polyhedral convex cone and there exist

$$M_I \in R^{(n+m) \times (n+m)} \quad and \quad q_I \in R^{n+m}$$

such that

$$\varphi_{KKT}^I(c, b) = \frac{1}{2}\begin{pmatrix} c \\ b \end{pmatrix}^T M_I \begin{pmatrix} c \\ b \end{pmatrix} + q_I^T \begin{pmatrix} c \\ b \end{pmatrix} \quad (16.16)$$

for every $(c, b) \in \mathrm{dom}\varphi_{KKT}^I$.

The following useful fact follows from Lemma 16.1.

Lemma 16.5. *For any* $(c, b) \in W$, *it holds*

$$\varphi(c, b) = \min\{\varphi_{KKT}^I(c, b) : I \subset \{1, 2, \ldots, m\}\}. \quad (16.17)$$

Proof. From (16.4), (16.13) and (16.15), we deduce that

$$\varphi_{KKT}(c, b) = \min\{\varphi_{KKT}^I(c, b) : I \subset \{1, 2, \ldots, m\}\}.$$

Combining this with (16.7) we obtain (16.17). $\quad \square$

Remark 16.1. From (16.17) it follows that, for for any $(c, b) \in W$ and for any $I \subset \{1, 2, \ldots, m\}$, we have $\varphi_{KKT}^I(c, b) > -\infty$.

Remark 16.2. It may happen that for some pairs $(c, b) \in W$ the function φ^I_{KKT} has the value $+\infty$. Note that $\varphi^I_{KKT}(c, b) = +\infty$ if and only if the solution set of (16.14) is empty. The example considered in Section 16.3 will illustrate this situation.

Remark 16.3. If D is a positive semidefinite matrix then (15.1) is a convex QP problem and the equality $\varphi^{I_1}_{KKT}(c, b) = \varphi^{I_2}_{KKT}(c, b)$ holds for any index sets $I_1, I_2 \subset \{1, 2, \ldots, m\}$ and for any point $(c, b) \in \mathrm{dom}\varphi^{I_1}_{KKT} \cap \mathrm{dom}\varphi^{I_2}_{KKT}$. The last equality is valid because any KKT point of a convex QP problem is a solution.

16.2   Directional Differentiability

In this section, we will prove that although φ is not a convex function but it enjoys the important property of convex functions of being directionally differentiable at any point in its effective domain. A formula for computing the directional derivative $\varphi'(\bar{w}; z)$ of φ at any $\bar{w} = (\bar{c}, \bar{b}) \in W$ in direction $z = (u, v) \in R^n \times R^m$ is also established.

Recall (Rockafellar (1970), p. 13) that a subset $K \subset R^p$ is called a *cone* if $tx \in K$ whenever $x \in K$ and $t > 0$. (The origin itself may or may not be included in K).

Proposition 16.1. *Let W be defined by* (16.1),

$$Z_1 = \{(c, b) \in R^n \times R^m : \varphi(c, b) = +\infty\},$$

$$Z_2 = \{(c, b) \in R^n \times R^m : \varphi(c, b) = -\infty\}, \quad and$$

$$L = \{b \in R^m : \Delta(A, b) \text{ is nonempty}\}.$$

Then Z_1 is an open cone, W is a closed cone, and Z_2 is a cone which is relatively open in the polyhedral convex cone $R^n \times L \subset R^n \times R^m$. Moreover, it holds

$$R^n \times L = W \cup Z_2, \ R^n \times R^m = W \cup Z_2 \cup Z_1, \ Z_1 = (R^n \times R^m) \backslash (R^n \times L).$$
$$(16.18)$$

The easy proof of this proposition is omitted.

Theorem 16.1. *The optimal value function φ defined in* (16.5) *is directionally differentiable on W, i.e., for any $\bar{w} = (\bar{c}, \bar{b}) \in W$ and for any $z = (u, v) \in R^n \times R^m$ there exists the directional derivative*

$$\varphi'(\bar{w}; z) := \lim_{t \downarrow 0} \frac{\varphi(\bar{w} + tz) - \varphi(\bar{w})}{t} \tag{16.19}$$

of φ at $\bar{w}$ in direction z.

Proof. Let $\bar{w} = (\bar{c}, \bar{b}) \in W$ and $z = (u, v) \in R^n \times R^m$ be given arbitrarily. If $z = 0$ then it is obvious that $\varphi'(\bar{w}; z) = 0$. Assume that $z \neq 0$. We first prove that one of the following three cases must occur:

(c1)   There exists $\bar{t} > 0$ such that $\bar{w} + tz \in Z_1$ for every $t \in (0, \bar{t}]$,

(c2)   There exists $\bar{t} > 0$ such that $\bar{w} + tz \in Z_2$ for every $t \in (0, \bar{t}]$,

(c3)   There exists $\bar{t} > 0$ such that $\bar{w} + tz \in W$ for every $t \in (0, \bar{t}]$.

For this purpose, suppose that (c3) fails to hold. We have to show that, in this case, (c1) or (c2) must occur. Since (c3) is not valid, we can find a decreasing sequence $t_k \to 0+$ such that $\bar{w}+t_k z \notin W$ for every $k \in N$. By (16.18), for each $k \in N$, we must have $\bar{w} + t_k z \in Z_1$ or $\bar{w} + t_k z \in Z_2$. Hence, there exists a subsequence $\{t_{k_i}\}$ of $\{t_k\}$ such that

$$\bar{w} + t_{k_i} z \in Z_1 \quad (\forall i \in N), \tag{16.20}$$

or

$$\bar{w} + t_{k_i} z \in Z_2 \quad (\forall i \in N). \tag{16.21}$$

Consider the case where (16.20) is fulfilled. If there exists an $\hat{t} \in (0, t_{k_1})$ such that

$$\bar{w} + \hat{t} z \in R^n \times L$$

then, by the convexity of $R^n \times L$,

$$\{\bar{w} + tz : t \in [0, \hat{t}]\} \subset R^n \times L.$$

By virtue of the first equality in (16.18), this yields $\varphi(\bar{w}+tz) \neq +\infty$ for every $t \in [0, \hat{t}]$, contradicting (16.20). Thus (16.20) implies that $\bar{w} + tz \notin R^n \times L$ for every $t \in (0, t_{k_1})$. Then, the third equality in (16.18) shows that $\bar{w} + tz \in Z_1$ for every $t \in (0, t_{k_1})$. Putting $\bar{t} = t_{k_1}$, we see at once that (c1) holds.

Consider the case where (16.21) is fulfilled. Since $\bar{w} \in W \subset R^n \times L$ and

$$\bar{w} + t_{k_1} z \in Z_2 \subset R^n \times L,$$

it follows that

$$\{\bar{w} + tz : t \in [0, t_{k_1}]\} \subset R^n \times L.$$

Therefore, we can deduce from the first equality in (16.18) that, for every $t \in (0, t_{k_1})$, $\bar{w} + tz \in Z_2$ or $\bar{w} + tz \in W$. If there exists $i \in N$ such that

$$\bar{w} + tz \in Z_2 \quad (\forall t \in (0, t_{k_i})) \tag{16.22}$$

then (c2) is satisfied if we choose $\bar{t} = t_{k_i}$. If there is no $i \in N$ such that (16.22) is valid, then for every $i \in N$ there must exist some $t'_{k_i} \in (0, t_{k_i})$ such that $\bar{w} + t'_{k_i} z \in W$. By (16.11), there is an index $j(k_i) \in \{1, \ldots, s\}$ such that

$$\bar{w} + t'_{k_i} z \in W_{j(k_i)}. \tag{16.23}$$

Without loss of generality, we can assume that

$$0 < t'_{k_{i+1}} < t_{k_{i+1}} < t'_{k_i} < t_{k_i} \quad (\forall i \in N). \tag{16.24}$$

Since $j(k_i) \in \{1, \ldots, s\}$, there must exist a pair (i, j) such that $j > i$ and $j(k_j) = j(k_i)$. By (16.23) and by the convexity of $W_{j(k_i)}$, we have

$$\{\bar{w} + tz : t'_{k_j} \leq t \leq t'_{k_i}\} \subset W_{j(k_i)} \subset W. \tag{16.25}$$

From (16.21) and (16.24) we get $\varphi(\bar{w} + t_{k_{i+1}}z) = -\infty$ and $t'_{k_j} < t_{k_{i+1}} < t'_{k_i}$, a contradiction to (16.25). We have thus proved that if (16.21) is valid then (c2) must occur.

Summarizing all the above, we conclude that one of the three cases (c1)–(c3) must occur.

If (c1) occurs then, by (16.19), we have $\varphi'(\bar{w}; z) = +\infty$. Similarly, if (c2) happens then $\varphi'(\bar{w}; z) = -\infty$.

Now assume that (c3) takes place. Denote by F the collection of the index sets $I \subset \{1, 2, \ldots, m\}$ for which there exists $t_I \in (0, \bar{t})$, where $\bar{t} > 0$ is given by (c3), such that

$$\{\bar{w} + tz : t \in [0, t_I]\} \subset \mathrm{dom}\varphi^I_{KKT}. \tag{16.26}$$

Recall that $\mathrm{dom}\varphi^I_{KKT}$ is a closed convex set (see Lemma 16.4). If $F = \emptyset$ then for any $I \subset \{1, 2, \ldots, m\}$ and for any $t \in (0, \bar{t}]$ one has $\varphi^I_{KKT}(\bar{w} + tz) = +\infty$. By (c3), $\bar{w} + tz \in W$ for all $t \in (0, \bar{t}]$. Then, according to (16.18) we have

$$\varphi(\bar{w} + tz) = \min\{\varphi^I_{KKT}(\bar{w} + tz) : I \subset \{1, 2, \ldots, m\}\} = +\infty$$

for all $t \in (0, \bar{t}]$, which is impossible. We have shown that $F \neq \emptyset$. Define

$$\hat{t} = \min\{t_I : I \in F\} > 0.$$

By virtue of (c3) and of (16.18), one has

$$\varphi(\bar{w} + tz) = \min\{\varphi_{KKT}^I(\bar{w} + tz) : I \in F\} \quad (\forall t \in [0, \hat{t}]). \quad (16.27)$$

It follows from (16.26) that

$$\bar{w} + tz \in \mathrm{dom}\varphi_{KKT}^I \quad (\forall I \in F, \forall t \in [0, \hat{t}]).$$

For each $I \in F$, let $M_I \in R^{(n+m) \times (n+m)}$ and $q_I \in R^{n+m}$ be such that the representation (16.16) holds for all $(c, b) \in \mathrm{dom}\varphi_{KKT}^I$. Setting

$$\tilde{\varphi}_{KKT}^I(c, b) = \frac{1}{2} \begin{pmatrix} c \\ b \end{pmatrix}^T M_I \begin{pmatrix} c \\ b \end{pmatrix} + q_I^T \begin{pmatrix} c \\ b \end{pmatrix} \quad (16.28)$$

for every $(c, b) \in R^n \times R^m$, we extend $\varphi_{KKT}^I(\cdot)$ from $\mathrm{dom}\varphi_{KKT}^I$ to the whole space $R^n \times R^m$. From (16.28) it follows that all the functions $\tilde{\varphi}_{KKT}^I(\cdot)$, $I \in F$, are smooth. According to Theorem 2.1 in Clarke (1975), the function

$$\tilde{\varphi}(c, b) = \min\{\tilde{\varphi}_{KKT}^I(c, b) : I \in F\}$$

is locally Lipschitz at $\bar{w} = (\bar{c}, \bar{b})$. Moreover, $\tilde{\varphi}$ is *Lipschitz regular* (see Definition 2.3.4 in Clarke (1983)) at $\bar{w}$, and

$$\tilde{\varphi}^0(\bar{w}; z) = \tilde{\varphi}'(\bar{w}; z) = \min\{(\tilde{\varphi}_{KKT}^I)'(\bar{w}; z) : I \in F\}, \quad (16.29)$$

where $\tilde{\varphi}^0(\bar{w}; z)$ (resp., $\tilde{\varphi}'(\bar{w}; z)$) denotes the Clarke generalized directional derivative (resp., the directional derivative) of $\tilde{\varphi}$ at $\bar{w}$ in direction z. Since

$$\tilde{\varphi}_{KKT}^I(c, b) = \varphi_{KKT}^I(c, b)$$

for all $(c, b) \in \mathrm{dom}\varphi_{KKT}^I$, from (16.27) and (16.29) it follows that the directional derivative $\varphi'(\bar{w}; z)$ exists, and we have

$$\varphi'(\bar{w}; z) = \min\{(\varphi_{KKT}^I)'(\bar{w}; z) : I \in F\}. \quad (16.30)$$

The proof is complete. $\quad \square$

In the course of the above proof we have obtained some explicit formulae for computing the directional derivative of the function φ. Namely, we have proved the following result.

Theorem 16.2. *Let $\bar{w} \in W$ and $z = (u, v) \in R^n \times R^m$. The following assertions hold:*

(i) *If there exists $\bar{t} > 0$ such that*

$$\bar{w} + tz \in Z_1 = \{(c, b) : \Delta(A, b) = \emptyset\}$$

for all $t \in (0, \bar{t}]$), then $\varphi'(\bar{w}; z) = +\infty$.

(ii) *If there exists $\bar{t} > 0$ such that*

$$\bar{w} + tz \in Z_2 = \{(c, b) : \Delta(A, b) \neq \emptyset, \ \varphi(c, b) = -\infty\}$$

for all $t \in (0, \bar{t}]$), then $\varphi'(\bar{w}; z) = -\infty$.

(iii) *If there exists $\bar{t} > 0$ such that*

$$\bar{w} + tz \in W = \{(c, b) : \Delta(A, b) \neq \emptyset, \ \varphi(c, b) > -\infty\}$$

for all $t \in (0, \bar{t}]$), then $\varphi'(\bar{w}; z)$ can be computed by formula (16.30), where F is the collection of all $I \subset \{1, 2, \ldots, m\}$ for which there exists some $t_I \in (0, \bar{t})$ satisfying condition (16.26).

At the end of the next section we shall use Theorem 16.2 for computing directional derivative of the optimal value function in a concrete nonconvex QP problem.

16.3   Piecewise Linear-Quadratic Property

The notion of *piecewise linear-quadratic function* (plq function, for brevity) was introduced in Rockafellar (1988).

Definition 16.1. (See Rockafellar and Wets (1998), p. 440) A function $\psi : R^l \to \bar{R}$ is piecewise linear-quadratic (plq) if the set

$$\text{dom}\psi = \{z \in R^l : -\infty < \psi(z) < +\infty\} \tag{16.31}$$

can be represented as the union of finitely many polyhedral convex sets, relative to each of which $\psi(z)$ is given by an expression of the form

$$\frac{1}{2}z^T Q z + d^T z + \alpha \tag{16.32}$$

for some $\alpha \in R$, $d \in R^l$, $Q \in R_S^{l \times l}$.

Note that in Rockafellar and Wets (1998) instead of (16.31) one has the following formula

$$\text{dom}\psi = \{z \in R^l : \psi(z) < +\infty\}. \tag{16.33}$$

If there exists some $\bar{z} \in R^l$ with $\psi(\bar{z}) = -\infty$ then, since $\bar{z}$ belongs to the set defined in (16.33), one cannot represent the latter as the union of finitely many polyhedral convex sets, relative to each of which $\psi(z)$ is given by an expression of the form (16.32). Hence ψ cannot be a plq function. This is the reason why we prefer (16.31) to (16.33).

If D is a positive semidefinite matrix then, by using the Eaves Theorem we can prove that W is a polyhedral convex cone. Using Lemmas 16.4, 16.5, and Remark 16.3, it is not difficult to show that *the optimal value function $\varphi(c,b) = \varphi(c,b)$ of a convex QP problem is plq.*

Example 16.1. (See Rockafellar and Wets (1998)) Consider the function

$$\psi(z) = |z_1^2 + z_2^2 - 1|, \quad z = (z_1, z_2) \in R^2.$$

We have $R^2 = \Omega_1 \cup \Omega_2$, where

$$\Omega_1 = \{z : z_1^2 + z_2^2 \leq 1\}, \quad \Omega_2 = \{z : z_1^2 + z_2^2 \geq 1\}.$$

The formulae

$$\psi(z) = -z_1^2 - z_2^2 + 1 \ (\forall z \in \Omega_1) \quad \text{and} \quad \psi(z) = z_1^2 + z_2^2 - 1 \ (\forall z \in \Omega_2)$$

show that ψ admits a representation of the form (16.33) on each domain Ω_i $(i = 1, 2)$. Meanwhile, it can be proved that ψ is *not* a plq function.

Note that if the function $\varphi(c,b)$ defined by (16.5) is plq then, for any $\bar{b} \in R^m$, the function $\varphi(\cdot, \bar{b})$ is also plq on its effective domain. Indeed, assume that $\varphi(c,b)$ is plq, that is W admits a representation of the form $W = \bigcup_{i=1}^s W_i$, where every W_i is a polyhedral convex set and there exist $Q_i \in R_S^{(n+m)\times(n+m)}$, $d_i \in R^{n+m}$ and $\alpha_i \in R$ such that

$$\varphi(c,b) = \frac{1}{2}\binom{c}{b}^T Q_i \binom{c}{b} + d_i^T \binom{c}{b} + \alpha_i \quad (\forall (c,b) \in W_i). \quad (16.34)$$

Let $\bar{b} \in R^m$ be given arbitrarily. Define

$$W' = \{c \in R^n : (c,\bar{b}) \in W\}, \quad W_i' = \{c \in R^n : (c,\bar{b}) \in W_i\}$$

for all $i = 1, \ldots, s$. It is obvious that $W' = \text{dom}\varphi(\cdot, \bar{b})$ and $W' = \bigcup_{i=1}^s W_i'$. Moreover, for every $i \in \{1, \ldots, s\}$, from (16.34) it follows that

$$\varphi(c,\bar{b}) = \frac{1}{2}\binom{c}{\bar{b}}^T Q_i \binom{c}{\bar{b}} + d_i^T \binom{c}{\bar{b}} + \alpha_i \quad (\forall c \in W_i').$$

Since the function in the right-hand-side of this formula is a linear-quadratic function of c and since each W_i' is a polyhedral convex set (maybe empty), we conclude that $\varphi(\cdot, \bar{b})$ is a plq function.

We are interested in solving the following question: *Whether the optimal value function in a general (indefinite) parametric quadratic programming problem is a plq function w.r.t. the linear parameters?*

It turns out that the plq property is not available in the general case.

Example 16.2. Consider the problem

$$\begin{cases} \text{Minimize } f(x, c) = \dfrac{1}{2}(x_1^2 + 2x_1 x_2 - x_2^2) + c_1 x_1 + c_2 x_2 \\ \text{subject to } \quad x = (x_1, x_2) \in R^2, \quad \dfrac{1}{2}x_1 + x_2 \geq 0, \\ \qquad\qquad x_2 - x_1 \geq 0, \quad -x_2 \geq -2, \end{cases} \tag{16.35}$$

and denote by $\varphi(c)$, $c = (c_1, c_2) \in R^2$, the optimal value of this nonconvex QP problem.

In the remainder of this section we will compute the values $\varphi(c)$, $c \in R^2$. In the next section it will be shown that the function $\varphi(c)$ is not plq. Then we can conclude that the optimal value function $\varphi(c, b)$, $c = (c_1, c_2) \in R^2$ and $b = (b_1, b_2, b_3) \in R^3$, of the following parametric QP problem is not plq:

$$\begin{cases} \text{Minimize } f(x, c) = \dfrac{1}{2}(x_1^2 + 2x_1 x_2 - x_2^2) + c_1 x_1 + c_2 x_2 \\ \text{subject to } \quad x = (x_1, x_2) \in R^2, \quad \dfrac{1}{2}x_1 + x_2 \geq b_1, \\ \qquad\qquad x_2 - x_1 \geq b_2, \quad -x_2 \geq b_3. \end{cases} \tag{16.36}$$

Indeed, if $\varphi(c, b)$ is plq then the arguments given after Example 16.1 show that $\varphi(c) = \varphi(c, \bar{b})$, where $\bar{b} = (0, 0, -2)$, is a plq function, which is impossible.

In order to write (16.35) in the form (15.1), we put

$$D = \begin{bmatrix} 1 & 1 \\ 1 & -1 \end{bmatrix}, \quad A = \begin{bmatrix} \dfrac{1}{2} & 1 \\ -1 & 1 \\ 0 & -1 \end{bmatrix}, \quad b = \bar{b} = \begin{pmatrix} 0 \\ 0 \\ -2 \end{pmatrix}, \quad c = \begin{pmatrix} c_1 \\ c_2 \end{pmatrix}.$$

Note that the feasible domain $\Delta(A, \bar{b})$ of (16.36) is a triangle with the vertexes $(0, 0)$, $(2, 2)$ and $(-4, 2)$. Since $\Delta(A, \bar{b})$ is compact, $\varphi(c, \bar{b}) \in R$ for every $c \in R^2$. In other words, $\text{dom}\varphi(\cdot, \bar{b}) = R^2$.

In agreement with (16.2) and (16.3), the auxiliary problem corresponding to (16.35) is the following one

$$
\begin{cases}
\text{Minimize } \dfrac{1}{2}(c^T x + b^T \lambda) = \dfrac{1}{2}(c_1 x_1 + c_2 x_2) - \lambda_3 \\
\text{subject to } \quad (x, \lambda) = (x_1, x_2, \lambda_1, \lambda_2, \lambda_3) \in R^2 \times R^3, \\
x_1 + x_2 - \dfrac{1}{2}\lambda_1 + \lambda_2 + c_1 = 0, \\
x_1 - x_2 - \lambda_1 - \lambda_2 + \lambda_3 + c_2 = 0, \\
\dfrac{1}{2}x_1 + x_2 \geq 0, \quad \lambda_1 \geq 0, \quad \lambda_1(\dfrac{1}{2}x_1 + x_2) = 0, \\
x_2 - x_1 \geq 0, \quad \lambda_2 \geq 0, \quad \lambda_2(x_2 - x_1) = 0, \\
x_2 \leq 2, \quad \lambda_3 \geq 0, \quad \lambda_3(2 - x_2) = 0.
\end{cases}
\tag{16.37}
$$

We shall apply formula (16.17) to compute the values $\varphi(c, b)$, $c \in R^2$, $b = \bar{b}$. To do so, we have to compute the optimal value $\varphi_{KKT}^I(c, b)$ defined by (16.15), where $I \subset \{1, 2, 3\}$ is an arbitrary subset. Since there are 8 possibilities to choose such index set I, we have to consider 8 linear subproblems of the problem (16.37).

Case 1. $I = I_1 = \{1, 2, 3\}$. In the corresponding subproblem we must have $\lambda_I = (\lambda_1, \lambda_2, \lambda_3) = (0, 0, 0)$. Taking account of (16.37), we can write that subproblem as follows

$$
\begin{cases}
\dfrac{1}{2}(c_1 x_1 + c_2 x_2) \to \min \\
x_1 + x_2 + c_1 = 0, \quad x_1 - x_2 + c_2 = 0, \\
\dfrac{1}{2}x_1 + x_2 \geq 0, \quad x_2 - x_1 \geq 0, \quad x_2 \leq 2.
\end{cases}
\tag{I_1}
$$

In accordance with (16.15), we denote the optimal value of (I_1) by $\varphi_{KKT}^{I_1}(c, \bar{b})$. An elementary investigation on (I_1) gives us the following result:

$$
\begin{cases}
\varphi_{KKT}^{I_1}(c, \bar{b}) = \dfrac{1}{4}(-c_1^2 - 2c_1 c_2 + c_2^2), \\
\mathrm{dom}\varphi_{KKT}^{I_1}(\cdot, \bar{b}) = \{c = (c_1, c_2) : -3c_1 + c_2 \geq 0, \\
\qquad\qquad\qquad\qquad\qquad c_2 \geq 0, \; -c_1 + c_2 \leq 4\}.
\end{cases}
\tag{16.38}
$$

The exact meaning of (16.38) is the following: We have

$$
\varphi_{KKT}^{I_1}(c, \bar{b}) = \dfrac{1}{4}(-c_1^2 - 2c_1 c_2 + c_2^2)
$$

for every $c \in \mathrm{dom}\varphi_{KKT}^{I_1}(\cdot, \bar{b})$ and

$$
\varphi_{KKT}^{I_1}(c, \bar{b}) = +\infty
$$

for every $c \notin \text{dom}\varphi_{KKT}^{I_1}(\cdot, \bar{b})$. A similar interpretation applies to the results of the forthcoming 7 cases.

Case 2. $I = I_2 = \{1, 2\}$. We have $\lambda_I = (\lambda_1, \lambda_2) = (0, 0)$, $\lambda_3 \geq 0$. The corresponding subproblem is

$$
\begin{cases}
\frac{1}{2}(c_1 x_1 + c_2 x_2) - \lambda_3 \to \min \\
x_1 + x_2 + c_1 = 0, \quad x_1 - x_2 + \lambda_3 + c_2 = 0, \\
\frac{1}{2}x_1 + x_2 \geq 0, \quad x_2 - x_1 \geq 0, \quad x_2 = 2, \quad \lambda_3 \geq 0.
\end{cases}
\quad (I_2)
$$

Then

$$
\begin{cases}
\varphi_{KKT}^{I_2}(c, \bar{b}) = -\frac{1}{2}c_1^2 - 2c_1 + 2c_2 - 4, \\
\text{dom}\varphi_{KKT}^{I_2}(\cdot, \bar{b}) = \{c = (c_1, c_2) : 2 - c_1 \geq 0, \\
\qquad\qquad 4 + c_1 \geq 0, \ 4 + c_1 - c_2 \geq 0\}.
\end{cases}
\quad (16.39)
$$

Case 3. $I = I_3 = \{2, 3\}$. We have $\lambda_I = (\lambda_2, \lambda_3) = (0, 0)$, $\lambda_1 \geq 0$. The corresponding subproblem is

$$
\begin{cases}
\frac{1}{2}(c_1 x_1 + c_2 x_2) \to \min \\
x_1 + x_2 - \frac{1}{2}\lambda_1 + c_1 = 0, \quad x_1 - x_2 - \lambda_1 + c_2 = 0, \\
\frac{1}{2}x_1 + x_2 = 0, \quad \lambda_1 \geq 0, \quad x_2 - x_1 \geq 0, \quad x_2 \leq 2.
\end{cases}
\quad (I_3)
$$

Then

$$
\begin{cases}
\varphi_{KKT}^{I_3}(c, \bar{b}) = 2c_1^2 + \frac{1}{2}c_2^2 - 2c_1 c_2, \\
\text{dom}\varphi_{KKT}^{I_3}(\cdot, \bar{b}) = \{c = (c_1, c_2) : c_2 \leq 3c_1, \\
\qquad\qquad c_2 - 2c_1 \geq 0, \quad c_2 - 2c_1 \leq 2\}.
\end{cases}
\quad (16.40)
$$

Case 4. $I = I_4 = \{1, 3\}$. We have $\lambda_I = (\lambda_1, \lambda_3) = (0, 0)$, $\lambda_2 \geq 0$. The corresponding subproblem is

$$
\begin{cases}
\frac{1}{2}(c_1 x_1 + c_2 x_2) \to \min \\
x_1 + x_2 + \lambda_2 + c_1 = 0, \quad x_1 - x_2 - \lambda_2 + c_2 = 0, \\
\frac{1}{2}x_1 + x_2 \geq 0, \quad x_2 - x_1 = 0, \quad \lambda_2 \geq 0, \quad x_2 \leq 2.
\end{cases}
\quad (I_4)
$$

Then

$$
\begin{cases}
\varphi_{KKT}^{I_4}(c, \bar{b}) = -\frac{1}{4}(c_1 + c_2)^2, \\
\text{dom}\varphi_{KKT}^{I_4}(\cdot, \bar{b}) = \{c = (c_1, c_2) : c_1 + c_2 \leq 0, \\
\qquad\qquad c_1 + c_2 \geq -4, \quad c_2 \geq 0\}.
\end{cases}
\quad (16.41)
$$

Case 5. $I = I_5 = \{1\}$. We have $\lambda_1 = 0$, $\lambda_2 \geq 0$, $\lambda_3 \geq 0$. The corresponding subproblem is

$$\begin{cases} \dfrac{1}{2}(c_1 x_1 + c_2 x_2) - \lambda_3 \to \min \\ x_1 + x_2 + \lambda_2 + c_1 = 0, \quad x_1 - x_2 - \lambda_2 + \lambda_3 + c_2 = 0, \qquad (I_5) \\ \dfrac{1}{2}x_1 + x_2 \geq 0, \quad x_2 - x_1 = 0, \quad \lambda_2 \geq 0, \quad x_2 = 2, \quad \lambda_3 \geq 0. \end{cases}$$

Then

$$\begin{cases} \varphi_{KKT}^{I_5}(c, \bar{b}) = 2c_1 + 2c_2 + 4, \\ \mathrm{dom}\varphi_{KKT}^{I_4}(\cdot, \bar{b}) = \{c = (c_1, c_2) : c_1 + 4 \leq 0, \quad c_1 + c_2 + 4 \leq 0\}. \end{cases}$$
$$(16.42)$$

Case 6. $I = I_6 = \{2\}$. We have $\lambda_2 = 0$, $\lambda_1 \geq 0$, $\lambda_3 \geq 0$. The corresponding subproblem is

$$\begin{cases} \dfrac{1}{2}(c_1 x_1 + c_2 x_2) - \lambda_3 \to \min \\ x_1 + x_2 - \dfrac{1}{2}\lambda_1 + c_1 = 0, \quad x_1 - x_2 - \lambda_1 + \lambda_3 + c_2 = 0, \qquad (I_6) \\ \dfrac{1}{2}x_1 + x_2 = 0, \quad \lambda_1 \geq 0, \quad x_2 - x_1 \geq 0, \quad x_2 = 2, \quad \lambda_3 \geq 0. \end{cases}$$

Then

$$\begin{cases} \varphi_{KKT}^{I_6}(c, \bar{b}) = -4c_1 + 2c_2 - 2, \\ \mathrm{dom}\varphi_{KKT}^{I_6}(\cdot, \bar{b}) = \{c = (c_1, c_2) : c_1 - 2 \geq 0, \\ \qquad\qquad\qquad\qquad\qquad 2 + 2c_1 - c_2 \geq 0\}. \end{cases}$$
$$(16.43)$$

Case 7. $I = I_7 = \{3\}$. We have $\lambda_3 = 0$, $\lambda_1 \geq 0$, $\lambda_2 \geq 0$. The corresponding subproblem is

$$\begin{cases} \dfrac{1}{2}(c_1 x_1 + c_2 x_2) \to \min \\ x_1 + x_2 - \dfrac{1}{2}\lambda_1 + \lambda_2 + c_1 = 0, \quad x_1 - x_2 - \lambda_1 - \lambda_2 + c_2 = 0, \\ \dfrac{1}{2}x_1 + x_2 = 0, \quad \lambda_1 \geq 0, \quad x_2 - x_1 = 0, \quad \lambda_2 \geq 0, \quad x_2 \leq 2. \end{cases}$$
$$(I_7)$$

Then

$$\begin{cases} \varphi_{KKT}^{I_7}(c, \bar{b}) = 0, \\ \mathrm{dom}\varphi_{KKT}^{I_7}(\cdot, \bar{b}) = \{c = (c_1, c_2) : c_1 + c_2 \geq 0, \quad c_2 - 2c_1 \geq 0\}. \end{cases}$$
$$(16.44)$$

Case 8. $I = I_8 = \emptyset$. We have $\lambda_1 \geq 0$, $\lambda_2 \geq 0$, $\lambda_3 \geq 0$. The corresponding subproblem is

$$
\begin{cases}
\dfrac{1}{2}(c_1 x_1 + c_2 x_2) - \lambda_3 \to \min \\[2mm]
x_1 + x_2 - \dfrac{1}{2}\lambda_1 + \lambda_2 + c_1 = 0, \quad x_1 - x_2 - \lambda_1 - \lambda_2 + \lambda_3 + c_2 = 0, \\[2mm]
\dfrac{1}{2}x_1 + x_2 = 0, \quad \lambda_1 \geq 0, \quad x_2 - x_1 = 0, \quad \lambda_2 \geq 0, \quad x_2 = 2, \quad \lambda_3 \geq 0.
\end{cases}
\tag{I_8}
$$

Then

$$
\begin{cases}
\varphi_{KKT}^{I_8}(c, \bar{b}) = +\infty \quad \text{for every } c \in R^2, \\
\mathrm{dom}\, \varphi_{KKT}^{I_8}(\cdot, \bar{b}) = \emptyset.
\end{cases}
\tag{16.45}
$$

Consider the following polyhedral convex subsets of R^2:

$$
\begin{aligned}
\Omega_1 &= \{c = (c_1, c_2) : c_2 \leq -c_1 - 4, \quad c_1 \leq -4\}, \\
\Omega_2 &= \{c = (c_1, c_2) : c_2 \geq -c_1 - 4, \quad c_2 \leq -c_1, \quad c_2 \geq c_1 + 4\}, \\
\Omega_3 &= \{c = (c_1, c_2) : c_2 \geq -c_1, \quad c_2 \geq c_1 + 4, \quad c_2 \geq 2c_1 + 2\}, \\
\Omega_4 &= \{c = (c_1, c_2) : c_2 \leq 2c_1 + 2, \quad c_2 \geq 2c_1 + 1, \quad c_1 \geq 2\}, \\
\Omega_5 &= \{c = (c_1, c_2) : c_2 \leq 2c_1 + 1, \quad c_1 \geq 2\}, \\
\Omega_6 &= \{c = (c_1, c_2) : c_1 \leq 2, \quad c_2 \leq 2c_1, \quad c_2 \geq 0\}, \\
\Omega_7 &= \{c = (c_1, c_2) : c_2 \leq 0, \quad c_1 \geq -4, \quad c_1 \leq 2\}, \\
\Omega_8 &= \{c = (c_1, c_2) : c_2 \geq 0, \quad c_2 \leq -c_1, \\
&\qquad\qquad\qquad c_2 \leq (\sqrt{2} - 1)(c_1 + 4)\}, \\
\Omega_9 &= \{c = (c_1, c_2) : c_2 \geq (\sqrt{2} - 1)(c_1 + 4), \\
&\qquad\qquad\qquad c_2 \leq -c_1, \quad c_2 \leq c_1 + 4\}, \\
\Omega_{10} &= \{c = (c_1, c_2) : c_2 \geq -c_1, \quad c_2 \leq c_1 + 4, \quad c_1 \leq 2, \quad c_2 \geq 2c_1\}.
\end{aligned}
$$

Using formulae (16.17) and (16.38)–(16.45), one can show that

$$
\begin{aligned}
\varphi(c, \bar{b}) &= \varphi_{KKT}^{I_5}(c, \bar{b}) = 2c_1 + 2c_2 + 4 \quad \text{for every } c \in \Omega_1, \\
\varphi(c, \bar{b}) &= \varphi_{KKT}^{I_4}(c, \bar{b}) = -\frac{1}{4}(c_1 + c_2)^2 \quad \text{for every } c \in \Omega_2 \cup \Omega_9, \\
\varphi(c, \bar{b}) &= \varphi_{KKT}^{I_7}(c, \bar{b}) = 0 \quad \text{for every } c \in \Omega_3 \cup \Omega_4, \\
\varphi(c, \bar{b}) &= \varphi_{KKT}^{I_6}(c, \bar{b}) = -4c_1 + 2c_2 - 2 \quad \text{for every } c \in \Omega_5,
\end{aligned}
$$

and

$$
\varphi(c, \bar{b}) = \varphi_{KKT}^{I_2}(c, \bar{b}) = -\frac{1}{2}c_1^2 - 2c_1 + 2c_2 - 4
$$

for every $c \in \Omega_6 \cup \Omega_7 \cup \Omega_8$.

We will pay a special attention to the behavior of $\varphi(\cdot, \bar{b})$ on the region Ω_{10}. In order to compute $\varphi(c, \bar{b})$ for $c \in \Omega_{10}$, we divide Ω_{10}

into two subsets:

$$\begin{aligned}
\Omega'_{10} &= \{c = (c_1, c_2) \in \Omega_{10} : c_2 \geq 3c_1\} \\
&= \{c = (c_1, c_2) : c_2 \geq -c_1, \quad c_2 \geq 3c_1, \quad c_2 \leq c_1 + 4\} \\
\Omega''_{10} &= \{c = (c_1, c_2) \in \Omega_{10} : c_2 \leq 3c_1\} \\
&= \{c = (c_1, c_2) : c_1 \leq 2, \quad c_2 \geq 2c_1, \quad c_2 \leq 3c_1\}.
\end{aligned}$$

For $c \in \Omega'_{10}$, by (16.17) and (16.38)–(16.45) we have

$$\varphi(c, \bar{b}) = \min\{\varphi_{KKT}^{I_1}(c, \bar{b}), \ \varphi_{KKT}^{I_2}(c, \bar{b}), \ \varphi_{KKT}^{I_7}(c, \bar{b})\}.$$

Since

$$\begin{aligned}
&\varphi_{KKT}^{I_1}(c, \bar{b}) - \varphi_{KKT}^{I_2}(c, \bar{b}) \\
&= \frac{1}{4}(-c_1^2 - 2c_1 c_2 + c_2^2) + \frac{1}{2}c_1^2 + 2c_1 - 2c_2 + 4 \\
&= \frac{1}{4}(c_2 - (c_1 + 4))^2 \geq 0
\end{aligned}$$

for every $c \in \Omega'_{10}$, we have

$$\varphi(c, \bar{b}) = \min\{\varphi_{KKT}^{I_2}(c, \bar{b}), \ \varphi_{KKT}^{I_7}(c, \bar{b})\} \quad (\forall c \in \Omega'_{10}). \tag{16.46}$$

For $c \in \Omega''_{10}$, by (16.17) and (16.38)–(16.45) we have

$$\varphi(c, \bar{b}) = \min\{\varphi_{KKT}^{I_2}(c, \bar{b}), \ \varphi_{KKT}^{I_3}(c, \bar{b}), \ \varphi_{KKT}^{I_7}(c, \bar{b})\}.$$

Since

$$\begin{aligned}
&\varphi_{KKT}^{I_3}(c, \bar{b}) - \varphi_{KKT}^{I_7}(c, \bar{b}) \\
&= 2c_1^2 + \frac{1}{2}c_2^2 - 2c_1 c_2 \\
&= \frac{1}{2}(2c_1 - c_2)^2 \geq 0
\end{aligned}$$

for every $c \in \Omega''_{10}$, we have

$$\varphi(c, \bar{b}) = \min\{\varphi_{KKT}^{I_2}(c, \bar{b}), \ \varphi_{KKT}^{I_7}(c, \bar{b})\} \quad (\forall c \in \Omega''_{10}). \tag{16.47}$$

From (16.46) and (16.47) it follows that

$$\begin{aligned}
\varphi(c, \bar{b}) &= \min\{\varphi_{KKT}^{I_2}(c, \bar{b}), \ \varphi_{KKT}^{I_7}(c, \bar{b})\} \\
&= \min\{-\frac{1}{2}c_1^2 - 2c_1 + 2c_2 - 4, 0\}
\end{aligned} \tag{16.48}$$

for all $c \in \Omega_{10} = \Omega'_{10} \cup \Omega''_{10}$. Consider the parabola

$$\Gamma = \{(c_1, c_2) \in R^2 : c_2 = \frac{1}{4}c_1^2 + c_1 + 2\}. \tag{16.49}$$

By (16.48), for each $c \in \Omega_{10}$ we have

$$\varphi(c, \bar{b}) = \begin{cases} 0 & \text{if } c_2 \geq \dfrac{1}{4}c_1^2 + c_1 + 2 \\ -\dfrac{1}{2}c_1^2 - 2c_1 + 2c_2 - 4 & \text{if } c_2 \leq \dfrac{1}{4}c_1^2 + c_1 + 2. \end{cases} \tag{16.50}$$

This amounts to saying that $\varphi(c, \bar{b}) = \varphi_{KKT}^{I_7}(c, \bar{b})$ for all the points $c \in \Omega_{10}$ lying above the parabola Γ, and $\varphi(c, \bar{b}) = \varphi_{KKT}^{I_2}(c, \bar{b})$ for all the points $c \in \Omega_{10}$ lying below the curve Γ.

We have thus computed the values $\varphi(c, \bar{b})$ for all $c \in R^2$. To have a better knowledge of the behavior of the function $\varphi(\cdot, \bar{b})$, the reader can draw a plane R^2 with the regions $\Omega_1, \ldots, \Omega_{10}$ and the parabola Γ.

Proposition 16.2. *The obtained optimal value function $\varphi(c, \bar{b})$ ($c \in R^2$) cannot be a piecewise linear-quadratic function.*

A detailed proof of this proposition will be given in the next section.

By virtue of Proposition 16.2 and the observation stated just after Example 16.1, we can conclude that the optimal value function $(c, b) \mapsto \varphi(c, b)$ of problem (16.36) cannot be a plq function. Thus, *if D is not assumed to be a positive semidefinite matrix then the optimal value function $\varphi(\cdot, \cdot)$ of (15.1) can fail to be piecewise linear-quadratic.*

We now apply formula (16.30) to compute directional derivative of the function $\varphi(\cdot, \bar{b})$ studied in this section.

Let $\bar{c} = \bar{c}(\mu) = (0, \mu)$, $\mu \in R$. Let $\varphi_1(c) := \varphi(c, \bar{b})$. For $\bar{w}(\mu) = (\bar{c}(\mu), \bar{b})$ and $\bar{z} = (\bar{u}, \bar{v})$, where $\bar{u} = (1, 0) \in R^2$ and $\bar{v} = (0, 0, 0) \in R^3$, we have $\varphi'(\bar{w}(\mu); \bar{z}) = \varphi_1'(\bar{c}(\mu); \bar{u})$. Using formulae (16.30) and (16.38)–(16.45), we obtain

$$\begin{aligned} &\varphi'(\bar{w}(\mu); \bar{z}) \\ &= \varphi_1'(\bar{c}(\mu); \bar{u}) \\ &= \begin{cases} (\varphi_{KKT}^{I_7})'(\bar{c}(\mu); \bar{u}) & \text{for } \mu > 2, \\ \min\{(\varphi_{KKT}^{I_7})'(\bar{c}(\mu); \bar{u}), (\varphi_{KKT}^{I_2})'(\bar{c}(\mu); \bar{u})\} & \text{for } \mu = 2, \\ (\varphi_{KKT}^{I_2})'(\bar{c}(\mu); \bar{u}) & \text{for } \mu < 2. \end{cases} \end{aligned}$$

Therefore

$$\varphi'(\bar{w}(\mu); \bar{z}) = \varphi_1'(\bar{c}(\mu); \bar{u}) = \begin{cases} 0 & \text{for } \mu > 2, \\ -2 & \text{for } \mu \leq 2. \end{cases}$$

By Lemma 16.3, the function $\varphi_1(\cdot) = \varphi(\cdot, \bar{b})$ is locally Lipschitz on R^2. From Theorems 16.1 and 16.2 it follows that $\varphi_1(\cdot)$ is directionally differentiable at every $c \in R^2$ and, for every $u \in R^2$, the directional derivative $\varphi_1'(c; u)$ is finite. One can expect that $\varphi_1(\cdot)$ is regular in the sense of Clarke (1983), i.e. for every $c \in R^2$ it holds $\varphi_1^0(c; u) = \varphi_1'(c; u)$, where

$$\varphi_1^0(c; u) := \limsup_{c' \to c, \, t \downarrow 0} \frac{\varphi_1(c' + tu) - \varphi_1(c')}{t}$$

denotes the generalized directional derivative of φ_1 at c in direction u. Unfortunately, *the function $\varphi_1(\cdot)$ is not Lipschitz regular.* Indeed, for $\bar{c} = (0, 2)$ and $\bar{u} = (0, 1)$, using (16.50) it is not difficult to show that

$$0 = \varphi_1^0(\bar{c}; \bar{u}) > \varphi_1'(\bar{c}; \bar{u}) = -2.$$

16.4   Proof of Proposition 16.2

Suppose, contrary to our claim, that the function $\varphi(\cdot, \bar{b})$ is plq. Then the set $\mathrm{dom}\varphi(\cdot, \bar{b}) = R^2$ can be represented in the form

$$R^2 = \bigcup_{j \in J} \Delta_j, \qquad (16.51)$$

where J is a finite index set and Δ_j $(j \in J)$ are polyhedral convex sets. Moreover, for every $j \in J$, one has

$$\varphi(c, \bar{b}) = \frac{1}{2} c^T Q_j c + d_j^T c + \alpha_j \qquad (16.52)$$

for all $c \in \Delta_j$, where $\alpha_j \in R$, $d_j \in R^2$, $Q_j \in R_S^{2 \times 2}$. Let

$$\Delta_j' = \Delta_j \cap \Omega_{10} \quad (j \in J).$$

Note that some of the sets Δ_j' can be empty. From (16.51) we deduce that

$$\Omega_{10} = \bigcup_{j \in J} \Delta_j'. \qquad (16.53)$$

Note also that on each set Δ_j' $(j \in J)$ the function $\varphi(\cdot, \bar{b})$ has the linear-quadratic representation (16.52). Define

$$\Omega_{10}^I = \{c = (c_1, c_2) \in \Omega_{10} : c_2 \geq \frac{1}{4} c_1^2 + c_1 + 2\},$$

$$\Omega_{10}^{II} = \{c = (c_1, c_2) \in \Omega_{10} : c_2 \leq \frac{1}{4}c_1^2 + c_1 + 2\}.$$

It is evident that Ω_{10}^I is a convex set. Note that Ω_{10}^I and Ω_{10}^{II} are compact sets which admit the curve $\Gamma \cap \Omega_{10}$, where Γ is the parabola defined by (16.49), as the common boundary. The set Ω_{10}^I (resp., Ω_{10}^{II}) has nonempty interior. Indeed, let $\hat{c} := (0, 3)$ and $\tilde{c} := (0, 1)$. Substituting the coordinates of these vectors into the inequalities defining Ω_{10}^I and Ω_{10}^{II}, one see at once that $\hat{c} \in \text{int}\Omega_{10}^I$ and $\tilde{c} \in \text{int}\Omega_{10}^{II}$.

Fix any index $j \in J$ for which $\Delta_j' \neq \emptyset$.

We first consider the case $\text{int}\Delta_j' \neq \emptyset$. If

$$\text{int}\Delta_j' \cap \text{int}\Omega_{10}^I \neq \emptyset \tag{16.54}$$

then we must have $\Delta_j' \subset \Omega_{10}^I$. Indeed, by (16.54) there must exist a ball $B \subset R^2$ of positive radius such that

$$B \subset \Delta_j' \cap \Omega_{10}^I.$$

By (16.50), $\varphi(c, \bar{b}) = 0$ for every $c \in \Omega_{10}^I$. Then, it follows from (16.52) that

$$\varphi(c, \bar{b}) = \frac{1}{2}c^T Q_j c + d_j^T c + \alpha_j = 0$$

for every $c \in B$. This implies that $Q_j = 0$, $d_j = 0$ and $\alpha_j = 0$. Consequently,

$$\varphi(c, \bar{b}) = 0 \quad (\forall c \in \Delta_j'). \tag{16.55}$$

We observe from (16.50) that $\varphi(c, \bar{b}) < 0$ for every $c \in \Omega_{10} \setminus \Omega_{10}^I$. Hence (16.55) clearly forces $\Delta_j' \subset \Omega_{10}^I$. If

$$\text{int}\Delta_j' \cap \text{int}\Omega_{10}^I = \emptyset$$

then we must have $\text{int}\Delta_j' \subset \Omega_{10}^{II}$. Since Ω_{10}^{II} is closed, we conclude that $\Delta_j' \subset \Omega_{10}^{II}$. Therefore, if $\Delta_j' \cap \Omega_{10}^I \neq \emptyset$ then $\Delta_j' \cap \Omega_{10}^I = \Delta_j' \cap \Gamma$. In this case, it is easy to show that $\Delta_j' \cap \Gamma$ is a singleton.

We now consider the case $\text{int}\Delta_j' = \emptyset$. Since Δ_j' is a compact polyhedral convex set in R^2, there are only two possibilities:

(i) Δ_j' is a singleton,

(ii) Δ_j' is a line segment.

In both situations, if $\Delta_j' \cap \Omega_{10}^I$ is nonempty then it is a compact polyhedral convex set (a point or a line segment).

From (16.53) and from the above discussion, we can conclude that Ω_{10}^I is the union of the following finite collection of polyhedral convex sets:

$$
\begin{array}{ll}
\Delta'_j & (j \in J \text{ is such that } \mathrm{int}\Delta'_j \cap \mathrm{int}\Omega^I_{10} \neq \emptyset), \\
\Delta'_j \cap \Gamma & (j \in J \text{ is such that } \mathrm{int}\Delta'_j \neq \emptyset, \ \mathrm{int}\Delta'_j \cap \mathrm{int}\Omega^I_{10} = \emptyset), \\
\Delta'_j \cap \Omega^I_{10} & (j \in J \text{ is such that } \mathrm{int}\Delta'_j = \emptyset, \ \Delta'_j \cap \Omega^I_{10} \neq \emptyset).
\end{array}
$$

As Ω^I_{10} is convex, it coincides with the convex hull of the above-named compact polyhedral convex sets. According to Theorem 19.1 in Rockafellar (1970), this convex hull is a compact polyhedral convex set. So it has only a finite number of extreme points (see Rockafellar (1970), p. 162). Meanwhile, it is a simple matter to show that *every point from the infinite set $\Gamma \cap \Omega_{10}$ is an extreme point of Ω^I_{10}.* We have arrived at a contradiction. The proof is complete. $\square$

16.5 Commentaries

The results presented in this chapter are taken from Lee et al. (2002a).

In this chapter we have studied a class of optimal value functions in parametric (nonconvex) quadratic programming. It has been shown that these functions are directionally differentiable at any point from their effective domains but, in general, they are not piecewise linear-quadratic and they may be not Lipschitz regular at some interior points in their effective domains.

The class of plq functions has been investigated systematically in Rockafellar and Wets (1998). In particular, the topics like sub-differential calculation, dualization, and optimization involving plq functions, are studied in the book.

The reader is referred to Gauvin and Tolle (1977), Gauvin and Dubeau (1982), Rockafellar (1982), Fiacco (1983), Clarke (1983), Janin (1984), Minchenko and Sakolchik (1996), Bonnans and Shapiro (1998, 2000), Ward and Lee (2001), and references therein, for different approaches in the study of differential properties of the optimal value functions in nonlinear optimization problems.

It would be desirable to find out what additional conditions one has to impose on the pair of matrices $(D, A) \in R_S^{n \times n} \times R^{m \times n}$, where D need not be a positive semidefinite matrix, so that the optimal value function

$$(c, b) \mapsto \varphi(c, b)$$

of the parametric problem (15.1) is piecewise linear-quadratic on $R^n \times R^m$.

Both referees of the paper Lee et al. (2001a) informed us that D. Klatte had constructed an example of an optimal value function in a linearly perturbed QP problem which is not plq. Being unaware of that (unpublished) example, we have constructed Example 16.2. One referee gave us some hints in detail on the example of Klatte. Namely, letting two components of the data perturbation of a QP problem considered by Klatte (1985) be fixed, one has the problem

$$\begin{cases} \text{Minimize } x_1 x_2 \\ \text{subject to } x = (x_1, x_2) \in R^2, \ -1 \le x_1 \le b_1, \ b_2 \le x_2 \le 1, \end{cases}$$

where $b = (b_1, b_2) \in R^2$, $b_1 \ge 0$ and $b_2 \le 0$, represents the perturbation of the feasible region. Denote by $\varphi(b_1, b_2)$ the optimal value function of this problem. It is easy to verify that

$$\varphi(b_1, b_2) = \begin{cases} -1 & \text{if } -1 \le b_1 b_2 \\ b_1 b_2 & \text{if } -1 > b_1 b_2. \end{cases}$$

If $b_1 < 0$ or $b_2 > 0$, then we put $\varphi(b_1, b_2) = +\infty$. Arguments similar to those of the proof of Proposition 16.2 show that $\varphi(b_1, b_2)$ is not a plq function. The main difference between this example and Example 16.2 is that here the feasible region is perturbed, while in Example 16.2 the objective function is perturbed.

Chapter 17

Quadratic Programming under Linear Perturbations: III. The Convex Case

The problem of finding the nearest point in a polyhedral convex set to a given point is a convex QP problem. That nearest point is called the *metric projection* of the given point onto the polyhedral convex set.

In this chapter we will see that the metric projection from a given point onto a moving polyhedral convex set is Lipschitz continuous with respect to the perturbations on the right-hand-sides of the linear inequalities defining the set. The property leads to a simple sufficient condition for Lipschitz continuity of a locally unique solution of parametric variational inequalities with a moving polyhedral constraint set. Applications of these results to traffic network equilibrium problems will be discussed in detail.

17.1   Preliminaries

We will study sensitivity of solutions to a parametric variational inequality (PVI, for brevity) with a parametric polyhedral constraint. Let

$$K(\lambda) = \{x \in R^n \ : \ Ax \geq \lambda, \ x \geq 0\}, \tag{17.1}$$

$$\Lambda = \{\lambda \in R^r \ : \ K(\lambda) \neq \emptyset\}, \tag{17.2}$$

where $A \in R^{r \times n}$ is a given matrix. Let $M \subset R^m$ be any subset and $f : R^n \times M \to R^n$ be a given function. Consider the following PVI depending on a pair of parameters $(\mu, \lambda) \in M \times \Lambda$:

$$\begin{cases} \text{Find} \quad x \in K(\lambda) \quad \text{such that} \\ \langle f(x, \mu), y - x \rangle \geq 0 \quad \text{for all} \quad y \in K(\lambda). \end{cases} \qquad (17.3)$$

Assume that $\bar{x}$ is a solution of the following problem

$$\begin{cases} \text{Find} \quad x \in K(\bar{\lambda}) \quad \text{such that} \\ \langle f(x, \bar{\mu}), y - x \rangle \geq 0 \quad \text{for all} \quad y \in K(\bar{\lambda}), \end{cases} \qquad (17.4)$$

where $(\bar{\mu}, \bar{\lambda}) \in M \times \Lambda$ are given parameters.

Our aim is to prove that under some appropriate conditions on f in a neighborhood of $(\bar{x}, \bar{\mu})$ and *no* conditions on the matrix A, there exist $k > 0$ and neighborhoods X, U, V of $\bar{x}, \bar{\mu}$, and $\bar{\lambda}$, respectively, such that

(i) *For every* $(\mu, \lambda) \in (M \cap U) \times (\Lambda \cap V)$ *there is a unique solution* $x = x(\mu, \lambda)$ *of* (17.3) *in* X;

(ii) *For every* $(\mu, \lambda), (\mu', \lambda') \in (M \cap U) \times (\Lambda \cap V)$,

$$\| x(\mu', \lambda') - x(\mu, \lambda) \| \leq k(\| \mu' - \mu \| + \| \lambda' - \lambda \|).$$

To this aim, in Section 17.2 we obtain a property of the metric projection onto a moving polyhedral convex set which can be stated simply, as follows: For a given a matrix $A \in R^{r \times n}$ there exists a constant $k_1 > 0$ such that for all $y \in R^n$ and $\lambda, \lambda' \in \Lambda$, we have

$$\| P_{K(\lambda')} y - P_{K(\lambda)} y \| \leq k_1 \| \lambda' - \lambda \|, \qquad (17.5)$$

where $K(\lambda)$ and Λ are defined by (17.1) and (17.2), $P_{K(\lambda)} y$ is the unique point in $K(\lambda)$ with the minimal distance to y. (The map $P_{K(\lambda)}(.)$ is said to be the *metric projection* onto $K(\lambda)$.)

Property (17.5) is established by using a result on linear complementarity problems in Mangasarian and Shiau (1987). Then the scheme for proving Lemma 2.4 in Dafermos (1988) enables us to get, in Section 17.3, the desired sensitivity result for PVI. The latter can be interpreted as a condition for Lipschitz continuity of the equilibrium flow in a traffic network with changing costs and demands. This fact is considered in Section 17.4.

17.2   Projection onto a Moving Polyhedral Convex Set

To establish property (17.5) we will consider $P_{K(\lambda)}y$ as the unique solution of a quadratic program with parameters (y, λ). By the standard procedure (see Murty (1976)) we reduce this program to an equivalent linear complementarity problem. Although the assumption on uniqueness of solutions of Theorem 3.2 in Mangasarian and Shiau (1987) is violated in our LCP problem, we will show that the partition method for obtaining that theorem is well adequate for our purpose.

So, let $y \in R^n$ and $\lambda \in \Lambda$ (see (17.2)) be given. From the definition it follows that $\xi := P_{K(\lambda)}y$ is the unique solution of the problem

$$\text{Minimize} \quad \|x - y\|^2 \quad \text{subject to} \quad Ax \geq \lambda, \ x \geq 0,$$

which is equivalent to the following one

$$\text{Minimize} \quad (-2y^T x + x^T x) \quad \text{subject to} \quad Ax \geq \lambda, \ x \geq 0. \quad (17.6)$$

It is clear that (17.6) is a particular case of the following convex QP problem

$$\text{Minimize} \quad \frac{1}{2}x^T Dx + c^T x \quad \text{subject to} \quad Ax \geq \lambda, \ x \geq 0, \quad (17.7)$$

where $c \in R^n$, D is a symmetric positive semidefinite matrix. Indeed, (17.7) becomes (17.6) if one takes $c = -2y$ and $D = 2E$, where E denotes the unit matrix of order n.

The next lemma follows easily from Corollary 3.1 and the convexity of problem (17.7).

Lemma 17.1. *Vector $\xi \in R^n$ is a solution of (17.7) if and only if there exists $\eta \in R^r$ such that*

$$z := \begin{pmatrix} \xi \\ \eta \end{pmatrix}$$

is a solution to the following LCP problem:

$$Mz + q \geq 0, \quad z \geq 0, \quad (Mz + q)^T z = 0, \quad (17.8)$$

where

$$M := \begin{pmatrix} D & -A^T \\ A & 0 \end{pmatrix} \quad \text{and} \quad q := \begin{pmatrix} c \\ -\lambda \end{pmatrix}. \quad (17.9)$$

We set $s = n + r$. For any subset $J \subset \{1, \cdots, s\}$, observe (see Mangasarian and Shiau (1987), p. 591) that every solution of the following system of $2s$ linear equalities and inequalities

$$\begin{cases} M_j z + q_j \geq 0, & z_j = 0, \quad j \in J, \\ M_j z + q_j = 0, & z_j \geq 0, \quad j \notin J, \end{cases} \qquad (17.10)$$

is a solution of (17.8). For every $J \subset \{1, \cdots, s\}$, symbol $Q(J)$ denotes the set of all vectors q such that (17.10) has a solution. Note that $Q(J)$ is a closed convex cone which is called a *complementary cone* of (M, q) (see Murty (1976), Mangasarian and Shiau (1987)). The union $\cup\{Q(J) : J \subset \{1, \cdots, s\}\}$ is the set of all q such that (17.8) is solvable.

For each subset $J \subset \{1, \cdots, s\}$, according to Corollary 7.3, we can find a constant $\theta = \theta_J > 0$ such that if z^1 is a solution of (17.10) at $q = q^1$ and the solution set of (17.10) at $q = q^2$ is nonempty, then there exists a solution z^2 of (17.10) at $q = q^2$ such that

$$\|z^2 - z^1\| \leq \theta_J \|q^2 - q^1\|.$$

Let us set

$$k_0 = \max\{\theta_J : J \subset \{1, \cdots, s\}\}. \qquad (17.11)$$

The next technical lemma is crucial for applying Corollary 7.3 to linear complementarity problems.

Lemma 17.2. (See Mangasarian and Shiau (1987), p. 591) *Let $q^1, q^2 \in R^s$ be two distinct vectors. Assume that for every $t \in [0, 1]$ system (17.8) is solvable for $q = q(t) := (1 - t)q^1 + tq^2$. Then there is a partition $0 = t_0 < t_1 < \cdots < t_\ell = 1$ such that for every $i \in \{1, \cdots, \ell\}$,*

$$q(t_{i-1}) \in Q(J_i), \quad q(t_i) \in Q(J_i) \quad \text{for some} \quad J_i \subset \{1, \cdots, s\}. \qquad (17.12)$$

The proof of this lemma is based on the observation that the intersection of each complementary cone of (17.8) with the segment $[q^1, q^2]$ is a closed interval (which may reduce to a single point or to the empty set). Since (17.8) is solvable for every $q \in [q^1, q^2]$, this segment is contained in the union of such intervals. Excluding some redundant intervals in that union and let t_i be 0, 1, or a point in the intersection of two neighbouring intervals, we get the desired partition.

The following theorem will be useful for obtaining the results in Section 17.3.

Theorem 17.1. *Given a matrix $A \in R^{r \times n}$, define the sets $K(\lambda)$ and Λ by (17.1) and (17.2). Then there exists a constant $k_1 > 0$ such that*

$$\|P_{K(\lambda')}y - P_{K(\lambda)}y\| \leq k_1 \|\lambda' - \lambda\|, \tag{17.13}$$

for all $y \in R^n$ and $\lambda, \lambda' \in \Lambda$, where $P_{K(\lambda)}y$ is the metric projection of y onto $K(\lambda)$.

From the discusion at the beginning of this section we see that Theorem 17.1 is a direct consequence of the next result.

Theorem 17.2. *(See Cottle et al. (1992), p. 696) Let $A \in R^{r \times n}$, $K(\lambda)$ and Λ be defined as in (17.1) and (17.2). Let $D \in R_S^{n \times n}$ be a positive definite matrix. Define M and q by (17.9), k_0 by (17.11). Then for every $\lambda, \lambda' \in \Lambda$ and $c, c' \in R^n$ we have*

$$\|x(c', \lambda') - x(c, \lambda)\| \leq k_0 \left(\|c' - c\| + \|\lambda' - \lambda\| \right), \tag{17.14}$$

where $x(c, \lambda)$ and $x(c', \lambda')$ are the unique solution of (17.7) at the parameters (c, λ) and (c', λ'), respectively.

Proof. We will follow the arguments for proving Theorem 3.2 in Mangasarian and Shiau (1987). Let there be given vectors $\lambda, \lambda' \in \Lambda$ and $c, c' \in R^n$. We set

$$q^1 = \begin{pmatrix} c \\ -\lambda \end{pmatrix}, \quad q^2 = \begin{pmatrix} c' \\ -\lambda' \end{pmatrix},$$

$$c(t) = (1-t)c + t\,c', \quad \lambda(t) = (1-t)\lambda + t\lambda',$$
$$q(t) = (1-t)q^1 + tq^2 \quad \text{for every} \quad t \in [0,1].$$

If $\lambda = \lambda'$ and $c = c'$, then (17.14) holds. Consider the other case where at least one of these equalities does not hold. Then we have $q^1 \neq q^2$. From the definition we see that Λ is a closed convex cone. Thus $\lambda(t) \in \Lambda$ for every $t \in [0,1]$. This means that $K(\lambda(t)) \neq \emptyset$ for every $t \in [0,1]$. Since D is assumed to be a symmetric positive definite matrix, for each $t \in [0,1]$ program (17.7), where $(c(t), \lambda(t))$ are in the place of (c, λ), must have a unique solution, denoted by $\xi(t)$. Using Lemma 17.1 we find a vector $\eta(t) \in R^r$ such that

$$z(t) := \begin{pmatrix} \xi(t) \\ \eta(t) \end{pmatrix}$$

is a solution of (17.8), where

$$q = \begin{pmatrix} c(t) \\ -\lambda(t) \end{pmatrix} = q(t).$$

(Note that vector $\eta(t)$ may not be uniquely defined.) Hence, according to Lemma 17.2 there is a partition $0 = t_0 < t_1 < \cdots < t_\ell = 1$ such that for every $1 \leq i \leq \ell$ condition (17.12) holds.

Consequently, for every $1 \leq i \leq \ell$ vectors $q(t_{i-1})$ and $q(t_i)$ belong to the cone $Q(J_i)$ for a subset $J_i \subset \{1, \cdots, s\}$. Hence the systems of linear equalities and inequalities

$$\begin{cases} M_j z + (q(t_i))_j \geq 0, & z_j = 0, \quad j \in J_i \\ M_j z + (q(t_i))_j = 0, & z_j \geq 0, \quad j \notin J_i \end{cases} \tag{17.15}$$

and

$$\begin{cases} M_j z + (q(t_{i-1}))_j \geq 0, & z_j = 0, \quad j \in J_i \\ M_j z + (q(t_{i-1}))_j = 0, & z_j \geq 0, \quad j \notin J_i \end{cases} \tag{17.16}$$

are solvable. Let

$$\bar{z}(t_i) = \begin{pmatrix} \bar{\xi}(t_i) \\ \bar{\eta}(t_i) \end{pmatrix}$$

be a solution of (17.15). According to Corollary 7.3 there exists a solution

$$\bar{z}(t_{i-1}) = \begin{pmatrix} \bar{\xi}(t_{i-1}) \\ \bar{\eta}(t_{i-1}) \end{pmatrix}$$

of (17.16) satisfying

$$\begin{aligned} \|\bar{z}(t_i) - \bar{z}(t_{i-1})\| &\leq \theta_J \|q(t_i) - q(t_{i-1})\| \\ &= \theta_J (t_i - t_{i-1}) \|q^1 - q^2\|. \end{aligned}$$

This implies

$$\|\bar{\xi}(t_i) - \bar{\xi}(t_{i-1})\| \leq k_0 (t_i - t_{i-1}) \|q^1 - q^2\|. \tag{17.17}$$

Since $\bar{z}(t_i)$ solves (17.15) it also solves (17.8) at $q = q(t_i)$. By virtue of Lemma 17.1, $\bar{\xi}(t_i)$ is a solution of (17.7), where $(c, \lambda) = (c(t_i), \lambda(t_i))$. As the latter problem has a unique solution, we have $\bar{\xi}(t_i) = \xi(t_i)$. Similarly, since $\bar{z}(t_{i-1})$ solves (17.16) it also solves (17.8) at $q = q(t_{i-1})$. Hence, $\bar{\xi}(t_{i-1}) = \xi(t_{i-1})$. These facts and (17.17) imply

$$\|\xi(t_i) - \xi(t_{i-1})\| \leq k_0 (t_i - t_{i-1}) \|q^1 - q^2\|.$$

Consequently,

$$\|\xi(t_\ell) - \xi(t_0)\| \leq \sum_{i=1}^{\ell} \|\xi(t_i) - \xi(t_{i-1})\| \leq k_0 \|q^1 - q^2\|.$$

Since $\xi(t_\ell) = \xi(1) = x(c', \lambda')$ and $\xi(t_0) = \xi(0) = x(c, \lambda)$, we obtain

$$\|x(c', \lambda') - x(c, \lambda)\| \leq k_0 (\|c' - c\| + \|\lambda' - \lambda\|).$$

The proof is complete. $\square$

17.3 Application to Variational Inequalities

Consider problem (17.3) and suppose that for a pair $(\bar{\mu}, \bar{\lambda}) \in M \times \Lambda$ vector $\bar{x}$ is a solution of (17.4). Following Dafermos (1988) we assume that there exist neighborhoods X of $\bar{x}$, U of $\bar{\mu}$, and two constants $\alpha > 0, l > 0$, such that

$$\|f(x', \mu') - f(x, \mu)\| \leq l(\|x' - x\| + \|\mu' - \mu\|) \qquad (17.18)$$

for all μ, μ' in $M \cap U$, x, x' in X, and

$$\langle f(x', \mu) - f(x, \mu), x' - x \rangle \geq \alpha \|x' - x\|^2 \qquad (17.19)$$

for all $\mu \in M \cap U$, x and x' in X. Without loss of generality we can assume that X is a polyhedral convex set and $\alpha < l$. Condition (17.18) means that f is locally Lipschitz at $(\bar{x}, \bar{\mu})$. Condition (17.19) means that $f(., \mu)$ is locally strongly monotone around $\bar{x}$ with a common coefficient for all $\mu \in M \cap U$. Using the notation of Dafermos (1988) we put

$$G(x, \mu, \lambda) = P_{K(\lambda) \cap X}[x - \rho f(x, \mu)] \quad \text{for all } (x, \mu, \lambda) \in R^n \times M \times \Lambda, \qquad (17.20)$$

where $\rho > 0$ is a fixed number and $P_{K(\lambda) \cap X} y$ is the metric projection of y onto $K(\lambda) \cap X$. Let us consider a number ρ satisfying

$$0 < \rho \leq \frac{\alpha}{l^2}. \qquad (17.21)$$

For every $\lambda \in \Lambda$ such that $K(\lambda) \cap X \neq \emptyset$, Lemma 2.2 from Dafermos (1988) shows that

$$\|G(x', \mu, \lambda) - G(x, \mu, \lambda)\| \leq \beta \|x' - x\| \qquad (17.22)$$

for all x and x' in X, $\mu \in M \cap U$, where

$$\beta := (1 - \rho\alpha)^{1/2} < 1. \qquad (17.23)$$

According to the Banach contractive mapping principle, there is a unique vector $x = x(\mu, \lambda) \in X$ satisfying

$$x(\mu, \lambda) = G(x(\mu, \lambda), \mu, \lambda). \qquad (17.24)$$

For the map $K(\lambda)$ defined by (17.1) we apply Corollary 7.3 to find an $\theta > 0$ such that if $\lambda, \lambda' \in \Lambda$ and $x \in K(\lambda)$, then there exists $x' \in K(\lambda')$ satisfying

$$\|x' - x\| \leq \theta\|\lambda' - \lambda\|. \qquad (17.25)$$

Since $\bar{x} \in K(\bar{\lambda})$, from (17.25) it follows that there is a neighborhood V_1 of $\bar{\lambda}$ such that

$$K(\lambda) \cap X \neq \emptyset \quad \text{for every} \quad \lambda \in \Lambda \cap V_1. \qquad (17.26)$$

Since X is a polyhedral convex set we can find a matrix C of order $r_1 \times n$ and a vector $b \in R^{r_1}$ such that $X = \{x \in R^n : Cx \geq b\}$. Therefore

$$K(\lambda) \cap X = \{x \in R^n : Ax \geq \lambda, \; Cx \geq b, \; x \geq 0\}. \qquad (17.27)$$

So, taking (17.26) into account we can apply Theorem 17.1 for system (17.27) to choose a constant $k_1 > 0$ such that

$$\|P_{K(\lambda') \cap X} y - P_{K(\lambda) \cap X} y\| \leq k_1\|\lambda' - \lambda\| \qquad (17.28)$$

for all $y \in R^n$, λ and λ' in $\Lambda \cap V_1$. (Note that k_1 depends not only on A but also on C, that is, on the neighborhood X.)

Lemma 17.3. *Let (17.18) and (17.19) be fulfilled. Assume that $k_1 > 0$ is a constant satisfying (17.28). Then for any $\rho > 0$ satisfying (17.21) there exist neighborhoods $\bar{U}$ and $\bar{V}$ of $\bar{\mu}$ and $\bar{\lambda}$, respectively, such that:*

(i) *For every $(\mu, \lambda) \in (M \cap \bar{U}) \times (\Lambda \cap \bar{V})$ vector $x(\mu, \lambda) \in X$ defined by (17.24) is the unique solution of (17.3) in X;*

(ii) *For all $\mu, \; \mu' \in M \cap \bar{U}$ and $\lambda, \; \lambda' \in \Lambda \cap \bar{V}$,*

$$\|x(\mu', \lambda') - x(\mu, \lambda)\| \leq \frac{1}{1 - \beta}(\rho l\|\mu' - \mu\| + k_1\|\lambda' - \lambda\|),$$

where β is defined in (17.23).

This lemma can be proved similarly as Lemma 2.1 in Yen (1995a). Note that the scheme given on p. 424 in Dafermos (1988) is our key argument.

Proof. Fixing any ρ satisfying (17.21), for every $(\mu, \lambda) \in (M \cap U) \times (\Lambda \cap V_1)$ we denote by $x(\mu, \lambda)$ the unique vector in X satisfying

(17.24). Let $(\mu, \lambda), (\mu', \lambda') \in (M \cap U) \times (\Lambda \cap V_1)$. Using (17.22) we have

$$
\begin{aligned}
\|x(\mu', \lambda') &- x(\mu, \lambda)\| \\
&= \|G(x(\mu', \lambda'), \mu', \lambda') - G(x(\mu, \lambda), \mu, \lambda)\| \\
&\leq \|G(x(\mu', \lambda'), \mu', \lambda') - G(x(\mu, \lambda), \mu', \lambda')\| \\
&\quad + \|G(x(\mu, \lambda), \mu', \lambda') - G(x(\mu, \lambda), \mu, \lambda)\| \\
&\leq \beta \|x(\mu', \lambda') - x(\mu, \lambda)\| \\
&\quad + \|G(x(\mu, \lambda), \mu', \lambda') - G(x(\mu, \lambda), \mu, \lambda)\|.
\end{aligned} \tag{17.29}
$$

Formula (17.20) and the fact that the metric projection onto a fixed closed convex set is a nonexpansive mapping yield

$$
\begin{aligned}
\|G(x(\mu, \lambda), \mu', \lambda') &- G(x(\mu, \lambda), \mu, \lambda)\| \\
&= \|P_{K(\lambda') \cap X}[x(\mu, \lambda) - \rho f(x(\mu, \lambda), \mu')] \\
&\quad - P_{K(\lambda) \cap X}[x(\mu, \lambda) - \rho f(x(\mu, \lambda), \mu)]\| \\
&\leq \|P_{K(\lambda') \cap X}[x(\mu, \lambda) - \rho f(x(\mu, \lambda), \mu')] \\
&\quad - P_{K(\lambda') \cap X}[x(\mu, \lambda) - \rho f(x(\mu, \lambda), \mu)]\| \\
&\quad + \|P_{K(\lambda') \cap X}[x(\mu, \lambda) - \rho f(x(\mu, \lambda), \mu)] \\
&\quad - P_{K(\lambda) \cap X}[x(\mu, \lambda) - \rho f(x(\mu, \lambda), \mu)]\| \\
&\leq \rho \|f(x(\mu, \lambda), \mu') - f(x(\mu, \lambda), \mu)\| \\
&\quad + \|P_{K(\lambda') \cap X} y(\mu, \lambda) - P_{K(\lambda) \cap X} y(\mu, \lambda)\|,
\end{aligned} \tag{17.30}
$$

where

$$
y(\mu, \lambda) := x(\mu, \lambda) - \rho f(x(\mu, \lambda), \mu).
$$

From (17.18), (17.29) and (17.30) it follows that

$$
\begin{aligned}
\|x(\mu', \lambda') &- x(\mu, \lambda)\| \\
&\leq \frac{1}{1 - \beta}(\rho l \|\mu' - \mu\| + \|P_{K(\lambda') \cap X} y(\mu, \lambda) - P_{K(\lambda) \cap X} y(\mu, \lambda)\|).
\end{aligned} \tag{17.31}
$$

Now we can find neighborhoods $\bar{U}$ and $\bar{V}$ of $\bar{\mu}$ and $\bar{\lambda}$ such that (i) and (ii) are fulfilled. Indeed, since $\bar{x}$ is a solution of (17.4), it is easy to show that

$$
\bar{x} = P_{K(\bar{\lambda})}[\bar{x} - \rho f(\bar{x}, \bar{\mu})].
$$

Therefore $\bar{x}$ is the unique fixed point in X of the map $G(\,\cdot\,, \bar{\mu}, \bar{\lambda})$ defined by (17.20). Hence $\bar{x} = x(\bar{\mu}, \bar{\lambda})$. Using this and putting $\bar{y} = \bar{x} - \rho f(\bar{x}, \bar{\mu})$, we substitute $(\mu, \lambda) = (\bar{\mu}, \bar{\lambda})$ into (17.31) to obtain

$$
\|x(\mu', \lambda') - \bar{x}\| \leq \frac{1}{1 - \beta}(\rho l \|\mu' - \bar{\mu}\| + \|P_{K(\lambda') \cap X} \bar{y} - P_{K(\bar{\lambda}) \cap X} \bar{y}\|).
$$

Taking account of (17.28) we have

$$\|x(\mu', \lambda') - \bar{x}\| \leq \frac{1}{1-\beta}(\rho l \|\mu' - \bar{\mu}\| + k_1 \|\lambda' - \bar{\lambda}\|, \qquad (17.32)$$

for all $(\mu', \lambda') \in (M \cap U) \times (\Lambda \cap V_1)$. Due to (17.32), there exist neighborhoods $\bar{U} \subset U$ of $\bar{\mu}$ and $\bar{V} \subset V_1$ of $\bar{\lambda}$ such that $x(\mu, \lambda)$ belongs to the interior of X for every $(\mu, \lambda) \in (M \cap \bar{U}) \times (\Lambda \cap \bar{V})$. For such a pair (μ, λ), since the vector $x(\mu, \lambda)$ satisfies (17.24) and belongs to the interior of X, Lemma 2.1 in Dafermos (1988) shows that $x(\mu, \lambda)$ is the unique solution of (17.3) in X. We have thus established the first assertion of the lemma. The second assertion follows easily from (17.31) and (17.28).   □

Now we can formulate the main result of this section.

Theorem 17.3. *Let $\bar{x}$ be a solution of (17.4). If conditions (17.18) and (17.19) are satisfied, then there exist constants $k_{\bar{\mu}} > 0$ and $k_{\bar{\lambda}} > 0$, neighborhoods $\bar{U}$ of $\bar{\mu}$ and $\bar{V}$ of $\bar{\lambda}$ such that:*

(i) *For every $(\mu, \lambda) \in (M \cap \bar{U}) \times (\Lambda \cap \bar{V})$ there exists a unique solution of (17.3) in X, denoted by $x(\mu, \lambda)$;*

(ii) *For all $(\mu', \lambda'), (\mu, \lambda) \in (M \cap \bar{U}) \times (\Lambda \cap \bar{V})$,*

$$\|x(\mu', \lambda') - x(\mu, \lambda)\| \leq k_{\bar{\mu}} \|\mu' - \mu\| + k_{\bar{\lambda}} \|\lambda' - \lambda\|.$$

Proof. It suffices to apply Lemma 17.3 with any ρ satisfying (17.21), and put

$$k_{\bar{\mu}} = \frac{1}{1-\beta} \rho l, \quad k_{\bar{\lambda}} = \frac{1}{1-\beta} k_1. \quad □$$

17.4   Application to a Network Equilibrium Problem

Let us consider problem (17.3) with $K(\lambda)$ defined in the following way:

$$K(\lambda) = \{x \in R^n : x = Zh, \ \Gamma h = \lambda, \ h \geq 0\}, \qquad (17.33)$$

where $h \in R^p$, $\lambda \in R^r$, Γ is an $r \times p$-matrix, Z is an $n \times p$-matrix. This is the variational inequality model for the traffic equilibrium problem (see Smith (1979), Dafermos (1980), De Luca and Maugeri

(1989), Qiu and Magnanti (1989)) which we have studied in Chapter 9. The matrices, the vectors, and the function $f(x, \mu)$ in the model have the following interpretations (see Qiu and Magnanti (1989), and Chapter 9 of this book):

> x = vector of flows on arcs, h = vector of flows on paths, Γ = the incidence matrix of the relation "paths - OD (origin-destination) pairs", Z = the incidence matrix of the relation "arcs - paths", λ = vector of demands for the OD pairs, $f(x, \mu)$ = vector of the costs on arcs when the network is loaded with flow x, μ = parameter of the perturbation of the costs on arcs.

For a given pair (μ, λ), solutions of (17.3) are interpreted as the equilibrium flows on the traffic network, corresponding to vector λ of demands an function $f(., \mu)$ of the costs on arcs.

Since $K(\lambda)$ is defined by (17.33) rather by (17.1), Theorem 17.1 cannot be applied directly. However, a property like the one in (17.13) is valid.

Lemma 17.4. *Assume that $K(\lambda)$ is given by (17.33), $H(\lambda) := \{h \in R^p : \Gamma h = \lambda, h \geq 0\}$, $\Lambda := \{\lambda \in R^r : H(\lambda) \neq \emptyset\}$. Then there exists a constant $k > 0$ such that*

$$\|P_{K(\lambda')}y - P_{K(\lambda)}y\| \leq k\|\lambda' - \lambda\|, \tag{17.34}$$

for every $y \in R^n$ and $\lambda, \lambda' \in \Lambda$.

Proof. Since $K(\lambda) = Z(H(\lambda)) := \{Zh : h \in H(\lambda)\}$, then $\{\lambda \in R^r : K(\lambda) \neq \emptyset\} = \{\lambda \in R^r : H(\lambda) \neq \emptyset\} = \Lambda$. For each $\lambda \in \Lambda$ and $y \in R^n$ we consider two quadratic programming problems:

> Minimize $\|y - x\|^2$ subject to $x - Zh = 0$, $\Gamma h = \lambda$, $h \geq 0$;
> $$\tag{17.35}$$

and

> Minimize $\|y - Zh\|^2$ subject to $\Gamma h = \lambda$, $h \geq 0$. $\tag{17.36}$

Observe that if h is a solution of (17.36) then $x := Zh$ is a solution of (17.35) and, hence, $x = P_{K(\lambda)}y$. Moreover, since (17.35) has a unique solution, for arbitrary two solutions h^1, h^2 of (17.36) we have $Zh^1 = Zh^2$, and $x := Zh^1 = Zh^2$ is the unique solution of (17.35). Also, note that (17.36) is solvable, because (17.35) is solvable.

Since $\|y - Zh\|^2 = \|y\|^2 - 2y^T Zh + h^T Z^T Zh$, (17.36) is equivalent to the following problem

$$\text{Minimize} \quad \frac{1}{2} h^T Dh + c^T h \quad \text{subject to} \quad Ah \geq \hat{\lambda}, \ h \geq 0, \quad (17.37)$$

where $c := -2Z^T y$, $D := 2Z^T Z$,

$$A := \begin{pmatrix} \Gamma \\ -\Gamma \end{pmatrix}, \quad \text{and} \quad \hat{\lambda} := \begin{pmatrix} \lambda \\ -\lambda \end{pmatrix}.$$

It is clear that D is a symmetric positive semidefinite matrix. Hence the scheme for reducing a convex quadratic programming problem to an equivalent LCP problem, recalled in Section 17.2, is applicable to (17.37). In particular, Lemma 17.1 asserts that $h \in R^p$ is a solution of (17.37) if and only if there exists $\eta \in R^{2r}$ such that

$$z := \begin{pmatrix} h \\ \eta \end{pmatrix}$$

is a solution of the LCP problem (17.8), where

$$M := \begin{pmatrix} D & -A^T \\ A & 0 \end{pmatrix} \quad \text{and} \quad q := \begin{pmatrix} c \\ -\hat{\lambda} \end{pmatrix}.$$

Let $s := p + 2r$ and k_0 be the constant defined by (17.11). We are going to prove that $k := \sqrt{2}k_0 \|Z\|$ is a constant satisfying (17.34). Indeed, let $y \in R^n$ and $\lambda, \lambda' \in \Lambda$ be given arbitrarily. For each $t \in [0, 1]$ we set

$$q(t) = (1 - t)q^1 + tq^2, \quad \lambda(t) = (1 - t)\lambda + t\lambda', \quad \hat{\lambda}(t) = (1 - t)\hat{\lambda} + t\hat{\lambda}',$$
$$(17.38)$$

where

$$q^1 := \begin{pmatrix} c \\ -\hat{\lambda} \end{pmatrix}, \quad q^2 := \begin{pmatrix} c \\ -\hat{\lambda}' \end{pmatrix}, \quad c := -2Z^T y, \quad \hat{\lambda} := \begin{pmatrix} \lambda \\ -\lambda \end{pmatrix},$$

$$\hat{\lambda}' := \begin{pmatrix} \lambda' \\ -\lambda' \end{pmatrix}.$$

Since (17.36) has a solution for every $\lambda \in \Lambda$, (17.37) is solvable for each

$$\hat{\lambda} = \begin{pmatrix} \lambda \\ -\lambda \end{pmatrix},$$

where $\lambda \in \Lambda$. Consequently, for every $t \in [0, 1]$ problem (17.8) is solvable for $q = q(t)$, where $q(t)$ is defined in (17.38). Applying Lemma 17.2 one can find a partition $0 = t_0 < t_1 < \cdots < t_\ell = 1$ such that for every $1 \leq i \leq \ell$ condition (17.12) holds. Therefore, for each $1 \leq i \leq \ell$ there is a subset $J_i \subset \{1, \cdots, s\}$ such that the vectors $q(t_{i-1})$ and $q(t_i)$ belong to the cone $Q(J_i)$. This implies that (17.15) and (17.16) are solvable. Let

$$z(t_i) = \begin{pmatrix} h(t_i) \\ \eta(t_i) \end{pmatrix}$$

be a solution of (17.15). According to Corollary 7.3, there is a solution

$$z(t_{i-1}) = \begin{pmatrix} h(t_{i-1}) \\ \eta(t_{i-1}) \end{pmatrix}$$

of (17.16) satisfying

$$\|z(t_i) - z(t_{i-1})\| \leq \theta_J \|q(t_i) - q(t_{i-1})\| = \theta_J(t_i - t_{i-1})\|q^1 - q^2\|. \tag{17.39}$$

Since $z(t_i)$ solves (17.15) it also solves (17.8) at $q = q(t_i)$. Hence $h(t_i)$ is a solution of (17.37) at $\hat{\lambda} = \hat{\lambda}(t_i)$ and of (17.36) at $\lambda = \lambda(t_i)$. Thus, as remarked before, $x := Zh(t_i)$ is the unique solution of (17.35) at $\lambda = \lambda(t_i)$. In our notation,

$$Zh(t_i) = P_{K(\lambda(t_i))} y. \tag{17.40}$$

Arguing similarly with the solution $z(t_{i-1})$ of (17.16) we conclude that

$$Zh(t_{i-1}) = P_{K(\lambda(t_{i-1}))} y. \tag{17.41}$$

As (17.39) implies

$$\|h(t_i) - h(t_{i-1})\| \leq \theta_J (t_i - t_{i-1})\|q^1 - q^2\|,$$

(17.40) and (17.41) yield

$$\|P_{K(\lambda(t_{i-1}))} y - P_{K(\lambda(t_i))} y\| = \|Zh(t_{i-1}) - Zh(t_i)\| \\ \leq \theta_J(t_i - t_{i-1})\|q^1 - q^2\| \|Z\|.$$

Therefore

$$\|P_{K(\lambda(t_0))} y - P_{K(\lambda(t_1))} y\| \\ \leq \sum_{i=1}^{\ell} \|P_{K(\lambda(t_{i-1}))} y - P_{K(\lambda(t_i))} y\| \\ \leq k_0 \|q^1 - q^2\| \|Z\| \\ = k_0 \|\hat{\lambda} - \hat{\lambda}'\| \|Z\| \\ = \sqrt{2} k_0 \|\lambda - \lambda'\| \|Z\|.$$

Since $\lambda(t_0) = \lambda(0) = \lambda, \lambda(t_\ell) = \lambda(1) = \lambda'$ and $k = \sqrt{2}k_0\|Z\|$, the above estimation implies (17.34). □

Now, let $K(\lambda)$ be defined by (17.33) and $\bar{x}$ be a solution of (17.4). Let the function $f(x, \mu)$ satisfy conditions (17.18) and (17.19). Again, we assume that X is a polyhedral convex set and $\alpha < l$. Let $\bar{h} \in H(\bar{\lambda})$ be a vector such that $\bar{x} = Z\bar{h}$. Then $(\bar{x}, \bar{h})$ is a solution at parameter $\lambda = \bar{\lambda}$ of the following system of linear inequalities and equalities

$$x - Zh = 0, \quad \Gamma h = \lambda, \quad h \geq 0. \tag{17.42}$$

Applying Corollary 7.3 we can find $\theta > 0$ such that for every $\lambda \in \Lambda$ there exists a solution (x, h) of (17.42) satisfying

$$\|(x, h) - (\bar{x}, \bar{h})\| \leq \theta\|\lambda - \bar{\lambda}\|.$$

This implies that $x \in K(\lambda)$ and $\|x - \bar{x}\| \leq \theta\|\lambda - \bar{\lambda}\|$. Consequently, there is a neighborhood V_1 of $\bar{\lambda}$ such that

$$K(\lambda) \cap X \neq \emptyset \quad \text{for every} \quad \lambda \in \Lambda \cap V_1. \tag{17.43}$$

Now, let C be a matrix of order $r_1 \times n$ and b be a vector from R^{r_1}, such that $X = \{x \in R^n : Cx \geq b\}$. We have

$$\begin{aligned} K(\lambda) \cap X &= \{x \in R^n : x = Zh, \; Cx \geq b, \; \Gamma h = \lambda, \; h \geq 0\} \\ &= \{x \in R^n : x = Zh, \; CZh \geq b, \; \Gamma h = \lambda, \; h \geq 0\}. \end{aligned}$$

Since this set has the same structure as the one in (17.33), taking account of (17.43) we can apply Lemma 17.4 (see also the arguments for proving it) to find a constant $k > 0$ such that

$$\|P_{K(\lambda')\cap X}y - P_{K(\lambda)\cap X}y\| \leq k\|\lambda' - \lambda\| \tag{17.44}$$

for all $y \in R^n$ and $\lambda, \lambda' \in \Lambda \cap V_1$.

Using property (17.44) instead of (17.28) one can see that Lemma 17.3 (with k_1 being replaced by k) holds for the case where $K(\lambda)$ is given by (17.33). This fact gives us the following result.

Theorem 17.4. *Let $K(\lambda)$ be defined by (17.33) and $\bar{x}$ be a solution of (17.4), where $\bar{\mu} \in M \times \Lambda$ is a given pair of parameters. If conditions (17.18) and (17.19) are satisfied, then there exist constants $k_{\bar{\mu}} > 0$ and $k_{\bar{\lambda}} > 0$, neighborhoods $\bar{U}$ of $\bar{\mu}$ and $\bar{V}$ of $\bar{\lambda}$ such that:*

(i) *For every $(\mu, \lambda) \in (M \cap \bar{U}) \times (\Lambda \cap \bar{V})$ there exists a unique solution of (17.3) in X, denoted by $x(\mu, \lambda)$;*

(ii) *For all* $(\mu', \lambda'), (\mu, \lambda) \in (M \cap \bar{U}) \times (\Lambda \cap \bar{V})$,

$$\|x(\mu', \lambda') - x(\mu, \lambda)\| \leq k_{\bar{\mu}}\|\mu' - \mu\| + k_{\bar{\lambda}}\|\lambda' - \lambda\|.$$

Theorem 17.4 can be interpreted by saying that: *"In a traffic network with locally Lipschitz, locally strongly monotone function of costs on arcs, the equilibrium arcs flow is locally unique and is a locally Lipschitz function of the perturbations of costs on arcs and of the vector of demands."*

17.5 Commentaries

The material of this chapter is taken from Yen (1995b).

Stability and sensitivity analysis is a central topic in the optimization theory (see Robinson (1979), Fiacco (1983), Bank et al. (1983), Malanowski (1987), Levitin (1994), Bonnans and Shapiro (2000), and references therein). Recently, much attention has been devoted to stability and sensitivity analysis of variational inequalities. Although the methods here resemble those used in nonlinear parametric mathematical programming, specific features of variational inequalities pose new problems. The case of PVI with a fixed constraint set is studied, for example, in Kyparisis (1988). The case of PVI whose constraint set depends on a parameter is considered, for example, in Tobin (1986), Dafermos (1988), Harker and Pang (1990), Kyparisis (1990), Yen (1995b), Domokos (1999), Kien (2001).

Chapter 18

Continuity of the Solution Map in Affine Variational Inequalities

This chapter presents a systematic study of the usc and lsc properties of the solution map in parametric AVI problems. We will follow some ideas of Chapters 10 and 11, where the usc and lsc properties of the Karush-Kuhn-Tucker point set map in parametric QP problems were investigated.

In Section 18.1 we obtain a necessary condition for the usc property of the solution map at a given point. We will show that the obtained necessary condition is not a sufficient one. Then, in the same section, we derive some sufficient conditions for the usc property of the solution map and consider several useful illustrative examples. The lsc property of the solution map is studied in Section 18.2.

18.1 USC Property of the Solution Map

Consider the following AVI problem

$$\begin{cases} \text{Find} \quad x \in \Delta(A, b) \quad \text{such that} \\ \langle Mx + q, y - x \rangle \geq 0 \quad \text{for all} \quad y \in \Delta(A, b), \end{cases} \tag{18.1}$$

which depends on the parameter $(M, A, q, b) \in R^{n \times n} \times R^{m \times n} \times R^n \times R^m$. Here $\Delta(A, b) := \{x \in R^n : Ax \geq b\}$. We will abbreviate (18.1) to AVI(M, A, q, b), and denote its solution set by Sol(M, A, q, b).

The multifunction $(M, A, q, b) \mapsto$ Sol(M, A, q, b) is called the *solution map* of (18.1) and is abbreviated to Sol$(\cdot)$. For a fixed pair $(q, b) \in R^n \times R^m$, the symbol $\mathrm{Sol}(\cdot, \cdot, q, b)$ stands for the multifunction $(M, A) \mapsto \mathrm{Sol}(M, A, q, b)$. Similarly, for a fixed pair $(M, A) \in R^{n \times n} \times R^{m \times n}$, the symbol $\mathrm{Sol}(M, A, \cdot, \cdot)$ stands for the multifunction $(q, b) \mapsto \mathrm{Sol}(M, A, q, b)$.

According to Theorem 5.3, solutions of an AVI problem can be characterized via Lagrange multipliers. Namely, $x \in R^n$ is a solution of $\mathrm{AVI}(M, A, q, b)$ if and only if there exists $\lambda \in R^m$ such that

$$Mx - A^T\lambda + q = 0, \quad Ax \geq b, \quad \lambda \geq 0, \quad \langle \lambda, Ax - b \rangle = 0. \quad (18.2)$$

Vector $\lambda \in R^m$ satisfying (18.2) is called a *Lagrange multiplier* corresponding to x.

The following theorem gives a necessary condition for the usc property of the multifunction $\mathrm{Sol}(\cdot, \cdot, q, b)$ and the solution map $\mathrm{Sol}(\cdot)$.

Theorem 18.1. *Let* $(M, A, q, b) \in R^{n \times n} \times R^{m \times n} \times R^n \times R^m$. *Suppose that the solution set* $\mathrm{Sol}(M, A, q, b)$ *is bounded. Then, the following statements are valid:*

(i) *If the multifunction* $\mathrm{Sol}(\cdot, \cdot, q, b)$ *is upper semicontinuous at* (M, A), *then*

$$\mathrm{Sol}(M, A, 0, 0) = \{0\}. \quad (18.3)$$

(ii) *If the solution map* $\mathrm{Sol}(\cdot)$ *is upper semicontinuous at* (M, A, q, b), *then* (18.3) *is valid.*

Proof. (We shall follow the proof scheme of Theorem 10.1). Clearly, if $\mathrm{Sol}(\cdot)$ is upper semicontinuous at (M, A, q, b) then the multifunction $\mathrm{Sol}(\cdot, \cdot, q, b)$ is upper semicontinuous at (M, A). Hence (i) implies (ii). It remains to prove (i).

To obtain a contradiction, suppose that $\mathrm{Sol}(M, A, q, b)$ is bounded, the multifunction $\mathrm{Sol}(\cdot, \cdot, q, b)$ is upper semicontinuous at (M, A), and

$$\mathrm{Sol}(M, A, 0, 0) \neq \{0\}. \quad (18.4)$$

Since $0 \in \mathrm{Sol}(M, A, 0, 0)$, (18.4) implies that there exists a nonzero vector $\bar{x} \in R^n$ such that $\bar{x} \in \mathrm{Sol}(M, A, 0, 0)$. Hence there exists $\bar{\lambda} \in R^m$ such that

$$M\bar{x} - A^T\bar{\lambda} = 0, \quad A\bar{x} \geq 0, \quad \bar{\lambda} \geq 0, \quad \langle \bar{\lambda}, A\bar{x} \rangle = 0. \quad (18.5)$$

For every $t \in (0, 1)$, we define

$$x_t = t^{-1}\bar{x}, \quad \lambda_t = t^{-1}\bar{\lambda}. \quad (18.6)$$

We shall show that for every $t \in (0,1)$ there exist $M_t \in R^{n \times n}$ and $A_t \in R^{m \times n}$ such that

$$M_t x_t - A_t^T \lambda_t + q = 0, \tag{18.7}$$

$$A_t x_t \geq b, \quad \lambda_t \geq 0, \tag{18.8}$$

$$\langle \lambda_t, A_t x_t - b \rangle = 0, \tag{18.9}$$

and $\|M_t - M\| \to 0$, $\|A_t - A\| \to 0$ as $t \to 0$.

We will find M_t and A_t in the following forms:

$$M_t = M + tM_0, \quad A_t = A + tA_0, \tag{18.10}$$

where the matrices $M_0 \in R^{n \times n}$ and $A_0 \in R^{m \times n}$ will be chosen so that they do not depend on t. If M_t and A_t are of the forms described in (18.10), then we have

$$
\begin{aligned}
M_t x_t - A_t^T \lambda_t + q &= t^{-1}(M + tM_0)\bar{x} - t^{-1}(A + tA_0)^T\bar{\lambda} + q \\
&= t^{-1}(M\bar{x} - A^T\bar{\lambda}) + (M_0\bar{x} - A_0^T\bar{\lambda} + q),
\end{aligned}
$$

$$A_t x_t = t^{-1}A\bar{x} + A_0\bar{x}, \quad \lambda_t = t^{-1}\bar{\lambda},$$

and

$$\langle \lambda_t, A_t x_t - b \rangle = \langle t^{-1}\bar{\lambda}, t^{-1}A\bar{x} + A_0\bar{x} - b \rangle.$$

On account of (18.5), we have

$$M_t x_t - A_t^T \lambda_t + q = M_0\bar{x} - A_0^T\bar{\lambda} + q,$$

$$A_t x_t - b = t^{-1}A\bar{x} + A_0\bar{x} - b \geq A_0\bar{x} - b, \quad \lambda_t \geq 0,$$

$$
\begin{aligned}
\langle \bar{\lambda}_t, A_t x_t - b \rangle &= \langle t^{-1}\bar{\lambda}, t^{-1}A\bar{x} + A_0\bar{x} - b \rangle \\
&= \langle t^{-1}\bar{\lambda}, A_0\bar{x} - b \rangle.
\end{aligned}
$$

It is clear that conditions (18.7)–(18.9) will be satisfied if we choose M_0 and A_0 so that

$$M_0\bar{x} - A_0^T\bar{\lambda} + q = 0, \tag{18.11}$$

$$A_0\bar{x} - b = 0. \tag{18.12}$$

Let $\bar{x} = (\bar{x}_1, \ldots, \bar{x}_n)$ and let $I = \{i : \bar{x}_i \neq 0\}$. Since $\bar{x} \neq 0$, I is nonempty. Let i_0 be any element in I. We define

$$A_0 = [c_1 \ldots c_n] \in R^{m \times n},$$

where each c_i $(1 \leq i \leq n)$ is a column with m components given by the following formula

$$c_i = \begin{cases} (\bar{x}_i)^{-1}b & \text{for } i = i_0 \\ 0 & \text{for } i \neq i_0. \end{cases}$$

We check at once that this A_0 satisfies (18.12). Similarly, we define

$$M_0 = [d_1 \dots d_n] \in R^{n \times n},$$

where each d_i $(1 \leq i \leq n)$ is a column with n components given by

$$d_i = \begin{cases} (\bar{x}_i)^{-1}(A_0^T \bar{\lambda} - q) & \text{for } i = i_0 \\ 0 & \text{for } i \neq i_0. \end{cases}$$

This M_0 satisfies (18.11). So, for these matrices M_0 and A_0, conditions (18.7)–(18.9) are satisfied. According to Theorem 5.3, we have

$$x_t \in \text{Sol}(M_t, A_t, q, b)$$

for every $t \in (0, 1)$. Since $\text{Sol}(M, A, q, b)$ is bounded, there exists a bounded open set $V \subset R^n$ such that $\text{Sol}(M, A, q, b) \subset V$. Since $\text{Sol}(\cdot, \cdot, q, b)$ is upper semicontinuous at (M, A), there exists $\delta > 0$ such that

$$\text{Sol}(M', A', q, b) \subset V$$

for all $(M', A') \in R^{n \times n} \times R^{m \times n}$ satisfying $\|(M', A') - (M, A)\| < \delta$. As $\|M_t - M\| < 2^{-1/2}\delta$ and $\|A_t - A\| < 2^{-1/2}\delta$ for all $t > 0$ small enough, we have $\text{Sol}(M_t, A_t, q, b) \subset V$ for all $t > 0$ small enough. Hence $x_t \in V$ for every $t > 0$ sufficiently small. This is impossible, because V is bounded and $\|x_t\| = t^{-1}\|\bar{x}\| \to +\infty$ as $t \to 0$. The proof is complete. □

It is easy to verify that the solution set $\text{Sol}(M, A, 0, 0)$ of the homogeneous AVI problem $\text{AVI}(M, A, 0, 0)$ is a closed cone. Condition (18.3) requires that this cone consists of only one element 0.

We can characterize condition (18.3) as follows.

Proposition 18.1 (cf. Proposition 3 in Gowda and Pang (1994a)). *Condition* (18.3) *holds if and only if for every* $(q, b) \in R^n \times R^m$ *the set* $\text{Sol}(M, A, q, b)$ *is bounded.*

Proof. Suppose that (18.3) holds. If there is a pair $(\bar{q}, \bar{b}) \in R^n \times R^m$ such that $\text{Sol}(M, A, \bar{q}, \bar{b})$ is unbounded, there exists an unbounded sequence $\{x^k\}$ such that $x^k \in \text{Sol}(M, A, \bar{q}, \bar{b})$ for all k. Without loss of generality we can assume that $\|x^k\| \neq 0$ for all k, $\|x^k\| \to \infty$ and

$\|x^k\|^{-1}x^k \rightarrow \bar{x}$ for some $\bar{x} \in R^n$ with $\|\bar{x}\| = 1$ as $k \rightarrow \infty$. Since $x^k \in \text{Sol}(M, A, \bar{q}, \bar{b})$, for any $y \in \Delta(A, \bar{b})$ we have

$$\langle Mx^k + \bar{q}, y - x^k \rangle \geq 0, \quad Ax^k \geq \bar{b}, \qquad (18.13)$$

for all k. Dividing the first inequality in (18.13) by $\|x^k\|^2$, the second inequality by $\|x^k\|$, and letting $k \rightarrow \infty$ we get

$$\langle M\bar{x}, -\bar{x} \rangle \geq 0, \qquad A\bar{x} \geq 0. \qquad (18.14)$$

Since $A(x^k + \bar{x}) = Ax^k + A\bar{x} \geq \bar{b}$, we have $x^k + \bar{x} \in \Delta(A, \bar{b})$. Substituting $x^k + \bar{x}$ for y in the first inequality in (18.13), we have $\langle Mx^k + \bar{q}, \bar{x} \rangle \geq 0$. Dividing this inequality by $\|x^k\|$ and letting $k \rightarrow \infty$ we get

$$\langle M\bar{x}, \bar{x} \rangle \geq 0. \qquad (18.15)$$

From (18.14) and (18.15) it follows that

$$\langle M\bar{x}, \bar{x} \rangle = 0. \qquad (18.16)$$

Let z be any point in $\Delta(A, 0)$. Clearly, $x^k + z \in \Delta(A, \bar{b})$. Substituting $x^k + z$ for y in the first inequality in (18.13), we have $\langle Mx^k + \bar{q}, z \rangle \geq 0$. Dividing this inequality by $\|x^k\|$ and letting $k \rightarrow \infty$ we get $\langle M\bar{x}, z \rangle \geq 0$. From this and (18.16) we deduce that $\langle M\bar{x}, z - \bar{x} \rangle \geq 0$. Since $z \in \Delta(A, 0)$ is arbitrary, we conclude that $\bar{x} \in \text{Sol}(M, A, 0, 0) \setminus \{0\}$. This contradicts our assumption that $\text{Sol}(M, A, 0, 0) = \{0\}$.

We now suppose that $\text{Sol}(M, A, q, b)$ is bounded for every $(q, b) \in R^n \times R^m$. Since the solution set $\text{Sol}(M, A, 0, 0)$ is a cone, from its boundedness we see that (18.3) is valid. □

The following example shows that condition (18.3) and the boundedness of $\text{Sol}(M, A, q, b)$ are not sufficient for the upper semicontinuity of $\text{Sol}(\cdot)$ at (M, A, q, b).

Example 18.1. Consider problem (18.1) with

$$M = \begin{bmatrix} -1 & 0 \\ 0 & -1 \end{bmatrix}, \quad A = \begin{bmatrix} 1 & 0 \\ 0 & 1 \\ 0 & -1 \end{bmatrix}, \quad q = \begin{pmatrix} 0 \\ 0 \end{pmatrix}, \quad b = \begin{pmatrix} 0 \\ 0 \\ -1 \end{pmatrix}.$$

For each $t \in (0, 1)$, we set

$$A_t = \begin{bmatrix} 1 & 0 \\ 0 & 1 \\ -t & -1 \end{bmatrix}.$$

Using Theorem 5.3, we find that

$$\text{Sol}(M, A, q, b) = \{(0,0), (0,1)\}, \quad \text{Sol}(M, A, 0, 0) = \{(0,0)\},$$

and

$$\text{Sol}(M, A_t, q, b) = \left\{ (0,0), (0,1), \left(\frac{1}{t}, 0\right), \left(\frac{t}{t^2+1}, \frac{1}{t^2+1}\right) \right\}.$$

Since $(t^{-1}, 0) \in \text{Sol}(M, A_t, q, b)$ for all $t \in (0,1)$, for any bounded open subset $V \subset R^2$ containing $\text{Sol}(M, A, q, b)$ there exists $\delta_V > 0$ such that

$$\text{Sol}(M, A_t, q, b) \setminus V \neq \emptyset$$

for every $t \in (0, \delta_V)$. Since $\|A_t - A\| \to 0$ as $t \to 0$, we conclude that $\text{Sol}(\cdot)$ is not upper semicontinuous at (M, A, q, b).

Our next goal is to find some sets of conditions which guarantee that the solution map $\text{Sol}(\cdot)$ is upper semicontinuous at a given point $(M, A, q, b) \in R^{n \times n} \times R^{m \times n} \times R^n \times R^m$.

In order to obtain some sufficient conditions for the usc property of $\text{Sol}(\cdot)$ to hold, we will pay attention to the behavior of the quadratic form $\langle Mv, v \rangle$ on the cone $\Delta(A, 0) = \{v \in R^n : Av \geq 0\}$ and to the regularity of the inequality system $Ax \geq b$.

The next proposition shows that, for a given pair $(M, A) \in R^{n \times n} \times R^{m \times n}$, for almost all $(q, b) \in R^n \times R^m$ the set $\text{Sol}(M, A, q, b)$ is bounded (may be empty).

Proposition 18.2 (cf. Lemma 1 in Oettli and Yen (1995)). *Let* $(M, A) \in R^{n \times n} \times R^{m \times n}$. *The set*

$$W = \{(q, b) \in R^n \times R^m : \text{Sol}(M, A, q, b) \text{ is bounded}\} \quad (18.17)$$

is of full Lebesgue measure in $R^n \times R^m$.

Proof. The set $\text{Sol}(M, A, q, b)$ is nonempty if and only if system (18.2) has a solution $(x, \lambda) \in R^n \times R^m$. We check at once that (18.2) has a solution if and only if there exists a subset $\alpha \subset I$, where $I := \{1, 2, \ldots, m\}$, such that the system

$$\begin{cases} Mx - A_\alpha^T \lambda_\alpha + q = 0, \\ A_\alpha x = b_\alpha, \quad \lambda_\alpha \geq 0, \\ A_{\bar\alpha} x \geq b_{\bar\alpha}, \quad \lambda_{\bar\alpha} = 0. \end{cases} \quad (18.18)$$

has a solution $(x, \lambda_\alpha, \lambda_{\bar\alpha})$, where $\bar\alpha = I \setminus \alpha$. If $\alpha = \emptyset$ (resp., $\bar\alpha = \emptyset$) then the terms indexed by α (resp., by $\bar\alpha$) are absent in (18.18). Hence

$$\text{Sol}(M, A, q, b) = \bigcup \{\text{Sol}(M, A, q, b)_\alpha : \alpha \subset I\},$$

where

$$\text{Sol}(M, A, q, b)_\alpha := \{x \in R^n : \text{ there exists } \lambda \in R^m \text{ such that}$$
$$(x, \lambda_\alpha, \lambda_{\bar{\alpha}}) \text{ is a solution of } (18.18)\}.$$

Note that the set $\text{Sol}(M, A, q, b)$ is unbounded if and only if there exists $\alpha \subset I$ such that $\text{Sol}(M, A, q, b)_\alpha$ is unbounded. We denote by $S(M, A, q, b)_\alpha$ the set of all $(x, \lambda_\alpha) \in R^n \times R^{|\alpha|}$ satisfying the system

$$Mx - A_\alpha^T \lambda_\alpha + q = 0, \quad A_\alpha x = b_\alpha,$$

where $|\alpha|$ is the number of elements of α. Let

$$\Omega_\alpha = \{(q, b) \in R^n \times R^m : S(M, A, q, b)_\alpha \text{ is unbounded}\}.$$

Obviously, if $\text{Sol}(M, A, q, b)_\alpha$ is unbounded then $S(M, A, q, b)_\alpha$ is unbounded. So, on account of (18.19), we have

$$\{(q, b) \in R^n \times R^m : \text{Sol}(M, A, q, b) \text{ is unbounded}\}$$
$$\subset \bigcup \{\Omega_\alpha : \alpha \subset I\}. \tag{18.20}$$

Clearly,

$$\Omega_\alpha = \left\{(q, b) \in R^n \times R^m : \det \widetilde{M}_\alpha = 0 \text{ and there exists}\right.$$
$$\left. (x, \lambda) \in R^n \times R^m \text{ such that } \widetilde{M}_\alpha \left(\begin{smallmatrix} x \\ \lambda_\alpha \end{smallmatrix}\right) = \left(\begin{smallmatrix} q \\ b_\alpha \end{smallmatrix}\right)\right\},$$

where

$$\widetilde{M}_\alpha = \begin{bmatrix} -M & A_\alpha^T \\ A_\alpha & 0 \end{bmatrix}.$$

If $\det \widetilde{M}_\alpha = 0$, then the image of the linear operator

$$\widetilde{M}_\alpha : R^n \times R^{|\alpha|} \longrightarrow R^n \times R^{|\alpha|}$$

corresponding to the matrix $\widetilde{M}_\alpha$ is a proper linear subspace of $R^n \times R^{|\alpha|}$. Hence Ω_α is a proper linear subspace of $R^n \times R^m$. So the set Ω_α is of Lebesgue measure 0 in $R^n \times R^m$. Therefore, from (18.20) we deduce that the set

$$\Omega := \{(q, b) \in R^n \times R^m : \text{Sol}(M, A, q, b) \text{ is unbounded}\}$$

is of Lebesgue measure 0 in $R^n \times R^m$. Since $W = (R^n \times R^m) \setminus \Omega$ by (18.17), the desired conclusion follows. $\square$

In Example 18.1, the system $Ax \geq 0$ is irregular and $\mathrm{Sol}(\cdot)$ is not upper semicontinuous (M, A, q, b). The following theorem shows that if the system $Ax \geq 0$ is regular, then condition (18.3) is necessary and sufficient for the usc property of $\mathrm{Sol}(\cdot)$.

Theorem 18.2. *Let $(M, A, q, b) \in R^{n \times n} \times R^{m \times n} \times R^n \times R^m$. Suppose that the system $Ax \geq 0$ is regular. Then, (18.3) holds if and only if for every $(q, b) \in R^n \times R^m$ the solution map $\mathrm{Sol}(\cdot)$ is upper semicontinuous at (M, A, q, b).*

Proof. Suppose that (18.3) holds. We have to prove that for every $(q, b) \in R^n \times R^m$ the solution map $\mathrm{Sol}(\cdot)$ is upper semicontinuous at (M, A, q, b). To obtain a contradiction, suppose that there is a pair $(\bar{q}, \bar{b}) \in R^n \times R^m$ such that $\mathrm{Sol}(\cdot)$ is not upper semicontinuous at $(M, A, \bar{q}, \bar{b})$. Then there exist an open set $V \subset R^n$ containing $\mathrm{Sol}(M, A, \bar{q}, \bar{b})$, a sequence $\{(M^k, A^k, q^k, b^k)\}$ converging to $(M, A, \bar{q}, \bar{b})$ in $R^{n \times n} \times R^{m \times n} \times R^n \times R^m$ and a sequence $\{x^k\}$ in R^n such that

$$x^k \in \mathrm{Sol}(M^k, A^k, q^k, b^k) \setminus V. \tag{18.21}$$

Let y be any point in $\Delta(A, \bar{b})$. Since the system $Ax \geq 0$ is regular, the system $Ax \geq \bar{b}$ is regular (see Lemma 13.2). By Lemma 13.1, there exists a subsequence $\{k'\}$ of $\{k\}$ and a sequence $\{y^{k'}\}$ in R^n converging to y such that

$$A^{k'} y^{k'} \geq b^{k'} \quad \text{for all} \ \ k'. \tag{18.22}$$

From (18.21) and (18.22) it follows that

$$A^{k'} x^{k'} \geq b^{k'}, \quad \langle M^{k'} x^{k'} + q^{k'}, y^{k'} - x^{k'} \rangle \geq 0. \tag{18.23}$$

We claim that $\{x^{k'}\}$ is bounded. If $\{x^{k'}\}$ is unbounded, then without loss of generality we can assume that $\|x^{k'}\| \neq 0$ for all k', $\|x^{k'}\| \to \infty$ and $\|x^{k'}\|^{-1} x^{k'} \to \bar{x}$ for some $\bar{x} \in R^n$ with $\|\bar{x}\| = 1$, as $k' \to \infty$.

Dividing the first inequality in (18.23) by $\|x^{k'}\|$, the second inequality in (18.23) by $\|x^{k'}\|^2$, and letting $k' \to \infty$, we get

$$A\bar{x} \geq 0, \quad \langle M\bar{x}, -\bar{x} \rangle \geq 0. \tag{18.24}$$

Let v be any point in $\Delta(A, 0)$. As $Ax \geq 0$ is regular, by Lemma 13.1 there exists a subsequence $\{k''\}$ of $\{k'\}$ and a sequence $\{v^{k''}\}$ converging to v such that

$$A^{k''} v^{k''} \geq 0 \quad \text{for all} \ \ k''. \tag{18.25}$$

By (18.23) and (18.25) we have

$$A^{k''}(x^{k''} + v^{k''}) \geq b^{k''} \quad \text{for all } k''.$$

From (18.21) it follows that

$$\langle M^{k''} x^{k''} + q^{k''}, y - x^{k''} \rangle \geq 0 \quad \text{for all} \quad y \in \Delta(A^{k''}, b^{k''}). \quad (18.26)$$

Substituting $x^{k''} + v^{k''}$ for y in (18.26), we obtain

$$\langle M^{k''} x^{k''} + q^{k''}, v^{k''} \rangle \geq 0. \quad (18.27)$$

Dividing (18.27) by $\|x^{k''}\|$ and letting $k'' \to \infty$ we get $\langle M\bar{x}, v \rangle \geq 0$. From the last inequality and (18.26) it follows that

$$\langle M\bar{x}, v - \bar{x} \rangle \geq 0, \quad A\bar{x} \geq 0. \quad (18.28)$$

Since v is arbitrary in $\Delta(A, 0)$, from (18.28) we deduce that $\bar{x} \in$ Sol$(M, A, 0, 0) \setminus \{0\}$, a contradiction. Thus the sequence $\{x^{k'}\}$ is bounded. There is no loss of generality in assuming that $x^{k'} \to \hat{x}$ as $k' \to \infty$. By (18.21),

$$\hat{x} \in \text{Sol}(M, A, \bar{q}, \bar{b}) \subset V.$$

Since $x^{k'} \in R^n \setminus V$ and V is open, we have $\hat{x} \in R^n \setminus V$, which is impossible. We have thus proved that for every $(q, b) \in R^n \times R^m$ the map Sol$(\cdot)$ is upper semicontinuous at (M, A, q, b).

Conversely, suppose that for every $(q, b) \in R^n \times R^m$ the solution map Sol$(\cdot)$ is upper semicontinuous at (M, A, q, b). By Proposition 18.1, there exists $(\bar{q}, \bar{b}) \in R^n \times R^m$ such that Sol$(M, A, \bar{q}, \bar{b})$ is bounded. Therefore, according to Theorem 18.1, condition (18.3) is satisfied. The proof is complete.   □

Theorem 18.3. *Let* $(M, A, b) \in R^{n \times n} \times R^{m \times n} \times R^m$. *Let*

$$K^- := \{v \in R^n \; : \; \langle Mv, v \rangle \leq 0, \; Av \geq 0\}.$$

If $K^- = \{0\}$ *and the system* $Ax \geq b$ *is regular then, for any* $q \in R^n$, *the solution map* Sol$(\cdot)$ *is upper semicontinuous at* (M, A, q, b).

Proof. Suppose that $K^- = \{0\}$ and the system $Ax \geq b$ is regular. Suppose that the assertion of the theorem is false. Then there exists $q \in R^n$ such that Sol$(\cdot)$ is not upper semicontinuous at (M, A, q, b). Thus there exist an open set $V \subset R^n$ containing Sol(M, A, q, b), a sequence $\{(M^k, A^k, q^k, b^k)\}$ converging to (M, A, q, b) in $R^{n \times n} \times R^{m \times n} \times R^n \times R^m$, and a sequence $\{x^k\}$ in R^n such that

$$x^k \in \text{Sol}(M^k, A^k, q^k, b^k) \setminus V \quad \text{for all } k. \tag{18.29}$$

Let $y \in \Delta(A, b)$. Since $Ax \geq b$ is regular, by Lemma 13.1 there exist a subsequence $\{k'\}$ of $\{k\}$ and a sequence $\{y^{k'}\}$ converging to y in R^n such that

$$A^{k'} y^{k'} \geq b^{k'} \quad \text{for all } k'. \tag{18.30}$$

From (18.29) and (18.30) it follows that

$$\langle M^{k'} x^{k'} + q^{k'}, y^{k'} - x^{k'} \rangle \geq 0 \quad \text{for all } k'. \tag{18.31}$$

We claim that the sequence $\{x^{k'}\}$ is bounded. Indeed, if $\{x^{k'}\}$ is unbounded then without loss of generality we can assume that $\|x^{k'}\| \neq 0$ for all k', $\|x^{k'}\| \to \infty$ and $\|x^{k'}\|^{-1} x^{k'} \to \bar{v}$ as $k' \to \infty$ for some $\bar{v} \in R^n$ with $\|\bar{v}\| = 1$. Dividing the inequalities in (18.31) by $\|x^{k'}\|^2$ and letting $k' \to \infty$, we get

$$\langle M\bar{v}, -\bar{v} \rangle \geq 0. \tag{18.32}$$

Since $Ax^{k'} \geq b^{k'}$, we have $A\bar{v} \geq 0$. Since $\bar{v} \neq 0$, from the last inequality and (18.32) we deduce that $K^- \neq \{0\}$, which is impossible. Thus the sequence $\{x^{k'}\}$ is bounded; so it has a convergent subsequence. Without loss of generality we can assume that $\{x^{k'}\}$ itself converges to some $\bar{x}$ in R^n. From (18.31) it follows that

$$\langle M\bar{x} + q, y - \bar{x} \rangle \geq 0. \tag{18.33}$$

Since y is arbitrary in $\Delta(A, b)$ and $\bar{x} \in \Delta(A, b)$, from (18.33) we deduce that $\bar{x} \in \text{Sol}(M, A, q, b) \subset V$. On the other hand, since $x^k \in \text{Sol}(M^k, A^k, q^k, b^k) \setminus V$ for every k and V is open, it follows that $\bar{x} \notin V$. We have thus arrived at a contradiction. The proof is complete. $\square$

Remark 18.1. It is a simple matter to show that if $K^- = \{0\}$ then (18.3) holds.

Corollary 18.1. *Let* $(A, b) \in R^{m \times n} \times R^m$. *Suppose that the system* $Ax \geq b$ *is regular and the set* $\Delta(A, b)$ *is bounded. Then, for any* $(M, q) \in R^{n \times n} \times R^n$, *the solution map* $\text{Sol}(\cdot)$ *is upper semicontinuous at* (M, A, q, b).

Proof. Since $Ax \geq b$ is regular, $\Delta(A, b) \neq \emptyset$. Since $\Delta(A, b)$ is bounded, we deduce that $\Delta(A, 0) = \{0\}$. Let there be given any $(M, q) \in R^{n \times n} \times R^n$. Since $K^- \subset \Delta(A, 0)$ and $\Delta(A, 0) = \{0\}$, we have $K^- = \{0\}$. By Theorem 18.2, $\text{Sol}(\cdot)$ is upper semicontinuous at (M, A, q, b). □

Corollary 18.2. *Let* $(M, A, b) \in R^{n \times n} \times R^{m \times n} \times R^m$. *Suppose that the matrix* M *is positive definite and the system* $Ax \geq b$ *is regular. Then, for every* $q \in R^n$, *the solution map* $\text{Sol}(\cdot)$ *is upper semicontinuous at* (M, A, q, b).

Proof. Since M is positive definite, we have $v^T M v > 0$ for every nonzero $v \in R^n$. Hence $K^- = \{0\}$. Applying Theorem 18.2 we see that, for every $q \in R^n$, $\text{Sol}(\cdot)$ is upper semicontinuous at (M, A, q, b). □

The next examples show that without the regularity condition imposed on the system $Ax \geq b$, the assertion of Theorem 18.2 may be false or may be true, as well.

Example 18.2. Consider problem (18.1), where $m = n = 1$, $M = [1]$, $A = [0]$, $q = 0$, and $b = 0$. It is easily seen that $Ax \geq b$ is irregular. For every $t \in (0, 1)$, let $A_t = [t^2]$ and $b_t = t$. We have

$$K^- = \{v \in R : \langle Mv, \ v \rangle \leq 0, \ Av \geq 0\} = \{v \in R : v^2 \leq 0\} = \{0\},$$

$$\text{Sol}(M, A, q, b) = \{0\}, \quad \text{Sol}(M, A_t, q, b_t) = \{t^{-1}\}.$$

Fix any bounded open set V satisfying $\text{Sol}(M, A, q, b) = \{0\} \subset V$. Since $t^{-1} \in \text{Sol}(M, A_t, q, b_t)$ for every $t \in (0, 1)$, the inclusion $\text{Sol}(M, A_t, q, b_t) \subset V$ does not hold for $t > 0$ small enough. Since $A_t \to A$, $b_t \to b$ as $t \to 0$, the multifunction $\text{Sol}(\cdot)$ cannot be upper semicontinuous at (M, A, q, b).

Example 18.3. Consider problem (18.1), where $n = 1$, $m = 2$,

$$M = [-1], \quad A = \begin{bmatrix} -1 \\ 1 \end{bmatrix}, \quad q = (0), \quad b = \begin{pmatrix} 1 \\ 0 \end{pmatrix}.$$

In this case we have

$$K^- = \{v \in R : \ -v^2 \leq 0, \ -v \geq 0, \ v \geq 0\} = \{0\},$$

$$\Delta(A, b) = \{x \in R : \ -x \geq 1, \ x \geq 0\} = \emptyset.$$

Obviously, the system $Ax \geq b$ is irregular. Since $\text{Sol}(M, A, q, b) \subset \Delta(A, b)$, we conclude that $\text{Sol}(M, A, q, b) = \emptyset$. Moreover, we can find $\delta > 0$ such that for any $(A', b') \in R^{m \times n} \times R^m$ with $\|(A', b') - (A, b)\| \leq \delta$, $\Delta(A', b') = \emptyset$. Hence the multifunction $\text{Sol}(\cdot)$ is upper semicontinuous at (M, A, q, b).

Lemma 18.1 (cf. Lemma 3 in Robinson (1977) and Lemma 10.2 in this book). *Let* $(A, b) \in R^{m \times n} \times R^m$. *Let*

$$\Lambda_0[A] = \{\lambda \in R^m \ : \ A^T \lambda = 0, \ \lambda \geq 0\}.$$

Then, the system $Ax \geq b$ *is regular if and only if* $\langle \lambda, b \rangle < 0$ *for every* $\lambda \in \Lambda_0[A] \setminus \{0\}$.

Proof. Suppose that $Ax \geq b$ is regular, that is there exists $x_0 \in R^n$ such that $Ax_0 > b$. Let $y_0 = Ax_0 - b > 0$. Let $\lambda \in \Lambda_0[A] \setminus \{0\}$, that is $A^T \lambda = 0$, $\lambda \geq 0$ and $\lambda \neq 0$. We have

$$\langle \lambda, b \rangle = \langle \lambda, Ax_0 \rangle - \langle \lambda, y_0 \rangle = \langle A^T \lambda, x_0 \rangle - \langle \lambda, y_0 \rangle = -\langle \lambda, y_0 \rangle < 0.$$

Conversely, suppose that $\langle \lambda, b \rangle < 0$ for every $\lambda \in \Lambda_0[A] \setminus \{0\}$. If $Ax \geq b$ is irregular then there exists a sequence $\{b^k\}$ converging to b in R^m such that the system $Ax \geq b^k$ has no solutions for all k. According to Theorem 22.1 from Rockafellar (1970), there exists a sequence $\{\lambda^k\}$ in R^m such that

$$A^T \lambda^k = 0, \quad \lambda^k \geq 0, \quad \langle \lambda^k, b^k \rangle > 0. \tag{18.34}$$

Since $\lambda^k \neq 0$, by the homogeneity of the inequalities in (18.34) we can assume that $\|\lambda^k\| = 1$ for every k. Thus the sequence $\{\lambda^k\}$ has a convergent subsequence. We can suppose that the sequence $\{\lambda^k\}$ itself converges to some $\bar{\lambda}$ with $\|\bar{\lambda}\| = 1$. Taking the limits in (18.34) as $k \to \infty$ we get

$$A^T \bar{\lambda} = 0, \quad \bar{\lambda} \geq 0, \quad \langle \bar{\lambda}, b \rangle \geq 0.$$

Hence $\bar{\lambda} \in \Lambda_0[A] \setminus \{0\}$ and $\langle \bar{\lambda}, b \rangle \geq 0$. We have arrived at contradiction. $\square$

Theorem 18.4. *Let* $(M, A, b) \in R^{n \times n} \times R^{m \times n} \times R^m$. *Let*

$$K^+ := \{v \in R^n \ : \ \langle Mv, v \rangle \geq 0, \ Av \geq 0\}.$$

If $K^+ = \{0\}$ *and the system* $Ax \geq -b$ *is regular then, for any* $q \in R^n$, *the solution map* $\text{Sol}(\cdot)$ *is upper semicontinuous at* (M, A, q, b).

Proof. Suppose that $K^+ = \{0\}$ and the system $Ax \geq -b$ is regular. Suppose that the assertion of the theorem is false. Then there exists

$q \in R^n$ such that $\mathrm{Sol}(\cdot)$ is not upper semicontinuous at (M, A, q, b). Thus there exist an open set $V \subset R^n$ containing $\mathrm{Sol}(M, A, q, b)$, a sequence $\{(M^k, A^k, q^k, b^k)\}$ converging to (M, A, q, b) in $R^{n \times n} \times R^{m \times n} \times R^n \times R^m$, and a sequence $\{x^k\}$ in R^n such that

$$x^k \in \mathrm{Sol}(M^k, A^k, q^k, b^k) \setminus V \quad \text{for all } k. \tag{18.35}$$

Since $x^k \in \mathrm{Sol}(M^k, A^k, q^k, b^k)$, there exists λ^k such that

$$M^k x^k - (A^k)^T \lambda^k + q^k = 0, \tag{18.36}$$

$$A^k x^k \geq b^k, \quad \lambda^k \geq 0, \quad \langle \lambda^k, A^k x^k - b^k \rangle = 0. \tag{18.37}$$

We claim that the sequence $\{(x^k, \lambda^k)\}$ is unbounded. Indeed, if $\{(x^k, \lambda^k)\}$ is bounded then $\{x^k\}$ and $\{\lambda^k\}$ are bounded sequences, so each of them has a convergent subsequence. Without loss of generality we can assume that $x^k \to \bar{x}$, $\lambda^k \to \bar{\lambda}$ as $k \to \infty$, where $\bar{x} \in R^n$ and $\bar{\lambda} \in R^m$. From (18.36) and (18.37) we deduce that

$$M\bar{x} - A^T \bar{\lambda} + q = 0,$$

$$A\bar{x} \geq b, \quad \bar{\lambda} \geq 0, \quad \langle \bar{\lambda}, A\bar{x} - b \rangle = 0.$$

Hence $\bar{x} \in \mathrm{Sol}(M, A, q, b) \subset V$. This is impossible because V is open and $x^k \in \mathrm{Sol}(M^k, A^k, q^k, b^k) \setminus V$ for all k. Thus $\{(x^k, \lambda^k)\}$ is unbounded.

There is no loss of generality in assuming that $\|(x^k, \lambda^k)\| \neq 0$ for all k, $\|(x^k, \lambda^k)\| \to \infty$,

$$\|(x^k, \lambda^k)\|^{-1}(x^k, \lambda^k) \to (\bar{v}, \bar{\lambda}) \in R^n \times R^m, \tag{18.38}$$

where $\|(\bar{v}, \bar{\lambda})\| = 1$. Dividing the equality in (18.36) and the first two inequalities in (18.37) by $\|(x^k, \lambda^k)\|$, the equality in (18.37) by $\|(x^k, \lambda^k)\|^2$ and letting $k \to \infty$ we get

$$M\bar{v} - A^T\bar{\lambda} = 0, \quad A\bar{v} \geq 0, \quad \bar{\lambda} \geq 0, \quad \langle \bar{\lambda}, A\bar{v} \rangle = 0. \tag{18.39}$$

From (18.39) it follows that $\langle M\bar{v}, \bar{v} \rangle = 0$ and $A\bar{v} \geq 0$. Thus $\bar{v} \in K^+$. As $K^+ = \{0\}$, we have $\bar{v} = 0$. By (18.38), $\|\bar{\lambda}\| = 1$. Since $v = 0$, from (18.39) we deduce that $\bar{\lambda} \in \Lambda_0[A]$. Since $Ax \geq -b$ is regular, by Lemma 18.1 we have $\langle \bar{\lambda}, -b \rangle < 0$. From (18.36) and (18.37) it follows that

$$\langle M^k x^k + q^k, x^k \rangle = \langle \lambda^k, b^k \rangle. \tag{18.40}$$

If there exists an integer k_0 such that $\langle \lambda^k, b^k \rangle \leq 0$ for all $k \geq k_0$ then without loss of generality we can suppose that $\langle \lambda^k, b^k \rangle \leq 0$ for all k. Dividing the last inequality by $\|(x^k, \lambda^k)\|$ and letting $k \to \infty$ we have

$$\langle \bar{\lambda}, b \rangle \leq 0,$$

which contradicts the fact that $\langle \bar{\lambda}, -b \rangle < 0$. So there exists a subsequence $\{k'\}$ of $\{k\}$ such that $\langle \lambda^{k'}, b^{k'} \rangle > 0$. From this and (18.40) we deduce that

$$\langle M^{k'} x^{k'} + q^{k'}, x^{k'} \rangle = \langle \lambda^{k'}, b^{k'} \rangle, \quad \langle M^{k'} x^{k'} + q^{k'}, x^{k'} \rangle > 0 \quad (18.41)$$

for all k'. If $\{x^{k'}\}$ is bounded then, dividing the equality in (18.41) by $\|(x^{k'}, \lambda^{k'})\|$ and letting $k' \to \infty$ we obtain $\langle \bar{\lambda}, b \rangle = 0$, which contradicts the fact that $\langle \bar{\lambda}, -b \rangle < 0$. Thus $\{x^{k'}\}$ is unbounded. Without loss of generality we can assume that $\|x^{k'}\| \neq 0$ for all k' and the sequence $\|x^{k'}\|^{-1} x^{k'}$ itself converges to some $\hat{v} \in R^n$ with $\|\hat{v}\| = 1$. Dividing the inequality in (18.41) by $\|x^{k'}\|^2$ and letting $k' \to \infty$ we get $\langle M\hat{v}, \hat{v} \rangle \geq 0$. By (18.37), we have $A^{k'} x^{k'} \geq b^{k'}$ for all k'. Dividing the last inequality by $\|x^{k'}\|$ and letting $k' \to \infty$ we get $A\hat{v} \geq 0$. From this and the inequality $\langle M\hat{v}, \hat{v} \rangle \geq 0$ we see that $\hat{v} \in K^+ \setminus \{0\}$, contrary to the assumption $K^+ = \{0\}$. The proof is complete. $\square$

Remark 18.2. It is easily verified that if $K^+ = \{0\}$ then (18.3) holds.

Corollary 18.3. Let $(A, b) \in R^{m \times n} \times R^m$. Suppose that the system $Ax \geq -b$ is regular and $\Delta(A, b)$ is nonempty and bounded. Then, for any $(M, q) \in R^{n \times n} \times R^n$ the solution map $\mathrm{Sol}(\cdot)$ is upper semicontinuous at (M, A, q, b).

Proof. Let $(M, q) \in R^{n \times n} \times R^n$ be given arbitrarily. Since $\Delta(A, b)$ is nonempty and bounded, $\Delta(A, 0) = \{0\}$. As $K^+ \subset \Delta(A, 0)$, we have $K^+ = \{0\}$. Applying Theorem 18.4 we conclude that the solution map $\mathrm{Sol}(\cdot)$ is upper semicontinuous at (M, A, q, b). $\square$

Corollary 18.4. Let $(M, A, b) \in R^{n \times n} \times R^{m \times n} \times R^m$. If the matrix M is negative definite and the system $Ax \geq -b$ is regular then, for any $q \in R^n$, the solution map $\mathrm{Sol}(\cdot)$ is upper semicontinuous at (M, A, q, b).

Proof. Since M is negative definite, we have $v^T M v < 0$ for any nonzero vector $v \in R^n$. Since $K^+ \subset \{v \in R^n : \langle Mv, v \rangle \geq 0\}$, we deduce that $K^+ = \{0\}$. Applying Theorem 18.4 we see that, for every $q \in R^n$, $\mathrm{Sol}(\cdot)$ is upper semicontinuous at (M, A, q, b). $\square$

Corollary 18.5. *Let* $(A, b) \in R^{m \times n} \times R^m$. *Suppose that* $\Delta(A, 0) = \{0\}$ *and* $\langle \lambda, b \rangle \neq 0$ *for every nonzero* λ *satisfying* $A^T \lambda = 0, \lambda \geq 0$. *Then, for any* $(M, q) \in R^{n \times n} \times R^n$, *Sol*$(\cdot)$ *is upper semicontinuous at* (M, A, q, b).

Proof. Let $(M, q) \in R^{n \times n} \times R^n$ be given arbitrarily. Since $\Delta(A, 0) = \{0\}$, we have $K^- = K^+ = \{0\}$. Since $\Lambda[A] = \{\lambda \in R^m : A^T \lambda = 0, \lambda \geq 0\}$, we see that $\Lambda[A]$ is a pointed convex cone. From the assumption that $\langle \lambda, b \rangle \neq 0$ for every nonzero $\lambda \in \Lambda[A]$ we deduce that one and only one of the following two cases occurs:

(i) $\langle \lambda, b \rangle > 0$ for every $\lambda \in \Lambda[A] \setminus \{0\}$;

(ii) $\langle \lambda, b \rangle < 0$ for every $\lambda \in \Lambda[A] \setminus \{0\}$.

In case (i), the system $Ax \geq b$ is regular by Lemma 18.1. Since $K^- = \{0\}$, the desired conclusion follows from Theorem 18.2. In case (ii), $Ax \geq -b$ is regular by Lemma 18.1. Since $K^+ = \{0\}$, the assertion follows from Theorem 18.4. $\square$

The following examples show that without the regularity of the system $Ax \geq -b$, the assertion in Theorem 18.4 may be true or may be false, as well.

Example 18.4. Consider problem (18.1), where $n = 1$, $m = 2$,

$$M = [1], \quad A = \begin{bmatrix} -1 \\ 1 \end{bmatrix}, \quad q = (0), \quad b = \begin{pmatrix} -1 \\ 0 \end{pmatrix}.$$

We can check at once that $K^- = K^+ = \{0\}$, $\Delta(A, b) = \{x \in R : 0 \leq x \leq 1\}$, and $\Delta(A, -b) = \emptyset$. Note that the system $Ax \geq b$ is regular and the system $Ax \geq -b$ is irregular. The usc property of Sol$(\cdot)$ at (M, A, q, b) follows from Theorem 18.2.

Example 18.5. Consider problem (18.1), where $n = 2$, $m = 3$ and (M, A, q, b) is defined as in Example 18.1. For this problem we find that $K^+ = \{0\}$ and $\Delta(A, -b) = \emptyset$. In particular, the system $Ax \geq -b$ is irregular. As it has been shown in Example 18.1, the solution map Sol$(\cdot)$ is not upper semicontinuous at (M, A, q, b).

18.2 LSC Property of the Solution Map

In this section we will find necessary and sufficient conditions for the lower semicontinuity of the solution map in parametric AVI problems.

Theorem 18.5. Let $(M, A, c, b) \in R^{n \times n} \times R^{m \times n} \times R^n \times R^m$. If the multifunction $\mathrm{Sol}(M, A, \cdot, \cdot)$ is lower semicontinuous at (q, b) then

(a) the set $\mathrm{Sol}(M, A, q, b)$ is finite, and

(b) the system $Ax \geq b$ is regular.

Proof. We will omit the easy proof of (b). To prove (a), for every index set $I \subset \{1, \cdots, m\}$ we define a matrix $S_I \in R^{(n+|I|) \times (n+|I|)}$, where $|I|$ is the number of elements of I, by setting

$$S_I := \begin{bmatrix} M & -A_I^T \\ A_I & O \end{bmatrix}$$

(If $I = \emptyset$ then we set $S_I = M$). Let

$$P_I = \Big\{ (u, v) \in R^n \times R^m : \binom{u}{v_I} = S_I \binom{x}{\lambda_I} \\ \text{for some } (x, \lambda) \in R^n \times R^m \Big\},$$

and

$$\mathcal{P} = \bigcup \{ \mathcal{P}_I : I \subset \{1, \cdots, m\}, \ \det S_I = 0 \}.$$

If $\det S_I = 0$ then $\mathcal{P}_I$ is a proper linear subspace of $R^n \times R^m$. Hence, by the Baire Lemma (see Brezis (1987)), $\mathcal{P}$ is nowhere dense in $R^n \times R^m$. Then there exists a sequence $\{(q^k, b^k)\}$ converging to (q, b) in $R^n \times R^m$ such that $(-q^k, b^k) \notin \mathcal{P}$ for all k.

Fix any $\bar{x} \in \mathrm{Sol}(M, A, q, b)$. As the multifunction $\mathrm{Sol}(M, A, \cdot, \cdot)$ is lower semicontinuous at (q, b), there exist a subsequence $\{(q^{k_l}, b^{k_l})\}$ of $\{(q^k, b^k)\}$ and a sequence $\{x^{k_l}\}$ converging in R^n to $\bar{x}$ such that $x^{k_l} \in \mathrm{Sol}(M, A, q^{k_l}, b^{k_l})$ for all k_l. Since $x^{k_l} \in \mathrm{Sol}(M, A, q^{k_l}, b^{k_l})$, there exists $\lambda^{k_l} \in R^m$ such that

$$\begin{cases} Mx^{k_l} - A^T \lambda^{k_l} + q^{k_l} = 0, \\ Ax^{k_l} \geq b^{k_l}, \quad \lambda^{k_l} \geq 0, \\ (\lambda^{k_l})^T (Ax^{k_l} - b^{k_l}) = 0. \end{cases} \tag{18.42}$$

For every k_l, let $I_{k_l} = \{i \in \{1 \ldots, m\} : \lambda_i^{k_l} > 0\}$. It is clear that there must exists an index set $I \subset \{1, \cdots, m\}$ such that $I_{k_l} = I$ for infinitely many k_l. Without loss of generality, we may assume that $I_{k_l} = I$ for all k_l. By (18.42), we have

$$Mx^{k_l} - A_I^T \lambda_I^{k_l} + q^{k_l} = 0, \quad A_I x^{k_l} = b_I^{k_l},$$

or, equivalently,

$$S_I \begin{pmatrix} x^{k_l} \\ \lambda_I^{k_l} \end{pmatrix} = \begin{pmatrix} -q^{k_l} \\ b_I^{k_l} \end{pmatrix}. \tag{18.43}$$

We claim that S_I is nonsingular. Indeed, if $\det S_I = 0$ then, by (18.43) and by the definitions of $\mathcal{P}_I$ and $\mathcal{P}$, we have

$$(-q^{k_l}, b^{k_l}) \in \mathcal{P}_I \subset \mathcal{P},$$

which is impossible because $(-q^k, b^k) \notin \mathcal{P}$ for all k. So S_I is non-singular. From (18.43) it follows that

$$\begin{pmatrix} x^{k_l} \\ \lambda_I^{k_l} \end{pmatrix} = S_I^{-1} \begin{pmatrix} -q^{k_l} \\ b_I^{k_l} \end{pmatrix}.$$

Letting $l \to \infty$, we get

$$\lim_{l \to \infty} \begin{pmatrix} x^{k_l} \\ \lambda_I^{k_l} \end{pmatrix} = S_I^{-1} \begin{pmatrix} -q \\ b_I \end{pmatrix}. \tag{18.44}$$

If $I = \emptyset$ then (18.44) has the following form

$$\lim_{l \to \infty} x^{k_l} = M^{-1}(-q).$$

By (18.44), the sequence $\{\lambda_I^{k_l}\}$ must converge to some $\lambda_I \geq 0$ in $R^{|I|}$. As $x^{k_l} \to \bar{x}$, from (18.44) we obtain

$$\begin{pmatrix} \bar{x} \\ \lambda_I \end{pmatrix} = S_I^{-1} \begin{pmatrix} -q \\ b_I \end{pmatrix}. \tag{18.45}$$

Let

$$Z = \{(x, \lambda) \in R^n \times R^m : \quad \text{there exists such } J \subset \{1, \cdots, m\} \\ \text{that } \det S_J \neq 0 \text{ and } \begin{pmatrix} x \\ \lambda_J \end{pmatrix} = S_J^{-1} \begin{pmatrix} -q \\ b_J \end{pmatrix}\}$$

and

$$X = \{x \in R^n : \text{ there exists } \lambda \in R^m \text{ such that } (x, \lambda) \in Z\}.$$

Similarly as in the proof of Theorem 11.3, X is a finite set. From (18.45) we conclude that $\bar{x} \in X$. We have thus proved that

$$\text{Sol}(M, A, q, b) \subset X.$$

In particular, $\text{Sol}(M, A, q, b)$ is a finite set. The proof is complete.
$\square$

The following example shows that, in general, the above conditions (a) and (b) are not sufficient for the lsc property of $\mathrm{Sol}(M, A, \cdot, \cdot)$ at (q, b).

Example 18.6. Consider the AVI problem (18.1) in which $n = 2$, $m = 3$, and

$$M = \begin{bmatrix} 1 & 0 \\ 0 & -2 \end{bmatrix}, \quad A = \begin{bmatrix} 1 & 0 \\ 0 & 1 \\ -1 & -1 \end{bmatrix}, \quad q = \begin{pmatrix} -1 \\ 0 \end{pmatrix}, \quad b = \begin{pmatrix} 0 \\ 0 \\ -2 \end{pmatrix}.$$

For every $\varepsilon > 0$, we set $q(\varepsilon) = (-1, -\varepsilon)$. We can perform some computations to show that

$$\mathrm{Sol}(M, A, q, b) = \{(0, 2),\ (1, 0)\}$$

and $\mathrm{Sol}(M, A, q(\varepsilon), b) = \{(0, 2)\}$ for every $\varepsilon > 0$. It is clear that the system $Ax \geq b$ is regular. Let

$$V = \left\{ (x_1, x_2) \in R^2 : \frac{1}{2} < x_1 < \frac{3}{2},\ -1 < x_2 < 1 \right\}.$$

Since $\mathrm{Sol}(M, A, q, b) \cap V = \{(1, 0)\}$ and $\mathrm{Sol}(M, A, q(\varepsilon), b) \cap V = \emptyset$ for every $\varepsilon > 0$, we conclude the multifunction $\mathrm{Sol}(M, A, \cdot, \cdot)$ is not lower semicontinuous at (q, b).

Let $(M, A, q, b) \in R^{n \times n} \times R^{m \times n} \times R^n \times R^m$. Let $x \in \mathrm{Sol}(M, A, q, b)$ and let $\lambda \in R^m$ be a Lagrange multiplier corresponding to x. Let $I = \{1, 2, \ldots, m\}$, and let K and J be defined, respectively, by (11.14) and (11.15). We set $I_0 = K \cup J$.

The following theorem gives a sufficient condition for the lsc property of the multifunction $\mathrm{Sol}(M, A, \cdot, \cdot)$ at a given point. By definition (see (Cottle et al. 1992)), a square matrix is called a *P-matrix* if the determinant of each of its principal submatrices is positive.

Theorem 18.7. *Let* $(M, A, q, b) \in R^{n \times n} \times R^{m \times n} \times R^n \times R^m$. *Suppose that*

(i) *the set* $\mathrm{Sol}(M, A, q, b)$ *is finite, nonempty,*

(ii) *the system* $Ax \geq b$ *is regular,*

and suppose that for every $x \in \mathrm{Sol}(M, A, q, b)$ *there exists a Lagrange multiplier* λ *corresponding to* x *such that at least one of the following conditions holds:*

(c1) $v^T M v \geq 0$ *for every* $v \in R^n$ *with* $A_{I_0} v \geq 0$ *and* $(Mx+q)^T v = 0$,

(c2) $J = K = \emptyset$,

(c3) $J = \emptyset$, $K \neq \emptyset$, *and the system* $\{A_i : i \in K\}$ *is linearly independent,*

(c4) $J \neq \emptyset$, $K = \emptyset$, M *is nonsingular and* $A_J M^{-1} A_J^T$ *is a P-matrix,*

where K and J are defined via (x, λ) by (11.14) and (11.15). Then, the multifunction $\text{Sol}(M, A, \cdot, \cdot)$ *is lower semicontinuous at* (q, b).

Proof. Since $\text{Sol}(M, A, q, b)$ is nonempty, in order to prove that $\text{Sol}(M, A, \cdot, \cdot)$ is lower semicontinuous at (q, b) we only need to show that, for any $x \in \text{Sol}(M, A, q, b)$ and for any open neighborhood V_x of x, there exists $\delta > 0$ such that

$$\text{Sol}(M, A, q', b') \cap V_x \neq \emptyset \qquad (18.46)$$

for every $(q', b') \in R^n \times R^m$ satisfying $\|(q', b') - (q, b)\| < \delta$.

Let $x \in \text{Sol}(M, A, q, b)$ and let V_x be an open neighborhood of x. By our assumptions, there exists a Lagrange multiplier λ corresponding to x such that at least one of the four conditions (c1)-(c4) holds.

Consider the case where (c1) holds. Since $\text{Sol}(M, A, q, b)$ is finite by (i), x is an isolated solution of (18.1). By Corollary 10 in Gowda and Pang (1994b), from our assumptions it follows that there exists $\delta > 0$ such that (18.46) is valid for every (q', b') satisfying $\|(q', b') - (q, b)\| < \delta$.

Analysis similar to that in the proof of Theorem 11.4 shows that if one of the conditions (c2)–(c4) holds then we can find $\delta > 0$ such that (18.46) is valid for every (q', b') satisfying $\|(q', b') - (q, b)\| < \delta$.

So we can conclude that the multifunction $\text{Sol}(M, A, \cdot, \cdot)$ is lower semicontinuous at (q, b). $\square$

It is interesting to see how the conditions (c1)–(c4) in Theorem 18.7 can be verified for concrete AVI problems.

Writing the necessary optimality conditions for the QP problems in Examples 11.3–11.5 as AVI problems we obtain the following examples.

Example 18.7. Consider problem (18.1) with $n = 2$, $m = 2$,

$$M = \begin{bmatrix} 1 & 0 \\ 0 & -1 \end{bmatrix}, \quad A = \begin{bmatrix} 1 & -2 \\ 1 & 2 \end{bmatrix}, \quad q = \begin{pmatrix} -1 \\ 0 \end{pmatrix}, \quad b = \begin{pmatrix} 0 \\ 0 \end{pmatrix}.$$

We can show that $\text{Sol}(M, A, q, b) = \{\bar{x}, \hat{x}, \tilde{x}\}$, where $\bar{x}$, $\hat{x}$, $\tilde{x}$ are the same as in Example 11.3. Note that $\tilde{\lambda} := (0, 0)$ is a Lagrange multiplier corresponding to $\tilde{x}$. We observe that conditions (i) and (ii) in Theorem 18.7 are satisfied and, for each point $x \in \text{Sol}(M, A, q, b)$, either (c1) or (c2) is satisfied. More precisely, if $x = \bar{x}$ or $x = \hat{x}$ then (c1) is satisfied; if $x = \tilde{x}$ then (c2) is satisfied. By Theorem 18.7, the multifunction $\text{Sol}(M, A, \cdot, \cdot)$ is lower semicontinuous at (q, b).

Example 18.8. Consider problem (18.1) with $n = 2$, $m = 3$,

$$M = \begin{bmatrix} 1 & 0 \\ 0 & -1 \end{bmatrix}, \quad A = \begin{bmatrix} 1 & -2 \\ 1 & 2 \\ 1 & 0 \end{bmatrix}, \quad q = \begin{pmatrix} -1 \\ 0 \end{pmatrix}, \quad b = \begin{pmatrix} 0 \\ 0 \\ 1 \end{pmatrix}.$$

It is easy to verify that $\text{Sol}(M, A, q, b) = \{\bar{x}, \hat{x}, \tilde{x}\}$, where $\bar{x}$, $\hat{x}$, $\tilde{x}$ are the same as in the preceding example. Note that $\tilde{\lambda} := (0, 0, 0)$ is a Lagrange multiplier corresponding to $\tilde{x}$. For $x = \bar{x}$ and $x = \hat{x}$, assumption (c1) is satisfied. For the pair $(\tilde{x}, \tilde{\lambda})$, we have $K = \emptyset$, $J = \{3\}$. Since $A_J = (1\ 0)$, $M^{-1} = M$, we get $A_J M^{-1} A_J^T = 1$. Thus (c4) is satisfied. By Theorem 18.7, $\text{Sol}(M, A, \cdot, \cdot)$ is lower semicontinuous at (q, b).

Example 18.9. Consider problem (18.1) with $n = 2$, $m = 3$,

$$M = \begin{bmatrix} 1 & 0 \\ 0 & -1 \end{bmatrix}, \quad A = \begin{bmatrix} 1 & -2 \\ 1 & 2 \\ 1 & 0 \end{bmatrix}, \quad q = \begin{pmatrix} -1 \\ 0 \end{pmatrix}, \quad b = \begin{pmatrix} 0 \\ 0 \\ 2 \end{pmatrix}.$$

We can show that $\text{Sol}(M, A, q, b) = \{\bar{x}, \hat{x}, \tilde{x}\}$, where $\bar{x} = (2, -1)$, $\hat{x} = (2, 1)$ and $\tilde{x} = (2, 0)$. Note that $\tilde{\lambda} := (0, 0, 1)$ is a Lagrange multiplier corresponding to $\tilde{x}$. For $x = \bar{x}$ and $x = \hat{x}$, condition (c1) is satisfied. For the pair $(\tilde{x}, \tilde{\lambda})$, we have $K = \{3\}$, $J = \emptyset$. Since

$$\{A_i : i \in K\} = \{A_3\} = \{(1\ 0)\},$$

assumption (c3) is satisfied. According to Theorem 18.7, $S(D, A, \cdot, \cdot)$ is lower semicontinuous at (q, b).

Let $(M, A, q, b) \in R^{n \times n} \times R^{m \times n} \times R^n \times R^m$. Let $x \in \text{Sol}(M, A, q, b)$ and let $\lambda \in R^m$ be a Lagrange multiplier corresponding to x. We define K and J by (11.14) and (11.15), respectively. Consider the case where both the sets K and J are nonempty. If the matrix

$$Q_K = \begin{bmatrix} M & -A_K^T \\ A_K & 0 \end{bmatrix} \in R^{(n+|K|) \times (n+|K|)}$$

is nonsingular, then we denote by S_J the Schur complement of Q_K in the following matrix

$$\begin{bmatrix} M & -A_K^T & -A_J^T \\ A_K & 0 & 0 \\ A_J & 0 & 0 \end{bmatrix} \in R^{(n+|K|+|J|)\times(n+|K|+|J|)}.$$

This means that

$$S_J = [A_J\ 0]Q_K^{-1}[A_J\ 0]^T.$$

Since M is not assumed to be symmetric, S_J is not necessarily a symmetric matrix. Consider the following condition:

(c5) $J \neq \emptyset$, $K \neq \emptyset$, the system $\{A_i : i \in K\}$ is linearly independent, $v^T M v \neq 0$ for every nonzero vector v satisfying $A_K v = 0$, and S_J is a $P-$matrix.

It can be shown that if $J \neq \emptyset$, $K \neq \emptyset$, the system $\{A_i : i \in K\}$ is linearly independent, and $v^T M v \neq 0$ for every nonzero vector v satisfying $A_K v = 0$, then Q_K is nonsingular.

It can be proved that the assertion of Theorem 18.7 remains valid if instead of (c1)–(c4) we use (c1)–(c3) and (c5).

18.3 Commentaries

The material of this chapter is taken from Lee et al. (2002b-d).

As it has been noted in Chapter 5, the affine variational inequality problem is a natural and important extension of the linear complementarity problem. Both of the problems are closely related to the Karush-Kuhn-Tucker conditions in quadratic programming.

Various continuity properties of the solution maps in parametric affine variational inequality problems and parametric linear complementarity problems have been investigated (see Robinson (1979, 1981), Bank et al. (1982), Jansen and Tijs (1987), Cottle et al. (1992), Gowda (1992), Gowda and Pang (1994a), Oettli and Yen (1995), Gowda and Sznajder (1996), and the references therein).

References

1. J.-P. Aubin (1984): Lipschitz behavior of solutions to convex minimization problems, *Mathematics of Operations Research*, **9**, 87–111.

2. J.-P. Aubin and H. Frankowska (1990): *Set-Valued Analysis*, Birkhäuser, Berlin.

3. A. Auslender and R. Cominetti (1990), First- and second-order sensitivity analysis of nonlinear programs under directional constraint qualification conditions, *Optimization*, **21**, 351–363.

4. A. Auslender and P. Coutat (1996): Sensitivity analysis for generalized linear-quadratic problems, *Journal of Optimization Theory and Applications*, **88**, 541–559.

5. A. Auslender and M. Teboulle (2003): *Asymptotic Cones and Functions in Optimization and Variational Inequalities*, Springer, New York, 2003.

6. H. Attouch, H. and R.J.-B. Wets (1993): Quantitative stability of variational systems, II. A Framework for nonlinear conditioning, *SIAM Journal on Optimization*, **3**, 359–381.

7. B. Bank, J. Guddat, D. Klatte, B. Kummer and K. Tammer (1982): *Non-Linear Parametric Optimization*, Akademie-Verlag, Berlin.

8. B. Bank and R. Hansel (1984): Stability of mixed-integer quadratic programming problems, *Mathematical Programming Study*, **21**, 1–17.

9. E. M. Bednarczuk (1995), Berge-type theorems for vector optimization problems, *Optimization*, **32**, 373–384.

10. J. Benoist (1998): Connectedness of the efficient set for strictly quasiconcave sets, *Journal of Optimization Theory and Applications*, **96**, 627–654.

11. J. Benoist (2001): Contractibility of the efficient set in strictly quasiconcave vector maximization, *Journal of Optimization Theory and Applications*, **110**, 325–336.

12. M. J. Best and N. Chakravarti (1990): Stability of linearly constrained convex quadratic programs, *Journal of Optimization Theory and Applications*, **64**, 43–53.

13. M. J. Best and B. Ding (1995): On the continuity of minimum in parametric quadratic programs, *Journal of Optimization Theory and Applications*, **86**, 245–250.

14. E. Blum and W. Oettli (1972): Direct proof of the existence theorem for quadratic programming, *Operations Research*, **20**, 165–167.

15. J. F. Bonnans and A. Shapiro (1998): Optimization problems with perturbations: A guided tour, *SIAM Reviews* **40**, 228–264.

16. J. F. Bonnans and A. Shapiro (2000): *Perturbation Analysis of Optimization Problems*, Springer-Verlag, New York.

17. H. Brezis (1987): *Analyse fonctionnelle*, 2^e tirage, Masson, Paris.

18. G.-Y. Chen and N. D. Yen (1993): On the variational inequality model for network equilibrium, Preprint 3.196, Gruppo di Ottimizzazione e Ricerca Operativa, University of Pisa.

19. E. U. Choo and D. R. Atkins (1982): Bicriteria linear fractional programming, *Journal of Optimization Theory and Applications*, **36**, 203–220.

20. E. U. Choo and D. R. Atkins (1983): Connectedness in multiple linear fractional programming, *Management Science*, **29**, 250–255.

21. E. U. Choo, S. Schaible and K. P. Chew (1985): Connectedness of the efficient set in three-criteria quasiconcave programming, *Cahiers du Centre d'Etudes de Recherche Opérationnelle*, **27**, 213–220.

22. F. H. Clarke (1975): Generalized gradients and applications, *Transactions of the Americal Mathematical Society*, **205**, 247–262.

23. F. H. Clarke (1983): *Optimization and Nonsmooth Analysis*, John Wiley & Sons, New York.

24. L. Contesse (1980): Une caractérisation complète des minima locaux en programmation quadratique, *Numerische Mathematik*, **34**, 315–332.

25. R. W. Cottle, J.-S. Pang and R. E. Stone (1992): *The Linear Complementarity Problem*, Academic Press, New York.

26. B. D. Craven (1988): *Fractional Programming*, Heldermann Verlag, Berlin, 1988.

27. S. Dafermos (1980): Traffic equilibrium and variational inequalities, *Transportation Science*, **14**, 42–54.

28. S. Dafermos (1988): Sensitivity analysis in variational inequalities, *Mathematics of Operations Research*, **13**, 421–434.

29. J. W. Daniel (1973): Stability of the solution of definite quadratic programs, *Mathematical Programming*, **5**, 41–53.

30. G. B. Dantzig (1963): *Linear Programming and Extensions*, Princeton University Press, Princeton.

31. M. De Luca and A. Maugeri (1989): Quasi-variational inequalities and applications to equilibrium problems with elastic demand, In *Nonsmooth Analysis and Related Topics* (F. H. Clarke, V. F. Demyanov and F. Giannessi, Eds.), Plenum Press.

32. P. H. Dien (1985): On the regularity condition for the extremal problem under locally Lipschitz inclusion constraints, *Applied Mathematics and Optimization*, **13**, 151–161.

33. A. Domokos (1999): Solution sensitivity of variational inequalities, *Journal of Mathematical Analysis Applications*, **230**, 381–389.

34. A. L. Donchev and R. T. Rockafellar (1996): Characterizations of strong regularity for variational inequalities over polyhedral convex sets, *SIAM Journal on Optimization*, **6**, 1087–1105.

35. B. C. Eaves (1971): On quadratic programming, *Management Science*, **17**, 698–711.

36. F. Facchinei and J.-S. Pang (2003): *Finite-Dimensional Variational Inequalities and Complementarity Problems*, Volumes I and II, Springer, New York.

37. A. V. Fiacco (1983): *Introduction to Sensitivity and Stability Analysis in Nonlinear Programming*, Academic Press, New York.

38. A. V. Fiacco and J. Liu (1995): On the stability of general convex programs under Slater's condition and primal solution boundedness, *Optimization*, **32**, 291–299.

39. M. Frank and P. Wolfe (1956): An algorithm for quadratic programming, *Naval Research Logistics Quarterly*, **3**, 95–110.

40. C. B. Garcia (1973): Some classes of matrices in linear complementarity theory, *Mathematical Programming*, **5**, 299–310.

41. J. Gauvin (1977): A Necessary and sufficient regularity condition to have bounded multipliers in nonconvex programming, *Mathematical Programming*, **12**, 136–138.

42. J. Gauvin and J. W. Tolle (1977): Differential stability in nonlinear programming, *SIAM Journal on Control and Optimization*, **15**, 294–311.

43. J. Gauvin and F. Dubeau (1982): Differential properties of the marginal function in mathematical programming, *Mathematical Programming Study*, **19**, 101–119.

44. F. Giannessi (1980): Theorems of alternative, quadratic programs and complementarity problems, In *Variational Inequality and Complementarity Problems* (R. W. Cottle, F. Giannessi and J.-L. Lions, Eds.), Wiley, New York, pp. 151–186.

45. M. S. Gowda (1992): On the continuity of the solution map in linear complementarity problems, *SIAM Journal on Optimization*, **2**, 619–634.

46. M. S. Gowda and J.-S. Pang (1992): On solution stability of the linear complementarity problem, *Mathematics of Operations Research*, **17**, 77–83.

47. M. S. Gowda and J.-S. Pang (1994a): On the boundedness and stability of solutions to the affine variational inequality problem, *SIAM Journal on Control and Optimization*, **32**, 421–441.

48. M. S. Gowda and J.-S. Pang (1994b): Stability analysis of variational inequalities and nonlinear complementarity problems, via the mixed linear complementarity problem and degree theory, *Mathematics of Operations Research*, **19**, 831–879.

49. M. S. Gowda and R. Sznajder (1996): On the Lipschitzian properties of polyhedral multifunctions, *Mathematical Programming*, **74**, 267–278.

50. J. Guddat (1976): Stability in convex quadratic parametric programming, *Mathematishe Operationsforschung und Statistik*, **7**, 223–245.

51. P. T. Harker and J.-S. Pang (1990): Finite-dimensional variational inequality and nonlinear complementarity problems: A survey of theory, algorithms and applications, *Mathematical Programming*, **48**, 161–220.

52. P. Hartman and G. Stampacchia (1966): On some non-linear elliptic differential-functional equations, *Acta Mathematica*, **115**, 271–310.

53. J.-B. Hiriart-Urruty (1985): Images of connected sets by semicontinuous multifunctions, *Journal of Mathematical Analysis and Applications*, **111**, 407–422.

54. T. N. Hoa, T. D. Phuong and N. D. Yen: Linear fractional vector optimization problems with many components in the solution sets, Preprint 2004/05, Institute of Mathematics, Hanoi. (Submitted)

55. R. Horst and H. Tuy (1990): *Global Optimization*, Springer-Verlag, Berlin, 1990.

56. Y. D. Hu and E. J. Sun (1993): Connectedness of the efficient set in strictly quasiconcave vector maximization, *Journal of Optimization Theory and Applications*, **78**, 613–622.

57. N. Q. Huy and N. D. Yen (2004a): Contractibility of the solution sets in strictly quasiconcave vector maximization on noncompact domains. Accepted for publication in *Journal of Optimization Theory and Applications*.

58. N. Q. Huy and N. D. Yen (2004b): Remarks on a conjecture of J. Benoist, Preprint 2004/06, Institute of Mathematics, Hanoi. (Submitted)

59. R. Janin (1984): Directional derivative of the marginal function in nonlinear programming, *Mathematical Programming Study*, **21**, 110–126.

60. M. J. M. Jansen and S. H. Tijs (1987): Robustness and non-degenerateness for linear complementarity problems, *Mathematical Programming*, **37**, 293–308.

61. B. T. Kien (2001): Solution sensitivity of a generalized variational inequality, *Vietnam Journal of Mathematics*, **29**, 97–113.

62. D. Kinderlehrer and G. Stampacchia (1980): *An Introduction to Variational Inequalities and Their Applications*, Academic Press, New York-London.

63. D. Klatte (1985): On the Lipschitz behavior of optimal solutions in parametric problems of quadratic optimization and linear complementarity, *Optimization*, **16**, 819–831.

64. S. Kum, G. M. Lee and N. D. Yen (2004): Remarks on the stability of linear fractional vector optimization problems. (Manuscript)

65. J. Kyparisis (1988): Perturbed solutions of variational problems over polyhedral sets, *Journal of Optimization Theory and Applications*, **57**, 295-305.

66. J. Kyparisis (1990): Sensitivity analysis for nonlinear programs and variational inequalities with nonunique multipliers, *Mathematics of Operations Research*, **15**, 286-298.

67. G. M. Lee, D. S. Kim, B. S. Lee and N. D. Yen (1998): Vector variational inequalities as a tool for studying vector optimization problems, *Nonlinear Analysis*, **34**, 745–765.

68. G. M. Lee, N. N. Tam and N. D. Yen (2002a): On a class of optimal value functions in quadratic programming. Accepted for publication in *Journal of Global Optimization*.

69. G. M. Lee, N. N. Tam and N. D. Yen (2002b): Continuity of the solution map in quadratic programs under linear perturbations. (Submitted)

70. G. M. Lee, N. N. Tam and N. D. Yen (2002c): Lower semicontinuity of the solution maps in quadratic programming under linear perturbations, Part I: Necessary conditions. (Submitted)

71. G. M. Lee, N. N. Tam and N. D. Yen (2002d): Lower semicontinuity of the solution maps in quadratic programming under linear perturbations, Part II: Sufficient conditions. (Submitted)

72. G. M. Lee and N. D. Yen (2001): A result on vector variational inequalities with polyhedral constraint sets, *Journal of Optimization Theory and Applications*, **109**, 193–197.

73. E. S. Levitin (1994): *Perturbation Theory in Mathematical Programming and Its Applications*, John Wiley & Sons, New York.

74. D. T. Luc (1987): Connectedness of the efficient set in quasi-concave vector maximization, *Journal of Mathematical Analysis and Applications*, **122**, 346–354.

75. D. T. Luc (1989): *Theory of Vector Optimization*, Springer-Verlag, Berlin.

76. A. Majthay (1971): Optimality conditions for quadratic programming, *Mathematical Programming*, **1**, 359–365.

77. K. Malanowski (1987): *Stability of Solutions to Convex Problems of Optimization*, Springer-Verlag, Berlin-Heidelberg.

78. C. Malivert (1995): Multicriteria fractional programming, In *Proceedings of the 2nd Catalan Days on Apllied Mathematics* (M. Sofonea and J. N. Corvellec, Eds.), Presses Universitaires de Perpinan, pp. 189–198.

79. C. Malivert and N. Popovici (2000): Bicriteria linear fractional optimization, In *Optimization*, Lecture Notes in Economic and Mathematical Systems, 481, Springer, Berlin, pp. 305–319.

80. O. L. Mangasarian (1969): *Nonlinear Programming*, McGraw-Hill Book Company, New York.

81. O. L. Mangasarian (1980): Locally unique solutions of quadratic programs, linear and nonlinear complementarity problems, *Mathematical Programming*, **19**, 200–212.

82. O. L. Mangasarian and T. H. Shiau (1987): Lipschitz continuity of solutions of linear inequalities, programs and complementarity problems, *SIAM Journal on Control and Optimization*, **25**, 583–595.

83. G. P. McCormick (1967): Second order conditions for constrained minima, *SIAM Journal on Applied Mathematics*, **15**, 641–652.

84. L. I. Minchenko and P. P. Sakolchik (1996): Hölder behavior of optimal solutions and directional differentiability of marginal functions in nonlinear programming, *Journal of Optimization Theory and Applications*, **90**, 555–580.

85. B. Mordukhovich (1988): *Approximation Methods in Problems of Optimization and Control*, Nauka, Moscow. (In Russian)

86. B. Mordukhovich (1993): Complete characterization of openness, metric regularity, and Lipschitzian properties of multifunctions, *Transactions of the American Mathematical Society*, **340**, 1–35.

87. B. Mordukhovich (1994): Generalized differential calculus for nonsmooth and set-valued mappings, *Journal of Mathematical Analysis and Applications*, **183**, 250–288.

88. K. G. Murty (1972): On the number of solutions to the complementarity problem and spanning properties of complementarity cones, *Linear Algebra and Applications*, **5**, 65–108.

87. B. Mordukhovich (1994): Generalized differential calculus for nonsmooth and set-valued mappings, *Journal of Mathematical Analysis and Applications*, **183**, 250–288.

88. K. G. Murty (1972): On the number of solutions to the complementarity problem and spanning properties of complementarity cones, *Linear Algebra and Applications*, **5**, 65–108.

89. K. G. Murty (1976): *Linear and Combinatorial Programming*, John Wiley, New York.

90. A. Nagurney (1993): *Network Economics: A Variational Inequality Approach*, Kluwer Academic Publishers, Dordrecht.

91. N. H. Nhan (1995): *On the Stability of Linear Complementarity Problems and Quadratic Programming Problems*, Master thesis, University of Hue, Hue, Vietnam.

92. W. Oettli and N. D. Yen (1995): Continuity of the solution set of homogeneous equilibrium problems and linear complementarity problems, In *Variational Inequalities and Network Equilibrium Problems* (F. Giannessi and A. Maugeri, Eds.), Plenum Press, New York, pp. 225–234.

93. W. Oettli and N. D. Yen (1996a): Quasicomplementarity problems of type R_0, *Journal of Optimization Theory and Applications*, **89**, 467–474.

94. W. Oettli and N. D. Yen (1996b): An example of a bad quasicomplementarity problem, *Journal of Optimization Theory and Applications*, **90**, 213–215.

95. M. Patriksson (1999): *Nonlinear Programming and Variational Inequality Problems. A Unified Approach,* Kluwer Academic Publishers, Dordrecht.

96. J.-P. Penot and A. Sterna-Karwat (1986): Parameterized multicriteria optimization: Continuity and closedness of optimal multifunctions, *Journal of Mathematical Analysis and Applications*, **120**, 150–168.

97. H. X. Phu and N. D. Yen (2001): On the stability of solutions to quadratic programming problems, *Mathematical Programming*, **89**, 385–394.

98. Y. Qiu and T. L. Magnanti (1989): Sensitivity analysis for variational inequalities defined on polyhedral sets, *Mathematics of Operations Research*, **14**, 410–432.

99. S. M. Robinson (1975): Stability theory for systems of inequalities, Part I: Linear systems, *SIAM Journal Numerical Analysis*, **12**, 754–769.

100. S. M. Robinson (1977): A characterization of stability in linear programming, *Operations Research*, **25**, 435–477.

101. S. M. Robinson (1979): Generalized equations and their solutions, Part I: Basic theory, *Mathematical Programming Study*, **10**, 128–141.

102. S. M. Robinson (1980): Strongly regular generalized equations, *Mathematics of Operations Research*, **5**, 43–62.

103. S. M. Robinson (1981): Some continuity properties of polyhedral multifunctions, *Mathematical Programming Study*, **14**, 206–214.

104. S. M. Robinson (1982): Generalized equations and their solutions, Part II: Applications to nonlinear programming, *Mathematical Programming Study*, **19**, 200–221.

105. R. T. Rockafellar (1970): *Convex Analysis,* Princeton University Press, Princeton, New Jersey.

106. R. T. Rockafellar (1982): Lagrange multipliers and subderivatives of optimal value functions in nonlinear programming, *Mathematical Programming Study*, **17**, 28–66.

107. R. T. Rockafellar (1988): First- and second-order epi-differentiability in nonlinear programming, *Transactions of the American Mathematical Society*, **307**, 75–108.

108. R. T. Rockafellar and R. J.-B. Wets (1998): *Variational Analysis,* Springer, Berlin-Heidelberg.

109. J.-F. Rodrigues (1987): *Obstacle Problems in Mathematical Physics*, North-Holland Publishing Co., Amsterdam.

110. S. Schaible (1983): Bicriteria quasiconcave programs, *Cahiers du Centre d'Etudes de Recherche Opérationnelle*, **25**, 93–101.

111. A. Seeger (1988): Second order directional derivatives in parametric optimization problems, *Mathematics of Operations Research*, **13**, 124–139.

112. M. J. Smith (1979): The existence, uniqueness and stability of traffic equilibrium, *Transportation Research*, **13B**, 295–304.

113. R. E. Steuer (1986): *Multiple Criteria Optimization: Theory, Computation and Application*, John Wiley & Sons, New York.

114. N. N. Tam (1999): On continuity of the solution map in quadratic programming, *Acta Mathematica Vietnamica*, **24**, 47–61.

115. N. N. Tam (2001a): Sufficient conditions for the stability of the Karush-Kuhn-Tucker point set in quadratic programming, *Optimization*, **50**, 45–60.

116. N. N. Tam (2001b): Directional differentiability of the optimal value function in indefinite quadratic programming, *Acta Mathematica Vietnamica*, **26**, 377–394.

117. N. N. Tam (2002): Continuity of the optimal value function in indefinite quadratic programming, *Journal of Global Optimization*, **23**, 43–61.

118. N. N. Tam and N. D. Yen (1999): Continuity properties of the Karush-Kuhn-Tucker point set in quadratic programming problems, *Mathematical Programming*, **85**, 193–206.

119. N. N. Tam and N. D. Yen (2000): Stability of the Karush-Kuhn-Tucker point set in a general quadratic programming problem, *Vietnam Journal of Mathematics*, **28**, 67–79.

120. R. L. Tobin (1986): Sensitivity analysis for variational inequalities, *Journal of Optimization Theory and Applications*, **48**, 191-204.

121. D. W. Walkup and R. J.-B. Wets (1969): A Lipschitzian characterization of convex polyhedra, *Proceedings of the American Mathematical Society*, **23**, 167–173.

122. F. Wantao and Z. Kunping (1993): Connectedness of the efficient solution sets for a strictly path quasiconvex programming problem, *Nonlinear Analysis: Theory, Methods & Applications*, **21**, 903–910.

123. A. R. Warburton (1983): Quasiconcave vector maximization: Connectedness of the sets of Pareto-optimal and weak Pareto-optimal alternatives, *Journal of Optimization Theory and Applications*, **40**, 537–557.

124. D. E. Ward and G. M. Lee (2001): Upper subderivatives and generalized gradients of the marginal function of a non-Lipschitzian program, *Annals of Operations Research*, **101**, 299–312.

125. J. G. Wardrop (1952): Some theoretical aspects of road traffic research, *Proceddings of the Institute of Civil Engineers*, Part II , 325–378.

126. E. W. Weisstein (1999): *CRC Concise Encyclopedia of Mathematics*, CRC Press, New York.

127. N. D. Yen (1987): Implicit function theorems for set-valued maps, *Acta Mathematica Vietnamica*, **12**, No. 2, 7–28.

128. N. D. Yen (1995a): Hölder continuity of solutions to a parametric variational inequality,*Applied Mathematics and Optimization*, **31**, 245–255.

129. N. D. Yen (1995b): Lipschitz continuity of solutions of variational inequalities with a parametric polyhedral constraint, *Mathematics of Operations Research*, **20**, 695–708.

130. N. D. Yen and N. X. Hung (2001): A criterion for the compactness of the solution set of a linear complementarity problem, In *Fixed Point Theory and Applications* Vol. 2 (Y. J. Cho, J. K. Kim and S. M. Kang, Eds.), Nova Science Publishers, New York.

131. N. D. Yen and T. D. Phuong (2000): Connectedness and stability of the solution set in linear fractional vector optimization problems, In *Vector Variational Inequalities and Vector Equilibria* (F. Giannessi, Ed.), Kluwer Academic Publishers, Dordrecht, pp. 479–489.

132. N. D. Yen and L. Zullo (1992): Networks with an equilibrium condition, Research report, Università di Pisa, Pisa, Italy.

133. E. Zeidler (1986): *Nonlinear Functional Analysis and its Applications. I: Fixed-Point Theorems*, Springer, New York.

Index